MICROBIAL ECOLOGY

MICROBIAL ECOLOGY

Martin Alexander
Cornell University

John Wiley & Sons, Inc.
New York • London • Sydney • Toronto

Library of Congress Catalogue Card Number: 71-137105

ISBN 0-471-02054-0

Printed in the United States of America

10 9 8 7 6 5 4 3 2 1

Preface

Ecology has been the subject of a remarkable surge in interest in the last few years. Many biologists have turned their attention to environmental problems, new concepts have been proposed, novel techniques have been devised, and a broad range of fascinating phenomena has been uncovered. The general public too has become increasingly aware of the significance of ecology and the importance of understanding how organisms behave and interact in nature.

The field of microbial ecology has expanded rapidly during the same period. Part of this rise in interest undoubtedly resulted from a greater appreciation of the critical role that microorganisms play in environments subject to pollution, but a significant part has come about because of the discovery of previously unknown microbial processes, the finding of novel organisms, the greater understanding of biological changes of geochemical importance, and the application of techniques not used heretofore. Nevertheless microbial ecology is far from a new discipline. Marine, freshwater, and soil microbiologists have devoted their attention for many years to the characterization of microorganisms and microbial activities in natural habitats. The medical and veterinary microbiologist and the plant pathologist likewise have long sought to understand the behavior of parasites in their host-regulated environments. Similarly, food and industrial scientists have for a long time investigated the relationship between microscopic living things and the products on which they grow—in effect, the habitats of the organisms.

The kinds of areas and the dimensions of the habitats wherein microorganisms reside differ from those of higher plants and animals. The practical problems attributable to microbial activity and inactivity in nature are quite distinct from those of higher organisms, and the techniques employed by the microbiologist differ from those

v

used in investigations of rooted plants and vertebrates. For these and other reasons, the approaches, needs, and even terminology may often appear to be unique to the microbiologist. Furthermore, conflicts in terminology even exist in the various facets of microbiology—bacteriology, mycology, phycology, protozoology, and virology—and sometimes the specialist in one type of habitat uses concepts foreign to the specialist in another. Faced with these difficulties and conflicts, I have endeavored to select not a middle course but rather one that should help the reader to understand the numerous ramifications of microbial ecology and introduce him to the many intriguing relationships and activities of microorganisms.

No approach to microbial ecology is universally accepted. Some biologists emphasize a single type of habitat; others, a general category of organisms. Some focus their attention on a descriptive approach to the topic, while others give greatest weight to the biochemical changes induced by the microbiota. The approach I have employed emphasizes the underlying phenomena in the hope that an awareness and understanding of these basic phenomena will introduce the reader in a broad way to microbial ecology. Examples are selected from diverse habitats and from different microbial groups to illustrate the phenomena, but it is my feeling that a thorough understanding of the way microorganisms behave and interact in their natural or alien environments provides the most adequate introduction to the subject. Of necessity, the examples cited are few in number and are briefly described, but it is hoped that they will illustrate the approach, define the phenomenon, or show the importance of the topic under discussion.

The book is not, and cannot be, a definitive monograph. The literature is far too vast, new information is being gathered much too quickly, and the diversity of locales and organisms is far too great. Many things will be left unsaid. The reader hopefully will become acquainted with one or more of the applied fields, such as pollution, medicine, public health, plant pathology, soil science, food production, limnology, or oceanography. The book, as an introduction, should open the door to both new and important problems in applied or basic fields, creating a foundation for the reader to delve more thoroughly or comprehend more fully those aspects of greatest interest to him.

In attempting to fuse microbiology with ecology, it is necessary to consider some elementary concepts. This discussion will be repetitious for some readers but likely will provide new information to others. I have attempted to keep this background material to a minimum, but I believe it is necessary so that the subsequent chapters of the book will be clear and understandable. Likewise, because a wide array of organisms is considered in the illustrations, it seems necessary to provide a brief taxonomic outline; this is presented at the end of the book. Though the taxonomic schemes are not accepted by all and some still are in a state of change, the outline should be adequate for the purposes of an introductory textbook.

At the end of each chapter are given the references cited in the preceding discussion. These references represent but a minute sampling of the enormous literature, but they are selected to provide the student and researcher with reports of individual investigations that delve in greater detail into the subject matter and present a few of the techniques used by the ecologist and microbiologist. Several reviews, not all of them cited in the text, are also given at the end of each chapter. These are included to allow the reader to explore the subject further or to find additional references.

Finally, I express my thanks to the United States Department of the Interior for the support they provided to allow for the preparation of the manuscript. To my wife, Renee, I also express my deep appreciation for her patience and encouragement.

MARTIN ALEXANDER

Ithaca, New York
July 1970

Contents

Part 3

**EFFECT OF MICROORGANISMS
ON THEIR SURROUNDINGS**

MICROBIAL ECOLOGY

Part 1

THE COMMUNITY
AND ITS DEVELOPMENT

1

The Microbial Community

Microbial ecology is concerned with the interrelationships between microorganisms and their environment. It relates the bacteria, fungi, algae, protozoa, actinomycetes, and viruses to the many environments in which these small organisms occur—aquatic or terrestrial regions, in or on animals and plants, in the atmosphere or deep below the surface of the earth, in marine sediments or on the surface of rocks located on remote islands. As the science dealing with the relations between such organisms and their sites of occurrence, ecology thus encompasses numerous habitats and many types of microorganisms. Because of the diversity of microorganisms and the regions in which they are situated, the modern ecologist is compelled to establish general principles governing the interactions in natural circumstances.

The relationships between the environment and the different organisms that make up a biological complex in a single locale are the concern of *synecology*. In synecology the various species, the complex of organisms, and the association of the assemblage of species with the biologically significant abiotic components of the environment are considered. *Autecology*, by contrast, is concerned with the ecology of a single species and the influence of environmental factors on that species. The attention of the autecologist may be focused on the effect of the individual species on its surroundings, the impact of the environment on the organism, or the basis for its adaptation to the features and stresses of the site of habitation. Microbiologists investigating human, animal, and plant pathogens are often involved in autecological investigations, for their attention is directed to the relationship of one microbial species to its environment, the environment in such instances frequently being the appropriate host. Certain microbiologists specialize in *habitat ecology*, a field of study in which individual habitats, the organisms residing in them, and the interac-

3

tions between the residents and the abiotic constituents are considered. The marine or soil microbiologist is in fact a habitat ecologist in large part, but so too are the medical or veterinary microbiologist and the plant pathologist, in that they examine the microorganisms of a particular habitat and attempt to analyze the effects of the organisms on their viable environment and the influence of the habitat on its residents.

ECOLOGICAL HIERARCHY

The complex of organisms in a specified environment and the abiotic surroundings with which the organisms are associated are known as an *ecosystem*. The ecosystem includes the assemblage of species and the organic and inorganic constituents characterizing the particular site. Each different ecosystem has a collection of organisms and abiotic components unique to it and it alone.

The organisms inhabiting a given site constitute a *community*. Often few components of a microbial community are known, and fewer have been well studied. This is particularly true of the members of marine, freshwater, and terrestrial communities. By contrast, communities containing species responsible for major public health or economic problems frequently have been explored in some detail, and the identities and behavior of many of the resident species have been characterized. Community and *biocoenosis* are commonly considered synonymous terms.

The assemblage of organisms constituting a community contains *populations* of individual microbial species or of distinguishable types. Each population may be viewed as a discrete entity within the community, and studies of the interactions of a population with its biotic and abiotic environment are of particular significance to the welfare of man when the species under inquiry is either beneficial or harmful to man, domestic animals, or plants. At a rank still lower than the population in the ecological hierarchy is the individual microorganism—the discrete cell, the hyphal strand, the resting body, the thallus, and so on. Thus the organism is at the bottom rung of the hierarchal ladder, the population of individuals of one type or of one species is at the second level, and the community encompassing all populations in a given locale occupies the third rung, while the ecosystem stands at the top of the hierarchy.

RELATION BETWEEN ORGANISM
AND ENVIRONMENT

Microorganisms are potentially everywhere on or near the surface of the earth. They are capable of being transported by currents of air and in rivers and streams. Some are moved about on the surfaces of animals and plants and on inanimate materials that are borne from place to place. A few are transported vertically in soil and through fissures and cracks, to enter sites remote from the zone of life of higher plants and animals. Many are ingested with the food and water consumed by man and animals. Others penetrate through the skin and outer layers of a variety of organisms.

In addition to their ready dispersal, microorganisms commonly grow rapidly. Certain bacteria reproduce with generation times of less than 10 minutes. The mass of many fungi, algae, and protozoa doubles in short periods of time. The capacity for rapid growth, exhibited by many microbial species, is evident in numerous environments provided that suitable nutrients are present. Moreover, the number of entirely different types of habitats readily colonizable by the multitude of easily transported microorganisms is vast.

Although organisms of a wide spectrum of morphological and physiological types penetrate a vast number of different environments they potentially could colonize, and despite the high rate of proliferation of many kinds of microorganisms, individual ecosystems have characteristic communities. The microbial communities of soil, oceans, fresh water, plant surfaces, roots, invaded tissues, human stools, the oral cavity, animal droppings, decaying organic materials, and sewage treatment plants are all unique. A typical example is presented in Table 1.1. A trained investigator seeking a certain bacterium, fungus, alga, or protozoan does not search into every possible habitat to find a representative of the genus he wishes to isolate. He has a reasonable idea that the species will be found in a particular surrounding and be absent from environments of different compositions. He may not know the reasons for the distribution, but he feels confident that the organism will be associated with a particular ecosystem and that the chosen ecosystem will have a distinct community composed of only certain genera and species. Given the tens of thousands of described species of fungi, protozoa, bacteria, and algae, it is quite apparent that microbial species are not randomly distributed, but rather exist in distinct patterns and assortments.

Table 1.1

Estimated Numbers of Bacterial Groups in the Oral Cavity of Man [a]

Bacterial Group	Saliva ($\times 10^7/ml$)	Gingival Crevice ($\times 10^6/mg$)
Total, anaerobic	11.0	36.0
Total, aerobic	4.0	15.0
Staphylococcus	0.0005	—
Streptococcus	2.9	5.5
Veillonella	1.7	—
Neisseriae	0.2	—
Fusobacteria	0.0056	0.043
Bacteroides	—	0.95
Spirochetes	—	0.8
Others	0.7	—

[a] From Rosebury (1962).

The environment selects. Each distinguishable ecosystem has associated with it certain physical, chemical, and biological determinants governing the composition of the community and dictating which invading organisms will be successfully established and which will not, which species will be dominant and which will be of lesser significance. A vast body of literature attests to the fact that certain environmental factors become expressed, in a biological sense, by communities reflecting these ecological determinants.

Few sets of conditions on or near the surface of the earth do not allow for the growth of at least some microbial groups. Among the areas devoid of microorganisms or having few viable cells per unit mass—but often potentially colonizable, sometimes with profound consequences—are the internal tissues of animals or plants, the surfaces of the newborn infant, the insides of rocks, and food products that are deliberately sterilized. If no organism in the vicinity is immediately capable of exploiting a newly created condition or a previously uncolonized locale, either a species with the requisite physiological potential will be transported to the colonizable region or an indigenous species will exploit the new situation following a change in its physiological characteristics.

The interactions between microorganisms and their habitat are reciprocal. The composition of natural communities is not haphazard; rather, each microbial species and genus has a certain distribution, a pattern determined by the physiological responses of the population to the environments into which it is introduced. The populations coexisting in the community are those selected from among the invaders of the locality as being the best adapted to the prevailing conditions—utilizing nutrients entering into or formed within the ecosystem, tolerating the prevailing physical and chemical conditions, and coping with or benefiting from associations with other components of the community. Populations not meeting the environmental stresses will have been eliminated during the early stages in the development of the community, or they will not survive in the presence of the established populations.

The environment, in effect, molds the composition of the community by selecting for populations with favorable nutritional, physiological, or even morphological characteristics and rejecting unsuited species. The organisms with traits that make them fit for life in that area will survive, reproduce, and assume major or minor importance; those not so endowed will die out, although a diminishing number of survivors may be detectable for some time. The community is thus in equilibrium with the environment as a result of the operation of natural selection.

The relations between the environment and the resident populations are not unidirectional, however. Although the ecological determinants of the environment, abiotic and biotic, govern what microorganisms can occupy a given site, the inhabitants in turn modify the composition of their surroundings. These minor or major changes are evident even to the casual observer, and many can be seen with the naked eye. The etching and weathering of rocks by lichen colonizers or lichen-bacterial associations, the modifications in the open sea or in impounded bodies of fresh water by algal blooms, the genesis of soils, the alterations in animals and plants resulting from microbial infection, the generation of simple compounds from the polysaccharides entering the bovine rumen, and the deterioration of foods or plant materials are but a few of the environmental changes induced by microorganisms. Some of these changes have a local impact only, but even these may possess great economic, medical, or biological importance. Others have global significance, affecting geochemical cycles or the very existence of life on the earth; for example, the regeneration of atmospheric CO_2, certain transformations of nitrogen, and the genesis of various forms of sulfur are biogeochemical processes resulting largely from the metabolic activities of microorganisms. Hence, in addition to the environment having a profound

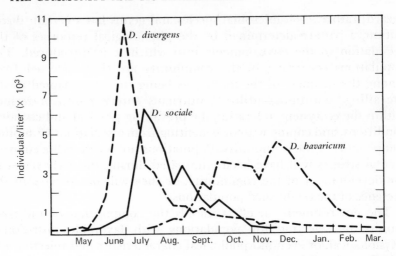

Figure 1.1   Seasonal succession of three species of *Dinobryon* in Lake Ohrid (After Stankovic, 1960).

influence in determining the composition of the community, the metabolism of the community in turn may dramatically alter the physical and chemical properties of its surroundings.

In an environment not subject to marked changes from without, the species composition of the community is remarkably stable. Not only the kinds of species but also the relative sizes of their populations often remain reasonably constant. On the other hand, a modification in an ecologically significant factor of the environment brings about a shift—sometimes minor, sometimes dramatic—in either the population densities of the indigenous species or the types of organisms present. Some groups are favored by the change in conditions and assume greater prominence, while others are not able to cope as effectively with the new circumstances or the newly significant inhabitants as with the old (Fig. 1.1). If the environmental modification is long-term, a different community becomes established, but if the modification is transitory the community readjusts and returns to its original composition. Examples of such readjustments can be seen in the seasonal fluctuations in the makeup of aquatic communities and in the temporary, antibiotic-induced changes in the communities inhabiting portions of the human or animal body.

The community finds in its characteristic ecosystem all things necessary for its maintenance—nutrients, cooperative relationships, and physical factors that permit growth of the assemblage of species.

The native microorganisms are able to multiply, or survive in the absence of conditions conducive to replication, because they are in harmony with their biotic and abiotic environment. The many relationships and biological pressures within the community naturally select for those species fit for life in that ecosystem and reject unsuited types. In effect, the collection of populations has come to be present because it contains those organisms that are suited to the environment and to the interactions occurring therein. The assemblage then indeed reflects its distinct surroundings, each resident species having succeeded in the struggle for life at that place and time.

INHABITANTS OF THE COMMUNITY

Within the community it is possible to distinguish several categories of microorganisms. The true inhabitants, often designated *indigenous* or *autochthonous* species, are native to the locale, and at one stage or another they grow, multiply, and contribute to the metabolism of the community. The mere presence of a particular species is itself no guarantee that it is a functional or permanent member of the community, inasmuch as microorganisms are readily disseminated, and many *invaders* or *allochthonous* forms originate elsewhere and are transported into the environment in a vegetative or a resting state. These transient types may maintain themselves for some time in an inactive form either as resistant structures morphologically different from the vegetative cells or filaments or as vegetative cells protected in some way from rapid destruction. Certain allochthonous species may even grow for short periods because they are deposited in their temporary domicile together with nutrient material or tissue derived from their old environment, and some may be able to exploit for short periods a portion of the resources of their new abode. Allochthonous forms on the human skin and leaf surfaces may fall from the air. Those in the oral cavity and gastrointestinal system may be derived from food and feed, certain ones in necrotic tissues may have come from adjacent healthy tissues, some in soil are introduced with diseased plant and animal tissue, and various types in the sea are traceable to rivers or polluted water. These are aliens, however, and are unable to cope successfully with the biological stresses or the abiotic factors in the new surroundings, and they are ultimately eliminated.

Certain microbial species are restricted to a single type of environment and are readily characterized as indigenous to that locality.

Thus some obligately parasitic bacteria and protozoa are unique to man or animals, several pathogenic fungi are strict parasites of plants, and a number of algae are situated only in the seas. Other species may grow best in one kind of ecosystem, but they are found and multiply in dissimilar types of environments, where they occupy a minor position. At the opposite pole are those ubiquitous bacteria, fungi, and other microorganisms that are widely distributed and presumably develop in a multitude of distinctly different habitats.

In view of the ease of dispersal of microorganisms and the finding of diverse genera in a host of circumstances, considerable attention has been given to establishing criteria of autochthony. Many investigators have attempted to establish which of the organisms present in one community or another are autochthonous and which are simply short-lived invaders; for example, what species are truly indigenous to man, which are involved in a specific necrosis, and which are the autochthonous components of the rumen, marine, or terrestrial environment? Some of the criteria or operational procedures proposed as a basis for considering a microorganism native to a particular habitat include the repeated isolation of the species from samples of the environment, its existence at a high population density, the ability of a pure culture of the organism to use nutrients normally entering that environment, and its ability to tolerate the environmental extremes typifying the site (Garrett, 1956; MacLeod, 1965; Rosebury, 1962; Winogradsky, 1949). Nevertheless, suitable or universal criteria for autochthony are not yet available for many microbial groups and for a number of types of ecosystems.

SPECIES DIVERSITY

Some habitats are thickly populated and show a high density of microbial cells and filaments, while other regions are sparsely populated at best. Sometimes the population density is greatest where the nutrient concentration is particularly abundant, as in polluted lakes, accumulations of plant debris, or root microsites from which organic materials are exuded. Frequently, however, a nutrient-rich area is poor in microorganisms because environmental conditions prevent extensive proliferation, as can be seen in acidified silage, in sugary syrups, and in plant and animal tissues containing antimicrobial agents. Marked differences in cell density may even be observed in what overtly seems to be the same environment; for example, a region of the skin, a minute portion of a plant root, a selected point in a lake, or a particular site in the soil may teem with life while a nearby region may have few individuals or appear to be sterile.

Not only do microbial communities vary from one another in the abundance of cells or filaments they contain, but they also differ in the number of microbial groups. Some possess a multitude of different species, others have no more than a few, while certain communities are monospecific and contain a population of but a single species. Thus, in many animal or plant infections, the responsible community is composed largely or entirely of one bacterial or fungal population. In other instances, although the community is heterogeneous, a broad taxonomic group may be represented by a solitary species, as in the monospecific algal blooms or fungal masses associated with spoiled foods.

Species diversity clearly varies from ecosystem to ecosystem; yet the reasons for the heterogeneity in microbial types remain unknown. The diversity is related to the habitat in question, the abundance of species gaining access to the particular region, and the numbers capable of coping with those environmental factors and host defense mechanisms affecting microbial activity. Low species diversities characterize areas in which the intensity of one or more ecological factors approaches the extremes capable of supporting life; thus, under the rigors of high temperature, high sugar or salt concentrations, or extremes of drought, few species are observed, presumably because the invaders were not able to survive long or to multiply under these stress circumstances. Conversely, species diversity is great in certain aquatic algal communities and in soil, and no one species is overly prominent (Fig. 1.2). In these instances the small population sizes and great species diversity seem to be associated with nutrient deficiencies, and when there is an inflow of nutrients species diversity frequently declines. Probably a limited few species are especially favored in these regions by the abundance of nutrients, whereas none has a particular advantage in nutrient-deficient environments that are otherwise suitable for microbial development (Alexander, 1961; Hulburt, 1963; Patrick, 1963).

Communities with high species diversity are often stated to be more resistant to change than those exhibiting low diversity. The statement seems valid for certain terrestrial communities of bacteria and fungi, for the microbial assemblage in the gastrointestinal system, and for those aquatic communities containing few algal types. In the first two instances modest environmental stresses do not appreciably modify the population sizes and the kinds of indigenous species. In the last instance even a minor ecological fluctuation may have an

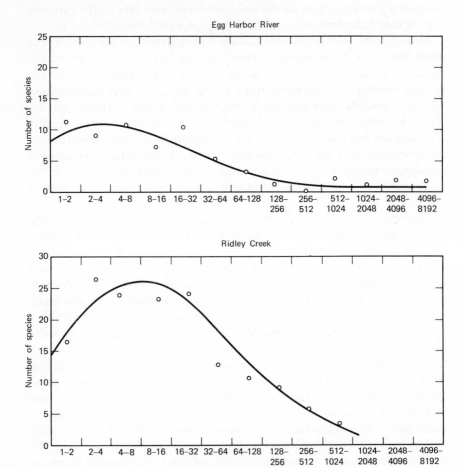

Figure 1.2   The structure of aquatic diatom communities. The number of species in a given set of water specimens is plotted against the number of individuals in those species. The community in the polluted Egg Harbor River is composed of fewer species, but the population sizes of some are quite large (After Patrick, 1963). By permission of The New York Academy of Sciences.

explosive effect, eliminating or reducing the abundance of a major species and allowing a minor alga to assume a dominant position. The statement may hold, too, for the relatively homogeneous communities colonizing or infecting higher animals or plants; which, when subjected to some host defense reaction or when exposed to an environmental stress in the form of a chemotherapeutic agent, are often catastrophically affected.

It is far from clear why so many types of microorganisms coexist indefinitely in a single ecosystem. The resources of the environment are clearly subdivided so that every resident population obtains what it needs to maintain its numerical relationship with its neighbors. It has been proposed that natural communities have a division of labor, each species specializing in one function (Elton, 1946), or that predators, and possibly parasites, serve to maintain species diversity by not allowing one population to monopolize the major environmental resources to the exclusion of others (Paine, 1966). Experimental confirmation of these views or alternative hypotheses to account for the coexistence of so many species in relatively homogeneous environments are largely lacking in microbiology.

By contrast with environments which are physically or chemically homogeneous, many regions where microorganisms reside are spatially heterogeneous or have characteristics whose intensities fluctuate in time, and species diversity is attributable to the lack of spatial homogeneity or the changes taking place in time. The spatial differences may reflect subtle or marked gradients or variations in some environmental factor—a mechanical barrier, a food particle, or an acid or anaerobic locality. Many environmental gradients are not detected simply because suitable microecological techniques have not been developed or are not widely employed. Because of the spatial heterogeneity, species diversity may be immense. Likewise, seasonal effects, variations in host behavior, or additional causes may be responsible for a fluctuation in the intensity of an ecological variable in time so that a rare organism is permitted to develop extensively, an organism whose population density may slowly decline after the disappearance of the particular conditions allowing for its proliferation.

COMPOSITION OF THE COMMUNITY

A *dominant* species is one exhibiting a particularly large population size or an abundance of filaments. Often the community contains not one dominant but rather two or more *codominants*. A *characteristic* species is especially abundant in a particular type of community, and it is either found only in that kind of community or, if located elsewhere, is present in small numbers. Although a list of species may be almost identical for slightly different sites, the quantitative proportions of the various populations may be markedly dissimilar, with the dominant in one place being of little consequence elsewhere.

Many communities have no firm boundaries, the edges of one overlapping a second to an extent that distinct lines cannot be drawn. Various organisms widely considered to be terrestrial probably multiply in rivers or in the sea, a coastal alga may be found and develop in oceanic communities, and a bacterium or fungus derived from the skin or plant surface may be detected in a deep wound.

Extensive lists showing the species composition of diverse communities have been prepared. In some instances the compilations present only representatives of one broad taxonomic category—bacteria, fungi, algae, protozoa, or actinomycetes. Often the investigator is concerned not even with a subphylum or class but with the representatives of a single family or genus, and many descriptions are available of the strains, varieties, or specialized forms of an individual species occurring in a particular ecosystem. Lists are available of microorganisms indigenous to man (Rosebury, 1962), the human skin (Marples, 1965), soil (Alexander, 1961), sewage beds (Hawkes, 1965), the atmosphere, oceans, estuaries, rivers, ponds, snow, rocks, rumen, plant surfaces, and foods, and predictions can be made of the types of organisms present in a site not yet studied but similar to an investigated habitat. Yet the species composition of many environments has been inadequately described, and evidence for entirely new and unusual microorganisms is still being obtained (Nikitin and Kuznetzov, 1967; Staley, 1968).

Similar assemblages of organisms are observed in related habitats because the regions have common physical and chemical properties and because certain populations modify local conditions to favor the occurrence of others. The fact that two organisms have similar distributions may reflect their capacity to grow or survive under comparable circumstances, or it may suggest that one is in some way dependent on the other. Species residing in close proximity or regularly occurring together make up an *interspecific association*. Interdependency and interactions are to be expected, since so many diverse groups are found in a small area, and many sorts of interactions have been described. The interactions may be harmful or beneficial to one or both of the interactants, and it is not rare for an organism to rely for its very existence in nature on its associate.

HABITAT AND NICHE

Of particular interest to the microbial ecologist is the *habitat* of an organism. The habitat is an area having a degree of uniformity in those characteristics deemed to be of ecological significance. Among the sorts of habitats occupied by microorganisms are surfaces of higher plants and animals, crevices of rocks, bodies of fresh water, the open sea, the soil of a fertile field or of the inhospitable desert sand, the bloodstream, the nasal cavity, or decaying plant remains. Few regions on the earth's surface are not populated by a variety of microbial species, so that the number of habitats is truly immense.

The size of the habitat varies considerably. The surface water of a large, clear lake may be considered to represent a single habitat, and it can be argued that large expanses of the open sea having a reasonably homogeneous composition constitute one habitat. On the other hand, the decaying portion of a live tree or a single wound in the tissue of an animal may also be taken as a distinct habitat. The word "habitat" denotes a certain set of conditions favorable or unfavorable to life and hence is more specific than the general term "environment." *Microhabitats* are of considerable interest to the microbiologist either because the functional area of the population is small or because the techniques at his disposal are uniquely suited for the examination of minute regions. Regardless of size, however, the habitat is the site at which the population or community grows, reproduces, or merely survives.

The composition of countless habitats is relatively constant. For example, the body fluids of warm-blooded animals, the bovine rumen, and ocean and lake bottom sediments provide reasonably constant conditions to microorganisms. Other habitats are modified at regular intervals, as by the diurnal changes in the intensity of light impinging on lakes or the slow seasonal drifts in temperature that affect organisms situated near the surface layers of soils or bodies of water. The makeup of still other environments changes dramatically in a short period of time, as when food enters an animal's mouth or when organic materials become suddenly available.

The ability of microorganisms to make use of the resources of their habitat is quite varied, but each must have an ecological *raison d'être*. It must participate in one way or another in the processes taking place at some time and place in that locality. The role of the species in its habitat is referred to as its *niche*. The niche is not a connotation of the physical position of an organism but rather is the designation of its unique function in the community. The activities performed by the species—its niche—are dictated by the biochemical, nutritional, and often morphological properties of that particular population. The activities of scores of heterotrophs and probably of all obligate chemoautotrophs are highly specialized; they occupy a narrow niche. Other species carry out many processes in the environments where they occur; these occupy a broad niche. Organisms having a narrow niche—the specialists—often develop explosively when their special growth requirements are satisfied, but they likewise may be acutely affected when their particular needs are no longer met. Species that have a wider niche are frequently not as subject to marked changes in population size, although many of these generalists never appear in overly large numbers.

Species with the same function in different habitats occupy identical ecological niches. Two organisms may be classified in different genera or classes and have no morphological similarities; yet they can function in their respective habitats in comparable ways. On the other hand, one species may possess quite different niches in dissimilar habitats because its activities are governed by local conditions.

Habitat refers to the location of an organism, essentially its address. By contrast, niche implies the function of the organism in a particular habitat, its occupation. It is commonly observed that a species introduced into an established community does not persist. Presumably it dies out because it finds no role for itself: there are no unoccupied niches and no occupations the alien can fill, and the microorganism has no way of maintaining itself.

THE FOOD CHAIN

For a microorganism to grow in an environment, it must be provided with a suitable source of energy, of carbon, and of additional nutrient elements and growth factors in utilizable forms. The energy source for algae and photosynthetic bacteria may be light, for the chemoautotrophic bacteria it may be H_2 or a reduced state of sulfur, nitrogen, or iron, while for heterotrophs energy is obtained by the oxidation of organic matter. These energy and nutrient sources may enter the habitat from the outside—as sunlight, as food consumed by the animal host, or as plant exudates—or they may be the

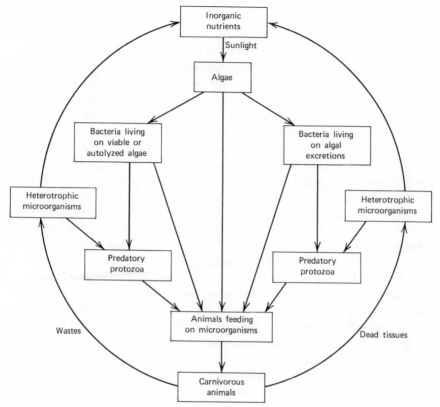

Figure 1.3   Some simplified food chains in water.

tissues of higher plants and animals. Organisms using the original energy sources of the ecosystem represent the *primary feeders;* they are at the first *trophic level.* Other populations, *secondary feeders,* obtain energy by utilizing products excreted by the primary species or live upon their cells by parasitism, predation, or digestion of constituents of the dead cells of the primary organisms; these species are at the second trophic level. The secondary feeders thus proliferate by oxidizing organic or inorganic compounds generated microbiologically within the ecosystem (Alexander, 1964).

Aquatic algae are typically primary organisms in that their energy is obtained from sunlight. These photosynthetic organisms form the base of a *food chain,* supplying the food that sustains the more nutritionally fastidious aquatic heterotrophs (Fig. 1.3). Similarly, cellulose-decomposing anaerobes of the rumen, bacteria metabolizing the simple organic compounds excreted by plant roots, and lignin-destroying fungi participating in the initial stages of wood decay serve as starting points for a concatenation of organisms constituting a food chain. In such habitats the secondary feeders include those hetero-trophs utilizing algal excretions or cells, rumen inhabitants making use of the simple materials liberated during cellulose decomposition, and protozoa preying on the bacteria decomposing root exudates. A well-known example of the relationship between primary and second-ary feeders is that involving a cellulose-decomposing anaerobe and specialized methane-forming bacteria. In the absence of O_2 copious quantities of methane are formed during the decomposition of cellu-losic materials; the primary organisms, cellulolytic species of *Clostridium*, degrade the polysaccharide to organic acids, alcohols, and other products, but are unable to generate methane. Bacteria like *Methanobacterium* cannot use cellulose as an energy source, but they grow well by oxidizing the simpler substances formed by the primary feeders, and it is these secondary feeders that make the methane.

The food chain is initiated by species relying for their growth on the energy sources entering the ecosystem. In aquatic environments these primary forms may be photosynthetic microorganisms—fre-quently algae, sometimes bacteria. In terrestrial environments the autotrophs generating the organic carbon from CO_2 commonly are not microorganisms but rather are green plants, which by their root exudates, leaf litter, or other tissues sustain heterotrophs. The photo-synthetic organisms—higher plants, algae, or photosynthetic bacteria —are *producers*, forming the organic molecules necessary for their more fastidious heterotrophic neighbors, the *consumers*. The con-sumers of the excretions, cells, and tissues of the producers are in turn attacked by other microorganisms, which in their turn become food for a third type of organism and thereby give rise to a food chain composed of a multitude of complex links. The chain is unquestion-ably not a simple linear interlinkage between species from bottom to top, but rather involves an intricate web of nutritional interrelation-ships. Even the introduction of a simple energy source into a hetero-geneous community almost invariably leads to the proliferation of a host of species with a wide range of feeding habits and physiological characteristics. Ultimately the energy that has entered the ecosystem and is utilized by the indigenous community will be dissipated, so that the system can only be maintained if there is a steady input of energy in the form of sunlight or biologically oxidizable substrates.

APPLICATION OF IN VITRO INFORMATION
TO CONDITIONS IN VIVO

The science of microbiology has to a large extent blossomed because of careful investigations of isolates obtained from nature and nurtured in pure culture, in which microbial behavior could be studied in carefully controlled, albeit highly artificial, conditions. The nutrition, genetics, and metabolism of at least selected species are, as a result, now reasonably well understood. But almost all the information has come from axenic cultures maintained in artificial media, the term *axenic* signifying the absence of strangers in the culture vessels. Research on the fundamental metabolic processes of living things and on the chemistry of organisms has necessarily been carried out in axenic cultures kept in carefully controlled environments. In nature, however, strangers are nearly always present, and the environmental conditions are frequently highly variable.

To what extent can in vitro nutritional, genetic, and biochemical information be applied in vivo, in natural ecosystems where microorganisms live, proliferate, and die? Can the voluminous literature provide a basis for understanding the behavior, distribution, activity, and interrelationships of microorganisms in environments not limited by the confining glass walls of a culture vessel? It is easy to determine the temperature, pH, or osmotic limits of a selected species in axenic culture in the laboratory. Such data are abundant. Nevertheless results of this type are often of little ecological value because the data suggest that the organism will be found in a wide variety of environments, where it obviously does not exist. The chief value of these studies is that they tell where an organism is not likely to occur. The species unquestionably does not grow when its tolerance for some factor is exceeded. Similarly, biochemical roles in nature have been ascribed to individual bacterial or fungal groups as a consequence of painstaking laboratory studies in artificial media, but equally careful investigations of the natural habitat of those organisms have shown that these biochemical transformations do not indeed take place.

Clearly the data obtained in the initial inquiry are insufficient. More information is required than simple tolerance limits or assessments of the biochemical potential of an organism. Other biological characteristics determine which of the environments suggested by the first inquiry are truly suitable and which of the biochemical traits are expressed in natural surroundings. A condition apparently ideal for an organism in axenic culture might not be optimum and may be highly deleterious in a heterogeneous microbial community, and the metabolic process so prominent in a culture pampered and nourished with choice nutrients designed to maximize growth may never be expressed in the reality of a harsh environment. The disciplines of ecology and physiology are intimately interrelated, but it is ludicrous to attempt to understand complex ecological interrelationships by applying no more than the most primitive of physiological principles.

Characterization of the relationships operating among microorganisms or between microorganisms and their habitat is made difficult by the fact that the individual species frequently does not act independently. It is affected by and, in turn, affects its neighbors in nature. It is part of a complex web of interrelationships and interactions that is hard to disentangle. Yet many of the problems and pitfalls inherent in a study of the ecology of microorganisms, at least from a physiological and biochemical viewpoint, are also inherent in investigations attempting to understand the physiology of an animal or plant by examining only single enzymes, polynucleotides, or polysaccharides. No one would dispute the fact that biology has come a long way in understanding the functioning of higher organisms by examination of distinct and discrete processes and components. On the one hand are the enzymes, cells, and tissues making up the organism; on the other hand are individuals, populations, and communities that constitute an ecosystem. In both instances the removal of a component and its study in isolation make the work more simple and the results less equivocal; yet in both instances the discrete pieces of information must be joined together judiciously to facilitate an understanding of the whole.

Granting that the interactions among microorganisms are complex, it should be possible nevertheless to establish, under controlled or recorded conditions, ecological models that represent or mimic nature. The models can be designed so that they approximate the normal surroundings of the organisms and their common interactions with neighbors. In this way extrapolations from the prototype to nature become increasingly less difficult or tenuous. By the use of such models it should be not excessively difficult to test ecological hypotheses, generate meaningful questions that are answerable in terms of still more realistic models, and ultimately establish theories that can be evaluated in vivo.

References

REVIEWS

Alexander, M. 1961. *Introduction to Soil Microbiology.* Wiley, New York.

Alexander, M. 1964. Biochemical ecology of soil microorganisms. *Ann. Rev. Microbiol.,* **18**, 217–252.

Garrett, S. D. 1956. *Biology of Root-Infecting Fungi.* Cambridge University Press, London.

Hawkes, H. A. 1965. The ecology of sewage bacteria beds. *In* G. T. Goodman, R. W. Edwards, and J. M. Lambert, Eds., *Ecology and the Industrial Society.* Wiley, New York. pp. 119–148.

MacLeod, R. A. 1965. The question of the existence of specific marine bacteria. *Bacteriol. Rev.,* **29**, 9–23.

Marples, M. J. 1965. *The Ecology of the Human Skin.* Thomas, Springfield, Ill.

Rosebury, T. 1962. *Microorganisms Indigenous to Man.* McGraw-Hill, New York.

Winogradsky, S. 1949. *Microbiologie du Sol.* Masson, Paris.

OTHER LITERATURE CITED

Elton, C. S. 1946. *J. Animal Ecol.,* **15**, 54–68.

Hulburt, E. M. 1963. *J. Marine Res.,* **21**, 81–93.

Nikitin, D. I., and Kuznetzov, S. I. 1967. *Mikrobiologiya,* **36**, 938–941.

Paine, R. T. 1966. *Am. Naturalist,* **100**, 65–75.

Patrick, R. 1963. *Ann. N.Y. Acad. Sci.,* **108**, 359–365.

Staley, J. T. 1968. *J. Bacteriol.,* **95**, 1921–1942.

Stankovic, S. 1960. *Cited in* G. E. Hutchinson, 1967. *A Treatise on Limnology,* Vol. 2. Wiley, New York.

2

Dispersal

The surface of the earth abounds in environments potentially colonizable and resources potentially exploitable by microorganisms. Nutrients become available as a result of vertical or horizontal movements of water, the introduction of organic pollutants into streams or soil, and the erection of man-made dams or ponds. Erosion exposes new surfaces for invasion, and the materials washed away by erosion often are deposited in a manner permitting microbial colonization. Seedlings emerging from seeds internally free of microorganisms, roots extending and exposing new surfaces, fruits damaged and thereby presenting a nutrient-rich substratum, and food and many manufactured products all offer hospitable conditions for microscopic life. The newborn infant or animal, too, is uninhabited. The healthy adult animal or plant also is potentially colonizable when its disease resistance declines, its tissues are injured, or the higher organism is exposed to a microbial pathogen with which it has had no prior contact.

The concept that microorganisms are everywhere is enormously valuable and serves many pragmatic functions, but it also represents a truism that glosses over problems of immense ecological, economic, or public health significance. No one will dispute that invasion and the successful establishment of one or another species is inevitable whenever a region containing adequate nutrients is exposed or created, provided that conditions are not overly harsh, but often the relevant question is not whether colonization will occur but rather what colonists will appear. Certain sequences of ecological events, the deterioration of an edible or manufactured product, and the onset of disease are usually not determined by the mere arrival and subsequent growth of any invader, but rather by a distinctive population with given physiological and biochemical attributes. In this light the processes of dissemination and the ability of individual species to spread by unique dispersal mechanisms assume particular significance.

Dissemination may occur by virtue of active movement or growth, or it may result solely from the passive transport of the microorganism by means not under its own control. Active or passive transport may require the presence of some structural feature or morphological property, but often none is evident. Moreover, an individual species may be disseminated in only one way, or its movement from place to place may entail several different mechanisms of dissemination.

Dispersal is essential for the continued existence of many species. This is true for free-living organisms residing in environments in which conditions regularly become adverse or where the nutrient supply is exhausted. It is particularly true for the obligate parasite or for a parasite that, though capable of independent existence in vitro, is restricted in nature to life in association with a suitable host. In such instances individuals of the species must escape from the locally detrimental or depleted environment and find a new habitat conducive to the continued existence of the species. For these organisms dispersal represents an escape in space, from an inhospitable region to one where multiplication is possible. The lack of a means of dissemination could mean the elimination of the species. On the other hand, certain microorganisms are not able to escape in space from unfavorable locales, but they have a mechanism for an escape in time: a structure or stage in the life cycle allowing the species to endure adversity.

For the species relying on some means of dissemination in order to survive, the term *imperative dispersal* is applied (Hirst, 1965). Inasmuch as new habitats must be found constantly, such dissemination must be repeated over and over again in time. Many free-living microorganisms, by contrast, do not have to be moved or transported to new surroundings, and the kind of dissemination involved here is termed *capricious dispersal*. Species subject to capricious dispersal are spread from place to place, but they are able to survive even in the absence of a means of dissemination.

Various terms have been proposed for the unit of dispersal, but *propagule* seems to be the one most applicable to microorganisms and has gained general acceptance. For some free-living organisms dissemination of the propagules proceeds in such a manner that the population gradually encompasses an ever widening but largely continuous region occupying a small area. Many organisms have a wide and discontinuous distribution, however, possibly as a result of sporadic dispersal occasionally coupled with locally unfavorable sites for colonization. Still others are readily transported and eminently successful in establishment; they are widely distributed and may be considered truly cosmopolitan.

CENTER AND DURATION OF DISPERSAL

Migration is frequently very hazardous, and many propagules do not survive the journey. The dispersal of some parasites is extremely tenuous, necessitating, for example, one or more hosts, movement from one tissue to another within a host, a morphogenetic sequence on the part of the parasite, and occasionally even a free-living stage. The species may lose immense numbers of vegetative cells, spores, or cysts in the process of locating a new habitat. The potential benefits of the migration, however, warrant the risk and the loss of propagules. By migrating, the species can find conditions where it once again has a selective advantage over nonmigrants—a new host individual, unexploited organic nutrients, or an incompletely colonized surface or body of water.

Microorganisms subject to either entirely local or widespread dissemination show *centers of dispersal*, regions from which the species are spreading or have spread. At these loci conditions generally are proper for extensive development, the population density becomes high, and the site serves as a point from which the generated propagules can be dispersed. This center serves essentially as a *reservoir* for the species, and for pathogenic microorganisms it is designated the *reservoir of infection*.

A species residing in one of the many habitats that are discontinuous in both space and time must have some means to bridge the spatial and temporal discontinuities. Discontinuities arise, for example, because of a depletion of nutrients essential for a free-living organism or from the death of the host for a parasite. To overcome problems of spatial or temporal discontinuities, many species have dispersal mechanisms or phases permitting their transport to a new, inanimate site or uninfected host where growth is again possible. Some are endowed with a resistant stage to tide them over a lean period, while still others are able to replicate in two different kinds of environments. Organisms in the latter category make the best of two worlds, protecting themselves from extinction by proliferating in dissimilar circumstances; in this instance the reservoir is in a sense merely a means of bridging the gap between discontinuous habitats.

Table 2.1

Reservoirs of Pathogenic Microorganisms

Microorganism	Reservoir
Bacteria	
Bacillus anthracis	Swine, cattle, sheep
Brucella spp.	Cattle, goats, swine
Erwinia amylovora	Infected plant tissues
Neisseria meningitidis	Man
Pasteurella pestis	Rodents
Pseudomonas solanacearum	Soil
Salmonella spp.	Man
Xanthomonas campestris	Crop residues
Rickettsiae	
Rickettsia prowazekii	Rats, man
Rickettsia rickettsii	Rodents
Protozoa	
Plasmodium spp.	Man
Fungi	
Histoplasma capsulatum	Soil
Puccinia graminis	Wheat
Pythium spp.	Soil

Noncosmopolitan organisms have distinct centers of dispersal. These centers may occupy areas of no more than microscopic size or they may be quite large. Some are continuous in space and time, but many are discontinuous. Because of their importance in epidemiology and disease control, the reservoirs of pathogens have received considerable attention. Many such centers of dispersal are known, a single type often harboring from a few to numerous pathogens. One species, moreover, may exist in a number of reservoirs, from which it spreads at appropriate times. Typical reservoirs are presented in Table 2.1. *E. amylovora, P. solanacearum, X. campestris, P. graminis* and *Pythium* are plant pathogens, while the remaining organisms listed cause animal or human disease. Viable hosts constitute the centers of dispersal for a range of pathogens, but soil and other nonviable habitats may play an identical role. For the obligate parasite the reservoir is an individual of the host species, within or upon which the microorganism multiplies and produces the propagules that subsequently colonize previously uninfected individuals.

Various centers of dispersal are difficult to detect. This is particularly true where a *carrier* is involved. Many potential incitants of animal or human disease survive in susceptible individuals, or in those who have recovered from the disease, without producing discernible ill effects. For example, humans who have recovered from typhoid frequently carry *Salmonella typhosa* and can be responsible for the typhoid bacilli found in food and water. A healthy person also may harbor *Neisseria meningitidis* in his nasopharynx, a center from which the bacterium may be transmitted to unexposed hosts. These *cryptic reservoirs* are of especial importance in disease transmission and control inasmuch as they constitute totally unrecognized sources of infectious agents.

Many and varied are the reservoirs of plant pathogens. Some, like species of *Rhizoctonia* and *Pythium*, grow with no difficulty in soil, but they are also capable of attacking tissues of higher plants. Others, like *Phymatotrichum*, *Ophiobolus*, and *Armillaria* species, also possess a subterranean reservoir but fare poorly when exposed to the indigenous soil microflora and must maintain themselves instead on the tissue remains of the host they previously infected (Muskett, 1960).

The *duration of escape* from a reservoir is often critical in determining success in dissemination. Among microorganisms having complex life cycles, frequently only one stage is suited for dispersal, and the duration of escape is limited to the period when this stage is in existence. Certain species reside in habitats subject to diurnal, seasonal, or irregular fluctuations, and the duration of their escape period often is directly related to the fluctuation in environmental conditions. Species inhabiting internal or external portions of animals or plants quite frequently have periods of escape entirely dependent on the behavior of the host, and a number of ecologically fascinating associations between microbial dissemination and host physiology have been described. Among the agents of disease the duration may be either quite short or extend for the life of the host; thus a human suffering from influenza is infectious for but a few days and diphtheria may be transmitted for several weeks, whereas leprosy may be passed on for the person's lifetime (Anderson, 1965).

Seasonal or diurnal migration is a widespread phenomenon. An excellent illustration is the downward migration of two species of *Ceratium* during the day and the movement of these marine dinoflagellates to the surface at night; by contrast, the dinoflagellate *Gonyaulax polyedra* rises during the day and migrates downward through the water at night (Hasle, 1950). Populations of airborne spores of certain fungi also exhibit a marked diurnal periodicity

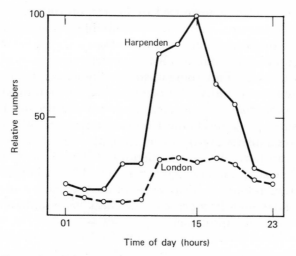

Figure 2.1   Diurnal periodicity in the abundance of *Cladosporium* in the air over London and Harpenden, England (Hamilton, 1959).

(Fig. 2.1). The discharges of fungal spores over agricultural land display a variety of dissemination rhythms; the peak of spore discharge for some populations is in the early morning, for others it is the forenoon, while other fungi favor the afternoon for the release of their spores into the atmosphere.

Infectious diseases caused by fungi, bacteria, and viruses have seasonal patterns of incidence and severity, and the patterns of dispersal of microbial agents of plant and animal disease thus assume particular importance. *Puccinia* rust spores, which survive neither the hot, dry summers of the southern portion nor the cold winters of the northern regions of the wheat-growing areas of North America, migrate north from the infested autumn-planted wheat of Texas and Mexico and shower down upon the wheat sown in the spring in Canada and northern United States, whereas the spores migrate southward in the autumn and rain down upon the winter-planted wheat of the south (Gregory, 1961). The spread of *Coccidioides immitis* is favored in months when there is little rainfall and abundant dust in the arid regions where this dust-borne human pathogen is prevalent, and the incidence of *C. immitis* infection among newcomers to the region is highest at this time of year (Conant, 1965). The basis for the seasonal patterns of dissemination of the causative agents of diphtheria, pertussis, scarlet fever, and measles remains somewhat obscure.

With microorganisms transmitted by insects, on the other hand, the seasonal variations are commonly attributable to the climatological influences on the insect, as is evident in the cyclic appearance of yellow fever, equine encephalomyelitis, and dengue.

The transmission of microorganisms by higher organisms is a common occurrence. An individual of one species that transmits propagules of another from place to place is known as a *vector*. The vector may serve simply as a mechanical transmitter of the microorganism, or it may have a distinct physiological or pathological association with the cells it transports. When the identity of the vector is known, the means of dispersal is largely established, but this kind of information may have tremendous practical value in disease control inasmuch as the vector is frequently more easy to check or eliminate than the pathogenic microorganism itself.

EFFICIENCY OF DISPERSAL

For many organisms dispersal is a chance phenomenon. The propagules are distributed widely and randomly, and undoubtedly they alight on inhospitable areas or hosts far more frequently than on localities favorable for the outgrowth or proliferation of the structure or cell that is disseminated. The risk and randomness of migration may be lessened in any one of a number of ways. Transmission by a vector usually increases greatly the likelihood of individuals of a population attaining a new, favorable habitat. Motile cells commonly are able to find hospitable and reject inhospitable environments through directed motions. Growth responses and tropisms of filamentous algae and fungi also contribute to lessening the hazards of transport. In species still exposed to a great risk, however, other mechanisms exist to enable at least a portion of the disseminating propagules to reach and establish themselves in a potentially colonizable habitat. These include the formation of resistant propagules and the release of spores in numbers so vast that the chances of a few locating a suitable substratum are notably increased.

The greater the efficiency of the dispersal mechanism, the smaller is the number of propagules necessary for successful dissemination. Species possessing no efficient way for transmitting their daughter cells have survived because they produce daughters in enormous numbers or because their propagules are repeatedly introduced in large numbers into an area from which they can make contact with a new habitat or new host. This sort of compensation for inefficiency is particularly evident in components of the airborne flora. Thus bacteria causing respiratory infections are expelled from the body in vast quantities during sneezing and coughing, and in a similar fashion an aerially transported plant-pathogenic fungus like *Puccinia graminis* releases millions or uredospores per square meter in a field of diseased wheat. Fungi dispersed through the air, whether they are free-living or parasitic, typically liberate multitudes of spores, the release lasting for either a short period or as long as 6 months. As the probability of finding a colonizable locale is improved by the existence of a vector or efficient vehicle of transfer, no need exists for the widespread scattering of hordes of propagules, so that the density of propagules responsible for the transport of bacteria and fungi transmitted by insect vectors and fungi carried on the surfaces of seeds is, not surprisingly, usually small.

Protozoan flagellates of termites exhibit a unique as well as efficient means of transmission. The parent termites feed their young with partially digested cellulosic foods populated with the flagellates, nutrients derived from the protozoan-rich alimentary canal of the adult. All the young become colonized shortly after their birth, therefore, with a microbiota essential in the digestion of the cellulosic materials on which the termites invariably feed (Hegner, 1938).

Many species reduce the risk of dispersal not by reason of their release in copious numbers or by virtue of their effectiveness in attaining a new locality, but rather because at one stage in their life cycle most or all individuals in the population are resistant, to some extent, to deleterious conditions in the environment in which the cells are deposited or through which they pass. Dissemination is typically associated with significant losses in viability of a population, and frequently the loss of viability as governed by local circumstances is probably more important in determining the sites a species will colonize than is the kind or efficiency of the migration and transport mechanisms. Aerial transport is extremely deleterious to continued survival, in that the cells are exposed to hazards of radiation and desiccation as well as extremes of temperature. The airborne biota must be resistant to drying and ultraviolet radiation, but even prolonged exposure to visible light can be lethal to cells suspended in the atmosphere. Fungal and bacterial components of the air biota are indeed able to overcome or tolerate these hazards; thus *Alternaria*, a fungus abundant in the air microflora, is notably resistant and maintains its viability, while thin-walled basidiospores, probably many other kinds of fungal spores, and the vegetative cells of bacteria rapidly die out (Kramer and Pady, 1968).

The transmission of many microorganisms is initiated as a result of some physical contact between a source of propagules and the potential habitat. Numerous soil-borne pathogens, for example, are unable to develop in soil but persist as resistant bodies until the cells make contact with the host or, for plant parasites, root exudates. The endospore of *Bacillus anthracis* or of *Clostridium* spp. causing gas gangrene or tetanus and spores, resting hyphae, and sclerotia of fungi are well known for their capacity to endure prolonged drought, nutrient depletions, or potential parasites, remaining inactive until a suitable habitat in the form of host tissue becomes once again available. The vegetative or active stages of the same species could not tolerate these extremes.

Cysts function in the dispersal of protozoa from one habitat or host to another. Whereas the vegetative stage of these protozoa dies quickly when the cells are separated from their native surroundings, the cysts maintain their viability, often for several years, in soil, sewage, water, and food, all habitats where one or another of the protozoa fail to replicate. A few of the aquatic algae that are transported from one body of water to a second in the digestive tract of birds possess a stage resistant to the harshness of the transporting fluid; *Chara zeylanica* is an ideal illustration, for its oospores pass through the avian digestive tract, and by the time they emerge the bird has migrated some distance from the original milieu of the alga (Proctor et al., 1967). On the other hand, the vegetative cells of several species of algae and bacteria—such as *Mycobacterium tuberculosis*—and viruses like that causing foot-and-mouth disease are resistant to desiccation and may be disseminated while in a dry condition.

The duration of the dispersal phase is of critical importance in determining whether the propagule will still be viable when it arrives. Species forming durable structures are not as subject to the time factor in dispersal as are those possessing only fragile propagules. Microbial inhabitants of the animal body that possess no resistant stage are notable in their fragility and inability to tolerate exposure to alien surroundings; populations of the syphilis spirochete *Treponema pallidum* and several rumen anaerobes are completely inactivated when apart from the hospitality of the animal ecosystem for more than a few minutes. Although the duration of life apart from the indigenous habitat might be brief for a fragile species, the viability of a few of the propagules nevertheless may be sufficiently long to allow for a successful migration; for example, although the bacteria causing typhoid, paratyphoid, cholera, and dysentery are not aquatic, if the period of their transmission through water is short before a suitable new environment is found the hazard of dissemination is tolerable. The same is true of many free-living and parasitic organisms disseminated through water or the atmosphere, on seed, planting stock, and foods, or on the surfaces of animals and inanimate objects. Nevertheless, for many species, even this short sojourn is too long, and they do not survive the trip.

Successful dispersal is never an end in itself: migration must be consummated with the establishment of the propagule. In order that establishment be effected, the locality that is reached must be receptive, and the organism must be capable of initiating growth therein. Air and water in tropical or temperate climatic areas contain a diversity of species neither able to grow in these fluids nor likely to find a suitable substratum for colonization. Even the Arctic or Antarctic air is inhabited by microorganisms, but these do not represent species likely to grow in the cold regions. Hence, granted that dissemination is a prelude to establishment, success in dispersal coupled with failure in colonization still represents no more than an ecological *cul-de-sac*.

The occupation of a new habitat does not necessarily entail an organism's moving or being transported to the new environment; the potential habitat may come to the microorganism. The locomotion of animals and their contact with one another and with inanimate objects may result in the transmission of microbial species indigenous to the animal. As they migrate through soil, plant roots come upon cells and filaments of fungi, bacteria, and actinomycetes suited to proliferation at the expense of root excretions. The release of organic pollutants into water, the fall of leaves to the ground, and the mixing of crop remains with soil during plowing all bring substrates or favorable conditions to the population instead of the species itself migrating in search of a habitat.

Dispersal mechanisms are selective for the microorganisms that can be transported or involved, and any one mechanism may be utilized by a few or a reasonably large number of species. The microbiotas of air, water, soil, insects, fecal matter, seeds, or droplets commonly show a sizable collection of genera, but the various communities of these ecosystems account for only a few of the dominant groups in nature. Conversely, every species is characterized by one or a select few mechanisms of dissemination, but rarely is the dispersal of a particular species effected by a variety of different means.

ACTIVE DISPERSAL

Transport from locality to locality results either from a physiological process under control of the organism itself or from factors not directly associated with the activities of the cell or structure that is translocated. The former is termed *active* and the latter *passive dispersal*. The linear extension of a growing filament and the swimming or gliding about of a nonfilamentous individual are two of the more usual means of active dispersal.

Dispersal by growth is evident among the soil and plant-pathogenic fungi, filamentous aquatic algae, and actinomycetes proliferating on decaying organic materials. Invasion of a nutrient-rich site from a depleted environment or escape from harsh surroundings is possible by mere linear extension of the filament. Development of an algal filament into uncolonized water, growth of fungal hyphae from the root to the aerial portion of a plant, or the spread of mycelium along the surfaces of food products allows for the occupation of incompletely exploited regions. The extent of spread by this kind of dispersal is frequently not great, but because immediately adjacent sites may contain vastly different quantities of nutrients and exhibit markedly diverse physical and chemical properties, the escape may be a real one. Spread over such apparently minute distances is typical of soil fungi and actinomycetes in general. The area covered by occasional soil fungi can be astonishingly large, however; *Armillaria mellea* has been reported to develop outward from a food base, in the form of colonized wood, to a distance of about 20 meters (Muskett, 1960). Similarly, the filaments of aquatic algae have been noted to ramify considerably and extend over an appreciable area, and the hyphae of root-invading fungi may be found at considerable heights within the plants they infect.

Microbial locomotion takes place either by swimming or by a type of gliding by cells in contact with a solid surface. Swimming is known to occur among the protozoa, algae, bacteria, and certain motile fungal forms. Protozoan locomotion may result from the presence of flagella in Mastigophora, cilia in Ciliata, pseudopodia in the Sarcodina, or polar filaments that serve to propel the infective stage of Cnidosporidia into new hosts or uncolonized tissues. Flagella are found also attached to the cells of motile bacteria, and a flagellated stage, the zoospore, is characteristic of the life cycles of certain fungi. In organisms undergoing a series of morphogenetic alterations or having a complex life cycle, only one stage may be motile; this stage, however, is frequently the one implicated in dispersal, as in the flagellated daughter cells of the stalk-forming bacterium *Caulobacter*, which swim to fresh sites to set the stage for the formation of new stalks (Poindexter, 1946), the flagellated fungal zoospores, and those motile protozoan forms important in the initiation of infection (Hickman and Ho, 1966; Garnham, 1966). The speed of movement is at times quite impressive and probably has selective value; thus ciliates may move as fast as 2.6 mm or as much as 20 times their own length in 1 second (Garnham, 1966). A slow, gliding motion, not apparently dependent on specialized external organelles of locomotion, typifies selected blue-green algae, fruiting myxobacteria, *Beggiatoa* and related bacteria, diatoms, species of Euglenophyta, and the aplanospores of red algae, to mention a few (Jahn and Bovee, 1965).

Typical of the behavior of motile organisms is their ability to move in response to external stimuli. These *taxes* are considered to be positive if the motion is toward, and negative if away from, the source of the stimulus. Phototaxis, chemotaxis, thigmotaxis, thermotaxis, aerotaxis, and geotaxis have been noted in one or another microbial group, the terms referring respectively to a movement in response to light, the presence of soluble compounds, a mechanical stimulus, a temperature gradient, particular gases, or gravitational force. The cell's behavior may also be conditioned by the intensity of the stimulus, being negative at one and positive at another intensity.

Both photosynthetic and nonphotosynthetic populations exhibit phototaxis. The photosynthetic species include (a) photosynthetic purple bacteria like *Rhodospirillum* and flagellates such as *Euglena*, in which motility is associated with the presence of flagella, and (b) diatoms and filamentous blue-green algae, as exemplified by *Oscillatoria*, that move by gliding. The heterotrophs exhibiting phototactic responses include individual species of amebae, ciliates, and flagellated protozoa. By means of positive or negative phototaxis, the cells and filaments orient themselves so that they are either in or out of the sunlight, and in some instances the organisms place themselves in regions with specific light intensities. As a consequence of this sort of activity, the vertical distribution of populations in aquatic habitats may be determined. Figure 2.2 depicts how phototaxis governs the localization of populations of two dinoflagellates within the lighted, upper zone of bodies of water. Migrating algae vary considerably in their rate of locomotion, some moving slowly, others as fast as 5–10 meters in 12 hours (Hasle, 1950). The capacity to migrate vertically is not restricted to algae.

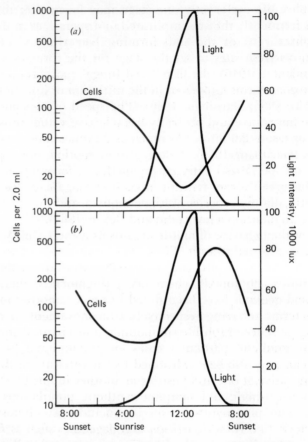

Figure 2.2 Vertical migration of two dinoflagellates in water. (*A*) *Ceratium fusus;* (*B*) *Gonyaulax polyedra* (Hasle, 1950). The changes in cell number in the surface water of the Oslo fjord, as shown here, reflect the vertical movement of the organisms.

Chemotaxis has been explored most extensively among the protozoa, in Phycomycetes possessing zoospores, and, to a lesser extent, in a small number of flagellated bacteria. Among the substances causing repulsion or attraction are a variety of organic and weak mineral acids, sugars, dyes, alcohols, esters, aldehydes, and ketones. The feeding of predatory protozoa is considered by some investigators to be initiated by chemotaxis, the finding of edible prey possibly being linked with a positive orientation toward a diffusible substance elaborated by the prey individual. Fungus zoospores are attracted to and accumulate on tissues of higher plants, hyphae of various fungi, and algal surfaces, and these motile spores may move up to 40μ/second under chemotactic influences. The zoospores of plant-pathogenic Phycomycetes have been seen moving to the stomata, where fungal germ tubes are formed, and they in turn enter the stomatal pores. Zoospores of occasional fungal genera show positive chemotaxis to wounds but no directed movement toward unwounded plant surfaces, while the zoospores of species of *Phytophthora* and *Pythium* are oriented toward distinct regions of the root, notably immediately behind the tip, in response to compounds exuded by the roots (Hickman and Ho, 1966).

Highly motile organisms may repeatedly come into contact with solid objects, and many protozoa react dramatically to such collisions and to the mechanical stimuli. The thigmotaxis may be negative or positive, the ultimate result being the avoidance or bringing together of the protozoan with the object it has encountered. Thermotactic responses are evident in the behavior of a number of microorganisms, including chlorophyll-containing flagellates, ciliates, amebae, and bacteria. Positive or negative geotaxis has been reported for *Volvox* and *Closterium* among the algae and *Paramecium* and *Metopus* among the protozoa (Noland and Gojdics, 1967). Individual genera exhibit both positive and negative aerotaxis, motile heterotrophs like *Pseudomonas* and photosynthetic bacteria such as *Chromatium* moving toward or away from the air-liquid interface.

The ecological significance of movements and orientations has received little attention, but enough information is available to warrant several suggestions as to their possible importance in nature. Because of the ability to respond to external stimuli, many motile organisms showing tactic responses undoubtedly move toward microlocalities suitable for their growth and reproduction while avoiding unfavorable regions. The extent of such migration is probably never great, but even the short path traversed could have a profound impact on the survival of the motile individual. The successful bridging of a short distance may be all that is necessary for an organism to avoid a harmful microsite, find suitable nutrients, or reach a colonizable locality on a potential host. A selective advantage is immediately apparent, moreover, in the movement toward a light source by the zoospores of fungi parasitic on algae, because such migrations serve to bring the parasite to a region of high host density. The protozoan *Metopus* likewise could not but benefit by its negative geotaxis, a behavioral pattern bringing it to the anaerobic habitats suited for its proliferation (Noland and Gojdics, 1967). Directed movements might also enable a population to reach soluble nutrients in sparse supply, and positive chemo- and thigmotaxis undoubtedly facilitate the contact of predator with prey and minimize random contacts with inedible food sources. Although zoospores of pathogenic species of *Pythium, Phytophthora, Aphanomyces,* and related fungi may be attracted to nonhost plants, instances of a specific host-induced chemotaxis also occur; however, whether the chemotaxis is host-specific or not, a species with motile zoospores must benefit if the spores it produces help to bridge the distance to the tissue surface of susceptible hosts or facilitate the movement toward a specific site from which infection may be initiated.

The capacity for movement and orientation toward chemical stimuli also confers an ecological advantage on the animal-parasitic protozoan possessing such traits. Penetration of host cells and tissues and the ability of the parasite to come upon and fix itself to appropriate sites outside or within the host's body are closely linked with locomotion and response to external stimuli. Animal parasites frequently require dispersion even within the confines of the animal; for example, the malaria parasite to locate red blood cells, *Toxoplasma* to find monocytes, *Leishmania* to the reticuloendothelium, and trypanosomes to muscle cells. Several of these organisms, or characteristic stages in species exhibiting simple or complex life cycles, move by virtue of their own locomotory organelles, but many are dispersed passively as a consequence of movements on the part of the host, alimentary functions of the metazoan, or the flow of the animal's body fluids. Pathogenic bacteria of a variety of genera also are transported through the host, although typically their movement is passive, as when they are carried about in the circulatory system.

PASSIVE DISPERSAL: AIR

Small or large numbers of microorganisms are transported passively in air, by water currents, through soil, on inanimate objects, or by means of biological vectors. Migrations of these kinds are independent of the presence of organelles of locomotion; yet many of the species involved have structural modifications or biochemical attributes specifically enabling them to be disseminated efficiently. The traits and structures linked with effective dispersal sometimes are readily apparent, but at times subtle morphological features or physiological attributes are involved. The medium of dispersal comes into contact with the environment serving as the reservoir of the species, picks

Table 2.2

Distance of Dispersal for Several Species

Microorganism	Distance Traversed	Probable Means of Transport	Reference
Asparagopsis armata	Mediterranean to Irish coast	Ship bottom	Elton (1958)
Biddulphia sinensis	Far East to North Sea	Ship bottom	Elton (1958)
Marine algae	65 km	Wind	Maynard (1968)
Pasteurella pestis	Transoceanic	Shipping (flea vector)	Meyer (1965)
Phytophthora infestans	Up to 64 km	Wind	Schrodter (1960)
Protozoa and algae	Possibly hundreds of kilometers	Dragonfly	Maguire (1963)
Puccinia graminis var. tritici	970 km	Wind	Stakman and Hamilton (1939)
Puccinia horiana	Japan to England	Aircraft	Zadoks (1967)

up a number of its propagules, and carries them either a short or a long distance, depending on the organism and the medium, to a new locality where establishment is possible. Although most species do not migrate far from their center of dispersal, the results of Table 2.2 show that surprisingly great distances are sometimes covered by the algae *Asparagopsis* and *Biddulphia,* the bacterium *Pasteurella,* and fungi like *Phytophthora* and *Puccinia.*

A large number of organisms are subject to dispersal by wind and air. These are forms that enter or are propelled into the air, by one means or another, only to settle out at a distance from their original habitat as they are carried down in rain showers or by the force of gravity. The susceptibility to aerial dissemination varies from group to group, but the proneness of a species to aerial dispersal can be a major contributing factor to its distribution pattern. Among the biological characteristics underlying an organism's success in airborne dissemination are the possession of a means of being picked up by or propelled into the air current and a resistance to desiccation, radiation, and temperature fluctuations.

Unique, frequently curious structural attributes are responsible for propelling the spores of many fungi into the air or merely for exposing them to the wind. These novel traits make the fungi ideally suited for aerial transport, and a high percentage of the air biota is thus made up of their spores. The aerial structures are, in addition, the agents of transmission of a variety of plant diseases and cause allergic reactions in man. Hyphal fragments are also observed above the surface of the ground, though not as frequently as fungal spores. Not as extensively studied but also present in the air, sometimes in large numbers, are bacteria, algae, protozoan cysts, and yeast cells.

A multitude of mechanisms are employed by the fungi to facilitate their aerial migration. In some instances the propagules are borne on specialized structures that extend some distance above the substratum where the propagules are liable to be caught by the wind. Other species residing on plant surfaces have dry, powdery spores, and these are released when the leaf or stem vibrates in a passing stream of air. Fascinating ballistic discharge mechanisms typify species of Ascomycetes, the spores being ejected violently into nearby layers of turbulent air. Current evidence suggests that airborne fungal spores are largely derived from species possessing explosive spore discharge mechanisms and inhabiting not the soil mass but rather surfaces of vegetation or growing out above the soil. This view is based on the significant differences in composition of the fungal flora of air and soil (Gregory, 1961). On the other hand, a few of the human and animal pathogens transmitted through the air unquestionably originate in the soil. The enormous scale of spore production contributes in no small way to the success of individual fungal populations in migrating through the air and in locating a colonizable site; for example, a field of wheat infected with *Puccinia graminis* might yield 25 million uredospores per square meter, and certain puffballs are reputed to liberate as many as 7×10^{12} spores (Gregory, 1961). Large sporophores, moreover, may release millions of spores per minute, a rate of liberation maintained for several months.

Several types of violent discharge are evident among the Ascomycetes and Basidiomycetes. The ascus of Ascomycetes may first swell and then, when its tip ruptures, the ascospores contained therein are propelled to distances ranging from a few to up to 50 cm. The basidiospore of the Basidiomycetes can also be expelled violently away from the source of its formation and then be caught by wind currents. Fruiting bodies of higher Basidiomycetes have been observed to discharge their spores continuously for prolonged periods of time. Depending on the fungus group, the projectile may be made up of single, discrete spores, or all the spores in an ascus may be cemented together into a unit that still is able to be propelled for some distance.

In *Pilobolus, Dasyobolus, Sphaerobolus,* and related fungi inhabiting dung, success in dispersal may be enhanced because the spore gun turns toward the light so that the spores have a good likelihood of finding a hospitable target area (Ingold, 1965).

The composition of the air flora varies according to altitude. Very near the ground, species dominance is regulated by local sources of origin. Spores of distinctive ubiquitous fungi, however, are prominent at a distance above ground level. *Cladosporium* and basidiospores of the Basidiomycetes are almost invariably noted, at least in the regions examined carefully to date, but *Alternaria* and ascospores are frequently abundant. Local disturbances, mowing, or other activities of man may introduce different dominants for short intervals, but immense numbers of species are wholly unsuited for aerial transmission, and their propagules are rarely if ever noted in the air.

Aerial spores or vegetative cells are commonly deposited near their centers of dispersal. Because the propagules are heavier than air, any ejected into the atmosphere fall to the ground or onto low-lying foliage in the absence of turbulence. It is the unusual rather than the common species that arrives and is capable of growth at a distant site. The distance of 970 km traversed by *P. graminis* (Table 2.2) is notable by the rarity rather than the frequency of such occurrences; yet the potential impact in plant pathology of wind-borne fungal propagules that traverse long distances makes this dispersal, though uncommon, of particular concern. Physical barriers and the fallout of spores in rain or under the influence of gravity contribute to the inability of many propagules to accomplish an extensive journey successfully, as does the loss of viability during transport. By means as yet unknown, the seas purify the air of the cells and spores being carried, and this too reduces the colonization range of microorganisms and minimizes or eliminates the problem of transoceanic transmission of infectious agents, except when carried by man, aircraft, or ships.

For the propagules that are both suited to this mode of transmission and able to maintain their viability during the trip, the local environmental variations typifying the distribution patterns of the aerial fungi and bacteria just above the ground disappear at high elevations. As a rule the number of propagules declines with increasing altitude (Fig. 2.3), although irregularities are sometimes noted, while the air several meters above sea level is especially poor in microorganisms because of the purifying effect of the ocean water on the biota of the lower atmosphere. The fungus isolates obtained from altitudes of 2000 meters and above tend to be dominated by *Cladosporium* and *Alternaria,* with lesser numbers of *Aspergillus* and *Peni-*

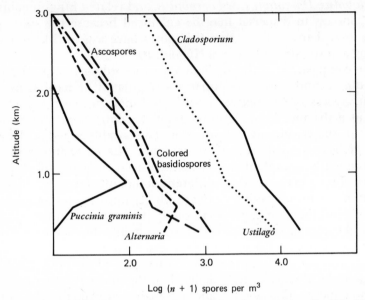

Figure 2.3   Spore types observed in air over the English Channel (Hirst et al., 1967).

cillium; Bacillus and *Micrococcus* often are the dominant bacterial genera, but cells of *Pseudomonas, Flavobacterium, Achromobacter, Corynebacterium,* and *Sarcina* are also present at reasonably high population densities.

In contrast with specialized mechanisms to eject or expose fungal spores for aerial dispersal, morphological devices for wind dissemination are not in evidence among the bacteria, protozoa, or algae. Also at variance with the airborne fungi, which originate largely from the surfaces of vegetation or structures emerging above the ground, the bacteria—and presumably also protozoa and algae—seem to be derived from windblown soil, spray, or water splashes, a fact attested to by the similarity in the airborne bacterial flora and the communities residing in soil and water. In the absence of specialized structures adventitious physical disturbances must be assumed to account for the presence of bacteria, protozoa, and algae in the atmosphere. To a large extent these organisms are borne aloft on soil particles to which they adhere, particles which when dry are easily raised from the ground by wind or by cultivation practices in agriculture. Ocean spray may also add its share of bacteria entrapped in minute droplets. Occasional fungi are also carried aloft and transmitted with windblown soil particles and dust; notorious in this group are *Coccidioides immitis* and *Histoplasma capsulatum*, organisms that attract attention because they are responsible for human and animal diseases. Infection by these two fungi usually is initiated after the inhalation of spore-laden dust, but occasionally the fungi lodge in and colonize other susceptible parts of the body. The view that *C. immitis* is subject to aerial transport is substantiated by the onset of epidemics of coccidioidomycosis, the disease brought about by *C. immitis,* following severe dust storms and by the appearance of the affliction in farm and construction workers, whose occupations entail disturbance of soil or the raising of dust (Kahrs, 1967). Psittacosis virus may also be transported through the air together with the dust raised by the flutter of parrot wings.

Algae and protozoa also are present in the air or on airborne particles. Representatives of Chlorophyta, Cyanophyta, and Chrysophyta among the algae and genera of amebae, ciliates, and flagellates of the phylum Protozoa have been observed in the air. Some species of these groups are quite abundant in the atmosphere, but several are noted infrequently. The algae may be derived from windblown soil particles or from wind-transported foam picked up from inland bodies of water or the open sea, and such algae can be recovered from altitudes of as much as 3000 meters above sea level. Wind action, moreover, may transport algae for distances of at least 65 km (Brown et al., 1964; Maynard, 1968). Wind dispersal has also been postulated to account for the similarities among the unique communities of thermophilic Cyanophyta inhabiting hot springs that are quite distant from one another; the remoteness of these ecosystems and the small likelihood of birds migrating from one locality containing hot springs to a second make this hypothesis attractive.

Air is an extremely important medium of microbial transport in enclosures of various sorts, as in hospital nurseries, operating rooms, barns, chicken coops, and the rooms of houses or public buildings. Microorganism-laden droplets emerging from the human or animal body in normal conversation, coughs, sneezes, or excreta generally settle out, but particles containing these organisms are resuspended in the air when the dust is raised. A variety of bacteria, including pathogens, retain their viability for several days or longer in floor dust or on bandages, and some of the propagules can be inhaled during sweeping, the making of beds, changing of bandages, or at other times when dust is disturbed. Virus and rickettsia infections have been traced to contaminated dust, and dust transmission of *Mycobacterium tuberculosis* has been observed experimentally (Cruickshank, 1957). Hides of anthrax-infected animals are invariably a potential source of the aerially dispersed *Bacillus anthracis* spores, notably for individuals working with the hides. The air of hospital wards also has been reported to contain *Staphylococcus aureus*, a bacterium often emerging as early colonizer of the newborn and of open wounds. The flight path for these organisms is short, in contrast with the air biota of nonenclosed areas, and they rapidly settle out once the disturbance is terminated. Nevertheless the frequency of their introduction into the air and the ready accessibility of a favorable habitat—in the form of a person with low disease resistance, a newborn infant, an open surgical wound, or an area of burned skin—aid a species with a short flight path in locating a new site where it can reinitiate multiplication.

Droplet infection is a common means for the transmission of pathogens between humans or two animals of the same species. The disease agents involved, generally bacteria and viruses, usually have relatively short survival times away from their typical and sometimes sole habitat, the host, and the capacity of these microorganisms to provoke their hosts to a droplet discharge is an ecological advantage of considerable consequence because it markedly increases the chance of successful migration of a population restricted to imperative dispersal. Many of the droplets discharged in sneezing, coughing, and talking fall quickly to the ground or adhere to clothing, but the liquid in minute droplets may evaporate readily so that the bacteria and virus particles remain in the air, whence they may come into contact with a nearby susceptible individual. Inasmuch as this type of dispersal mechanism never covers great distances, intimate association between the two humans or animals—the old and the new habitat—is required; yet, given the appropriate conditions, droplets of saliva or respiratory discharges teeming with microscopic life do serve to transmit pathogens effectively. Capable of being transported in this way, and in some instances largely limited to such a dispersal mechanism in nature, are the chicken pox, measles, mumps, influenza, and common cold viruses and *Neisseria meningitidis*, *Mycobacterium tuberculosis*, and *Bordetella pertussis*.

Another mode of aerial dissemination is *splash dispersal*. On colliding with a surface richly endowed with microbial cells or spores, a single large raindrop will break up into thousands of small, propagule-laden droplets. Thus rain impacting on leaves, soil, or other surfaces introduces an inoculum into the air that ultimately alights, depending on wind speed and height above the ground, on a nearby or remote site where growth may be reinitiated. Splash dispersal is an important means for the dissemination of vegetative cells of plant-pathogenic strains of *Xanthomonas* and *Pseudomonas* and of spores of diverse fungi. Although the extent of migration in both splash dispersal and droplet infection is never great and neither mechanism makes an appreciable contribution to the air biota at large, the survival of a variety of microbial species is nevertheless dependent on such processes to locate an as yet incompletely exploited habitat for continued development.

PASSIVE DISPERSAL: WATER AND SOIL

Populations of all major groups can be transported for considerable distances through or in water or soil until the organisms reach a region where they can grow. This constant horizontal and vertical migration of indigenous species entails a variety of modes of passive transport. The possibility of dissemination through water and soil is not restricted to indigenous populations, however, and many alien species, although often suffering high mortality rates, are carried about from place to place, some propagules ultimately arriving at new habitats wherein their proliferation can once again proceed.

Microorganisms may be transported in the sea or in fresh water by the lateral movement of currents, the vertical upwelling motions of water masses, and the mixing of surface waters under the influence of wind action. Fast-flowing or leisurely currents in oceans, rivers, streams, and estuaries carry not only populations residing in the water but also the heterotrophs and autotrophs adhering to floating plants, leaves, logs, and branches, and colonization of aquatic localities, shorelines, and islands at a distance from the center of dispersal of some species can be achieved through the movements of water. *Biddulphia sinensis* provides an excellent illustration. After this diatom initially accomplished the enormous feat of transfer from the Far East to European seas, apparently by adhering to the hulls of ships (Table 2.2), it moved along the coast of Denmark to the west coast of Norway and also through the English Channel, but this subsequent migration was achieved with currents serving as the motive force (Wimpenny, 1966). *Coccolithus huxleyii* also is carried along with flowing sea water, moving each year from an area in the North Atlantic out into the Norwegian Sea (Oppenheimer, 1966). Stream flow or the entry of one lake into another likewise may be the prelude to successful algal invasions of new aquatic ecosystems. In each instance the organism's own power of locomotion, if any, contributes little or not at all to the spread of cells or filaments.

Lateral movement through subterranean ground water is an important means for the translocation of various bacteria and viruses of public health significance. Ground water supports little heterotrophic and no photosynthetic proliferation, but it is an effective vehicle for the transport of organisms successfully penetrating to the ground water stratum underlying the surface soil. Wells, septic tanks, latrines, or polluted river water not uncommonly are sources for the subterranean bacteria and viruses, which, by moving along with the underground water, can contaminate potable water supplies and cause serious health problems. The extent of lateral movement is quite limited by the adsorption of organisms to soil particles and the rate of ground water flow, but bacteria and viruses are nevertheless known to migrate for some distance in the direction of ground water flow. Experiments under controlled conditions, for example, have shown that bacteria travel more than 60 meters, a span frequently great enough to lead to pollution of a drinking water supply, and outbreaks of disease have been reported where the lateral spread of pathogens through the ground water has been as much as 250 meters for the bacteria causing dysentery and typhoid and 15–25 meters for the infectious hepatitis virus (Mallmann and Mack, 1961).

Entire populations or significant numbers of cells or filaments may be translocated upward or downward in conjunction with disturbances in the water, a mixing that sometimes extends to depths of several hundred meters in the sea. Heterotrophs adhering to leaves or other floating matter sink when the object itself settles to the bottom, and the microorganisms may thereby become inhabitants of aquatic sediments. Many free-floating algae also sink because at some stage during their life the density of the cell contents exceeds that of the surrounding water. For algae and for heterotrophs as well, vertical migrations are often of no avail inasmuch as one or more of the essential nutrients, O_2, or sunlight is no longer available or all available niches are occupied; yet many species undoubtedly benefit as the nutrient supply near the surface is depleted while the underlying strata of water or the bottom sediments are rich in utilizable substances. Moreover, in the initial development of communities in newly constructed dams, reservoirs, and farm ponds, downward migration undoubtedly plays a role in determining the ultimate composition of the biota of the sediments.

Lateral translocation of microorganisms over the land is governed to a large degree by the movement of waterborne soil. Soil particles and cells adhering to them are transported downhill with water flowing from high-lying ground, and large numbers of soil residents are thereby introduced into streams and rivers and become deposited ultimately at some distance from the original center of dispersal. The spread of certain fungi in agricultural fields indeed sometimes follows the patterns of water drainage (Hickman, 1940), and instances have been described of the appearance of enormous quantities of presumably terrestrial microorganisms in aquatic communities following a heavy rainfall on the land. Storm water likewise introduces multitudes of bacteria from urban sources, sewage, and farm animal feces into reservoirs, rivers, springs, and wells, and because some of these bacteria are pathogenic to man, pollution of water supplies by surface runoff bearing passive migrants is a matter of continuing public health concern.

Flowing or quiescent water functions as a vehicle for the dissemination of a number of nonaquatic pathogens having human or animal reservoirs because, although the cells fail to proliferate and frequently their density declines rapidly, for a few cells in the population only a brief period elapses between the time of leaving the reservoir and the time when a new host-provided environment is located. Thus, when a source of human or animal fecal contamination is close to the place where drinking water is removed, sufficient propagules of the invader may have retained their viability to allow for establishment in a new host. The major microbial waterborne human diseases—typhoid, paratyphoid, cholera, dysentery, and infectious hepatitis—are caused by organisms unable to develop significantly in the water that serves as their vehicle of spread, and with the notable exception of the cyst-forming protozoan *Entamoeba histolytica,* they are not known to produce resistant stages.

Not easily accomplished, but still occurring for small distances, is vertical transport through soil. Water infiltrated through columns of soil or sand, a practice sometimes used for the purification of polluted water, is rapidly freed of viable cells, and few bacteria and protozoa move downward appreciably with the percolating liquid, most cells having been retained by particles near the surface. The sand or other soil particles present a barrier impenetrable to all but an extremely small number of cells, and it is the extremely rare individual that penetrates a homogeneous column to a depth of more than 1 meter. Yet channels are created in nature by plant roots or burrowing earthworms, insects, and larger animals, and these tunnels or other fractures in the surface or subsoil probably facilitate vertical transport. Algae are found at soil depths where photosynthesis is precluded, and it must be assumed that many of these, rather than grow-

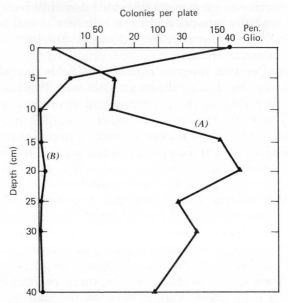

Figure 2.4   Downward movement of spores added to top of sand column. *(A) Gliomastix convoluta; (B) Penicillium cyclopium* (Burges, 1950).

ing heterotrophically in competition with the far more active bacteria and fungi, have been washed downward and exist in a largely dormant form (Alexander, 1961). Spores of several fungi, by contrast, have been shown unequivocally to be susceptible to vertical transport through sand (Fig. 2.4) and soil (Hepple, 1960). Nevertheless, vertical migrations of this sort generally lead to ecological dead ends, unless root constituents or excretions are available as nutrients, for the subsurface soil has little organic matter to sustain heterotrophic life.

PASSIVE DISPERSAL: INANIMATE OBJECTS

Imperative or capricious dispersal of many species occurs as a result of direct contact of some inanimate object bearing viable propagules with a new habitat. Such contacts take place in diverse ways and involve a variety of types of environments and inanimate objects. The juxtaposition of healthy vegetables with rotting produce, for example, frequently is a prelude to the initiation of decay of previously uncolonized tissue, and processed food products acquire spoilage organisms from inadequately cleaned equipment with which they come in contact. Similarly, the mere act of handling diseased animal carcasses by butchers, veterinarians, and hunters may result in infection of the human. Contaminated clothing, books, toys, pencils, dishes, door handles, toilet seats, agricultural implements, pruning knives, floors adjacent to swimming pools, and materials soiled with excreta are acknowledged means for the dissemination of pathogenic bacteria and fungi. Likewise *Clostridium tetani* and species of *Microsporum, Cryptococcus, Sporotrichum,* or other fungi implicated in mycotic infections gain entry into the human body by way of injured tissues that come into contact with infested soil (Ajello, 1962).

Foods and food products are sometimes transporters of species of public health interest. The food generally receives its inoculum from a host harboring the population or from an inanimate object passively bearing the propagules, but certain harmful microorganisms are indigenous to the consumable product. Some groups multiply in the foodstuff whereas others are unable to proliferate to a significant extent, and only the survivors from the original inoculum are associated with the transmission. *Salmonella, Shigella,* and *Entamoeba histolytica* find their way to new hospitable habitats by way of food. Milk and milk products also are vehicles for the introduction of species of *Streptococcus, Mycobacterium, Brucella, Corynebacterium, Salmonella,* and *Coxiella* into the human body, but sanitation practices and pasteurization have made dairy products a far less frequent medium for dissemination than heretofore.

Owing to their impact on man, human pathogens transmitted through food, milk, and water have been the focus of attention, but nonpathogens too are transported in foodstuffs. Cited previously is the fashion in which young termites gain their flagellates, the parents feeding the young with food derived from the protozoan-rich excreta of the adult termite, a transfer resulting in infection of all the offspring. The healthy newborn of many vertebrates, including man, emerge from the mother with no microbial colonists, and it is also likely that a considerable portion of their initial gastrointestinal communities is derived from the first feeding.

PASSIVE DISPERSAL BY BIOLOGICAL VECTORS

Animal vectors convey a multitude of populations from colonized environments to ones where the particular microorganisms are not yet present. Some of the microorganisms borne about lead a tenuous existence and would not survive in nature apart from their host or carrier, but others are quite hardy and have merely become linked

Table 2.3

Microorganisms Disseminated by Animal Vectors

Microorganism	Vector	Reference
Viruses		
Rabies virus	Dogs	Burnet (1962)
Soil-borne viruses	Nematodes	Hewitt and Grogan (1967)
Wheat streak mosaic virus	Mites	Broadbent (1960)
Yellow fever virus	Mosquitoes	Burnet (1962)
Rickettsiae		
Rickettsia prowazekii	Lice	Burnet (1962)
Rickettsia ricketsii	Ticks	Burnet (1962)
Bacteria		
Corynebacterium fascians	Nematodes	Broadbent (1960)
Erwinia carotovora	Cabbage root fly	Broadbent (1960)
Pasteurella pestis	Fleas	Burnet (1962)
Soil heterotrophs	Earthworms	Alexander (1961)
Fungi		
Botrytis anthophila	Bees	Broadbent (1960)
Ceratocystis ulmi	Bark beetles	Zadoks (1967)
Claviceps purpurea	Flies, beetles	Austwick (1957)
Endothia parasitica	Birds	Zadoks (1967)
Phytophthora palmivora	Snails	Turner (1967)
Protozoa		
Histomonas meleagridis	Nematodes	Gibbs (1962)
Leishmania	Mosquitoes	Garnham (1965)
Trypanosoma	Tsetse flies	Hagan and Bruner (1961)
Various protozoa	Birds, crustaceans	Dogiel et al. (1965)
Various protozoa	Raccoons	Maguire (1963)

with an efficient dispersal agent. A few of the microbial species not only are carried by the vector but also grow within or upon it, while in many instances the propagules receive no more from their carrier than transportation.

Animals of diverse phyla participate directly in microbial dispersal. A small sample of the known vectors and the species they bear is presented in Table 2.3. The vector may gain its passengers by touching a colonized surface or feeding on a viable host inhabited by the particular populations. As the animal moves to a new location, cells, spores, or other structures adhering to the animal's surfaces are rubbed off or are shed, while those carried within the animal's body may be released into the new environment with excreta or directly inserted into the locality or host by a boring, biting, or wounding mechanism. Many vectors bite, bore into, or cause wounds in potential hosts, and it is by means of these openings that the propagules make their way into the blood or underlying plant or animal tissues. Certain vectors themselves become not only colonized but seriously diseased by the microorganisms they bear, but many are totally unaffected or even benefit from the associates.

Vector transmission is not obligate for a variety of the populations that are carried by migrating animals, as the microorganisms have alternative means of dissemination, but other species rely entirely on the vector for transport. Where vectors function in bringing a parasite to a new host, the migrant bypasses the rigors of environmental stress and the possible detrimental effects induced by the heterogeneous communities of soil, water, and so on, and it also gains the enormous advantage of being borne, usually unerringly, from one host to another, relying on the animal to locate the next site for exploitation.

To accomplish their migration, occasional species have unique devices or distributions that lure the vector or facilitate a successful completion of the dispersal. Pollinating insects are attracted to flowers, and other insects go to the slime oozing from plant lesions, from which cells residing therein are readily conveyed to a second nutrient-rich site. *Botrytis anthophila* is particularly well suited for spread by pollinating insects because it sporulates abundantly in nature on the anthers of red clover. Representatives of the Diptera and Coleoptera are attracted by the odor or color associated with the growths of specific fungi, and while the insects feed on the fungi they pick up and subsequently transport the spores to a fresh area (Broadbent, 1960). *Ceratocystis ulmi,* the fungus causing Dutch elm disease, has a novel interrelationship with the bark beetle, its vector. The pathogen grows in infected trees in the egg galleries of the beetle, and the emerging adult insects become coated with a multitude of spores, which they transmit to healthy elms. The beetles thus set the stage for a new infection and derive benefit thereby because they subsequently are able to breed below the bark of trees that have since become diseased (Clinton and McCormick, 1936). Flies alighting on fecal matter can transport the bacteria, like *Shigella,* they acquire to food supplies, and insects also carry yeasts and other microbial types to honey or concentrated fruit products, providing an inoculum that may induce food spoilage (Mossel and Ingram, 1955). Similarly, insects are capable of disseminating unicellular and filamentous algae and protozoa, often for considerable distances, and such dispersal by aquatic insects could be a key factor in determining the composition of the community that appears in newly built reservoirs and farm ponds (Maguire, 1963; Stewart and Schlichting, 1966).

Successful passage through the alimentary tract is achieved by a few heterotrophs and algae not indigenous to this site of habitation, and the viable cells are in time deposited in the feces some distance from the point where the mobile animal gathered them. The spores of coprophilic fungi are thus swallowed and, upon emerging with the fecal matter, initiate extensive multiplication. Birds, mice, worms, centipedes, slugs, and larger animals all convey an assortment of fungi, and probably other groups, from place to place. Oospores produced by algae have been reported to survive passage through the digestive tract of birds, as have algae with no resting stage, and shore birds may transport these propagules internally and deposit them in new surroundings (Proctor et al., 1967). Birds also pick up hordes of algal cells on their feet, feathers, and bills, and these may well be translocated during the migration of the birds.

One microbial species on occasion may be a vector for a second type of microorganism. This is apparently a mode of transmission for at least three soil-borne plant viruses, the vector being *Olpidium brassicae*, but additional fungi may be revealed as vectors for plant viruses in the future (Hewitt and Grogan, 1967).

Man's activities contribute constantly to local, regional and global traffic in innocuous or harmful heterotrophs. Species initially of strictly local consequence are transported to new regions by human actions, and the invaders may become established with modest or at times disastrous consequences. Resting structures and vegetative cells or filaments have been transmitted in or on nursery trees, seeds, tubers, plant cuttings, shoes, clothing, farm implements, irrigation water, ships, aircraft, and dust raised in cultivating or construction activities. Indeed, scientists are not beyond suspicion; for example, *Peronospora tabacina*, the cause of tobacco blue mold, was introduced deliberately into England, with the consent of the British authorities, for controlled experiments to determine the effectiveness of fungicides, but the year of introduction into the experimental area was highlighted by an accidental release of the fungus and its spread to distant tobacco plants in England and the Netherlands. By the following year the fungus had migrated, without human intervention, to tobacco fields in Belgium (Zadoks, 1967). The migration of representatives of scores of bacterial, fungal, and algal genera has been facilitated in related manners involving human culpability.

References

REVIEWS

Anderson, G. W. 1965. The principles of epidemiology as applied to infectious diseases. *In* R. J. Dubos and J. G. Hirsch, Eds., *Bacterial and Mycotic Infections of Man.* Lippincott, Philadelphia. pp. 886–912.

Broadbent, L. 1960. Dispersal of inoculum by insects and other animals, including man. *In* J. G. Horsfall and A. E. Dimond, Eds., *Plant Pathology,* Vol. 3. Academic Press, New York. pp. 97–135.

Elton, C. S. 1958. *The Ecology of Invasions by Animals and Plants.* Methuen, London.

Garnham, P. C. C. 1966. Locomotion in the parasitic protozoa. *Biol. Rev.,* **41**, 561–586.

Gregory, P. H. 1961. *The Microbiology of the Atmosphere.* Hill, London.

Halldal, P. 1962. Taxes. *In* R. A. Lewin, Ed., *Physiology and Biochemistry of Algae.* Academic Press, New York. pp. 583–593.

Hewitt, W. B., and Grogan, R. G. 1967. Unusual vectors of plant viruses. *Ann. Rev. Microbiol.,* **21**, 205–224.

Hickman, C. J., and Ho, H. H. 1966. Behaviour of zoospores in plant-pathogenic Phycomycetes. *Ann. Rev. Phytopathol.,* **4**, 195–220.

Hirst, J. M. 1965. Dispersal of soil microorganisms. *In* K. F. Baker and W. C. Snyder, Eds., *Ecology of Soil-Borne Plant Pathogens.* University of California Press, Berkeley. pp. 69–81.

Ingold, C. T. 1965. *Spore Liberation.* Clarendon Press, Oxford.

Jahn, T. L., and Bovee, E. C. 1965. Movement and locomotion of microorganisms. *Ann. Rev. Microbiol.,* **19**, 21–58.

Muskett, A. E. 1960. Autonomous dispersal. *In* J. G. Horsfall and A. E. Dimond, Eds., *Plant Pathology,* Vol. 3. Academic Press, New York. pp. 57–96.

OTHER LITERATURE CITED

Ajello, L. 1962. *In* G. Dalldorf, Ed., *Fungi and Fungous Diseases.* Thomas, Springfield, Ill. pp. 69–83.

Alexander, M. 1961. *Introduction to Soil Microbiology*. Wiley, New York.

Austwick, P. K. C. 1957. *In* C. Horton-Smith, Ed., *Biological Aspects of the Transmission of Disease*. Oliver & Boyd, Edinburgh, pp. 73–79.

Brown, R. M., Larson, D. A., and Bold, H. C. 1964. *Science,* **143,** 583–585.

Burges, A. 1950. *Trans. Brit. Mycol. Soc.,* **33,** 142–147.

Burnet, F. M. 1962. *Natural History of Infectious Disease*. Cambridge University Press, London.

Clinton, G. P., and McCormick, F. A. 1936. *Conn. Agr. Expt. Sta. Bull., New Haven,* No. 389, 703–752.

Conant, N. F. 1965. *In* R. J. Dubos and J. G. Hirsch, Eds., *Bacterial and Mycotic Infections of Man*. Lippincott, Philadelphia. pp. 825–885.

Cruickshank, R. 1957. *In* C. Horton-Smith, Ed., *Biological Aspects of the Transmission of Disease*. Oliver & Boyd, Edinburgh. pp. 51–57.

Dogiel, V. A., Poljanskij, J. I., and Chejsin, E. M. 1965. *General Protozoology*. Oxford University Press, London.

Garnham, P. C. C. 1965. *Am. Zoologist,* **5,** 141–151.

Gibbs, B. J. 1962. *J. Protozool.,* **9,** 288–293.

Hagan, W. A., and Bruner, D. W. 1961. *The Infectious Diseases of Domestic Animals*. Cornell University Press, Ithaca, N.Y.

Hamilton, E. D. 1959. *Acta Allergol.,* **13,** 143–173.

Hasle, G. R. 1950. *Oikos,* **2,** 162–175.

Hegner, R. W. 1938. *Big Fleas Have Little Fleas*. Williams & Wilkins, Baltimore.

Hepple, S. 1960. *Trans. Brit. Mycol. Soc.,* **43,** 73–79.

Hickman, C. J. 1940. *J. Pomol.,* **18,** 89–118.

Hirst, J. M., Stedman, O. J., and Hogg, W. H. 1967. *J. Gen. Microbiol.,* **48,** 329–355.

Kahrs, R. F. 1967. *In* N. C. Brady, Ed., *Agriculture and the Quality of Our Environment*. American Association for the Advancement of Science, Washington, D.C. pp. 97–104.

Kramer, C. L., and Pady, S. M. 1968. *Mycologia,* **60,** 448–449.

Maguire, B. 1963. *Ecol. Monogr.,* **33,** 161–185.

Mallmann, W. L., and Mack, W. N. 1961. *In* Ground Water Contamination. *U.S. Publ. Health Serv. Tech. Rept.,* **W61-5,** 35–43.

Maynard, N. G. 1968. *Z. Allgem. Mikrobiol.,* **8,** 225–226.

Meyer, K. F. 1965. *In* R. J. Dubos and J. G. Hirsch, Eds. *Bacterial and Mycotic Infections of Man*. Lippincott, Philadelphia. pp. 659–697.

Mossel, D. A. A., and Ingram, M. 1955. *J. Appl. Bacteriol.,* **18,** 232–268.

Noland, L. E., and Gojdics, M. 1967. Ecology of free-living protozoa. *In* T.-T. Chen, Ed., *Research in Protozoology,* Vol. 2. Pergamon Press, New York. pp. 215–266.

Oppenheimer, C. H., Ed. 1966. *Marine Biology. II.* New York Academy of Sciences, New York.

Poindexter, J. S. 1964. *Bacteriol. Rev.,* **28,** 231–295.

Proctor, V. W., Malone, C. R., and DeVlaming, V. L. 1967. *Ecology,* **48,** 672–676.

Schrodter, H. 1960. *In* J. G. Horsfall and A. E. Dimond, Eds., *Plant Pathology,* Vol. 3. Academic Press, New York. pp. 169–227.

Stakman, E. C., and Hamilton, L. M. 1939. *Plant Dis. Reptr. Suppl.* **117,** 69–83.

Stewart, K. W., and Schlichting, H. E. 1966. *J. Ecol.,* **54,** 551–562.

Turner, G. J. 1967. *Trans. Brit. Mycol. Soc.,* **50,** 251–258.

Wimpenny, R. S. 1966. *The Plankton of the Sea.* Faber and Faber, London.

Zadoks, J. C. 1967. *Neth. J. Plant Pathol.,* **73** (Suppl. 1), 61–80.

3

Colonization

Changes in the species composition of environments invaded or inhabited by microorganisms are seen in innumerable places. The inoculation of a simple medium or of a complex infusion containing natural materials initiates a dramatic sequence of changes in the types of organisms present. Such modifications in species composition occur whether the inoculum was derived from sewage, fresh water, soil, rumen contents, animal droppings, or the exposed surfaces of animals or plants. Still more spectacular biological sequences are frequently observed in natural habitats as microorganisms respond to the availability of uninhabited sites, to seasonal or cataclysmic influences affecting their habitat, to the activities of animals or plants, or to processes brought about by their microbial neighbors. These changes and the reasons for them are considered as part of a discussion of microbial colonization and succession.

Succession is the replacement of one type of community by another in response to modifications in the habitat. It commonly involves a sequence of continuous replacements of the microbial communities located in a particular place. The changes are rarely abrupt, and individual species tend to appear and disappear slowly in the processional series, so that two successive communities may contain common components.

Clear-cut stages in certain successions are readily observable, as in certain sequences of freshwater or marine algae. In other instances no abrupt beginning or end to a stage is discernible, but the subtle changes in density of certain populations or in the kinds of species nevertheless show a directional sequence. As a result of careful study of a number of ecosystems, it is now possible to predict in a few instances the sorts of directional changes, despite a frequent inability to recognize discrete steps in the sequence. At each phase in the con-

54

tinuum the organisms present are those particularly suited for the conditions of the habitat as it exists at that point in time.

Populations appear in the gradual and continuous progression, reach a certain abundance, and then decline. In a region initially devoid of microbial life, a succession is initiated by one or more *pioneer* species, the organisms first proliferating in the particular circumstance. As they grow, these organisms, which together comprise the *pioneer community,* have an influence on their surroundings, selectively utilizing certain environmental resources and forming cellular constituents and extracellular products. The now modified circumstances are generally not as suitable for representatives of the pioneer groups as for later arrivals, so that the existence of the pioneer community is slowly but irrevocably terminated, and new species assume the forefront. The species that become established in the previously uninhabited zone occasionally may resist elimination for a short time, or they may have a reasonably long life in the ecosystem because of their ability to cope with the new neighbors and new conditions, but generally the pioneers have a transitory existence and are rapidly eliminated. Thus some Phycomycetes appearing in the initial stages of organic matter decay and *Clostridium perfringens,* which is abundant in the early life of several kinds of animals and the human baby (Dubos et al., 1963), become less abundant with the progress of the succession in which they participate.

COLONIZATION AND THE PIONEER

Environments free of microbial life are not uncommon, as stated previously, despite the truism that microorganisms are potentially everywhere. The newborn infant, newly exposed rock surfaces, and the tissues of animals and plants are entirely devoid of microorganisms until the pioneer arrives. The term pioneer is applied not only to the species first occupying the site but also to the initial member of a taxonomic group appearing in a region which, though not sterile, was largely or entirely free of vegetative forms of that particular microbial type; thus the first algae appearing in bodies of water, eroded land, or desert rain crusts and the first fungi showing hyphal development on plant materials undergoing decay are considered pioneers. These early arrivals possess *invasiveness,* the ability of an organism to migrate and establish itself in a location possessing resources as yet unexploited. The microbial propagule—that is, the cell, fragment of a filament, conidium, resting body, or any structural element capable of initiating growth—finds itself in a location with utilizable nutrients and an absence or paucity of competitors. Although the conditions may be far from suitable, the fact that the pioneer has the available nutrients largely to itself allows for its successful establishment. The few species of the pioneer community multiply, and their population densities rise until forces are encountered that tend to limit the population size. The limiting force may be a chemical or physical change that makes the environment unsuitable for further proliferation by the pioneers—a depletion of the supply of nutrients utilizable by those species, an accumulation of toxins, or a fall in pH—or the limiting factor may be a biologically imposed stress induced by late-coming parasites or predators. In the case of microorganisms proliferating in or on the tissues of a higher organism, the stress is not uncommonly a host response detrimental to the invader.

The identity of the pioneer species is governed in large part by the composition of the environment to be colonized. Chemically dissimilar bodies of water allow for the growth of different algae; plant materials having more or less lignin or cellulose support markedly different fungal colonists; and the identity of the bacteria establishing themselves in the gastrointestinal system of the newborn depends on whether the baby is breast-fed or receives cow's milk. The organisms initially appearing in sterilized food products also vary with the chemistry of the food. However, astonishingly little is known of the particular chemical constituents that determine whether a propagule alighting on a colonizable substrate will succeed in establishing itself as a member of the pioneer community.

Some locations exposed to colonizing species have little or no organic matter, and it is the absence of carbonaceous nutrients necessary to support heterotrophic life that determines the character of the pioneer. Typically such ecological successions are dominated by photosynthetic microorganisms, usually algae, that are able to gain a foothold because of their capacity, as autotrophs, to proliferate in the absence of organic matter. As the algae furnish organic carbon to the ecosystem, the stage becomes set for heterotrophs, which feed on the products or cell components of the pioneers and follow in the succession. With time, the amount of protoplasmic material present (the *biomass*), the variety of biochemicals, and the species diversity increase, while the relative significance in the community of autotrophy declines. Bare rock surfaces show an initial colonization of this sort, one including lichens (Haynes, 1964), diatoms (Blum, 1956; Chapman, 1957), and certain other algae. Blue-green algae such as *Schizothrix, Microcoleus,* and *Scytonema* may develop initially in

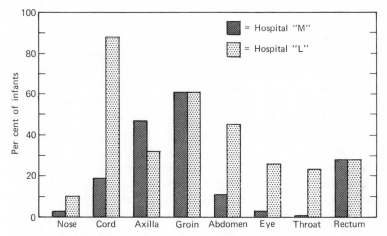

Figure 3.1 Percentage of infants initially colonized by *Staphylococcus aureus* at different sites on the skin. Many of the infants are initially colonized by these bacteria at more than one site (Marples, 1965). Courtesy of Charles C Thomas, Publishers, Springfield, Ill.

areas of badly eroded land to form a distinct algal crust (Booth, 1941), and blue-greens may also come to the fore in desert soils shortly after the infrequent rains but before the successional stage characterized by a variety of heterotrophic types (Fletcher and Martin, 1948). Similarly, the life cycle in land denuded by lava flows is started by lichens and blue-green algae.

Other environments are rapidly colonized by heterotrophs because the areas are well endowed with organic nutrients. The human fetus is free of microbial life, but immediately after birth a veritable biological horde alights on the sterile surfaces of the newborn infant through its contacts with humans or inanimate objects. Only a few of these organisms are able to establish themselves on the skin and mucous membranes to form either a dense or a sparse population, depending on the pioneer species and the site (Fig. 3.1). The oral cavity of the infant often shows distinct evidence of colonization between the eighth and twelfth hours following delivery. In the first day the streptococci are particularly prominent. The pioneer community is displaced as the infant ages, and lactobacilli, fusiform bacilli, and other bacteria assume positions in the community (Hoffman, 1966). Freshly exposed tissues of mature animals and growing plants, bruised portions of fruits and vegetables, and environments whose communities were deliberately destroyed by man, such as in the processes of food sterilization or in the fumigation of pathogen-infested soil, are also rapidly invaded by pioneer members of various bacterial and fungal genera.

One species occasionally dominates during the initial colonization and is undeniably the most prominent member of the entire pioneer community or of a single taxonomic group of early colonists. Dominance is characteristic, for example, of the fungi *Hypodermella mirabilis* and *Bifusella faullii* in the initial invasion of *Abies balsamea* (balsam fir) needles (Shigo, 1967), *Ceratocystis monoliforme* in the primary stages of attack on felled beech logs (Ueyama, 1965), an *Oscillatoria* species in the appearance of biological crusts in desert soils following a rain (Fletcher and Martin, 1948), *Streptococcus* in the mouth of the very young infant (McCarthy et al., 1965), *Lactobacillus bifidus* in the feces of young breast-fed infants (Rosebury, 1962), and *Trichoderma viride* in soils that have been steamed or treated with fungicidal agents. Each of these organisms has some uncommon physiological or morphological trait, possibly coupled with an efficient means of dispersal, allowing it to dominate the initial flora appearing in these circumstances.

In a number of ecosystems the new dominant arising to replace the dominant pioneer has been identified and characterized. The successor is not well suited to the pioneering role, but it does have characteristics permitting it to follow immediately after the ground-laying done by others. The pioneer community, moreover, frequently appears to determine the pattern of succession. The reason for the replacement of one dominant by another is readily determined in some habitats. Thus, in the fermentation of grapes and other sugar-rich materials, yeasts are prominent initially and are responsible for converting the sugar to ethyl alcohol, the latter in turn reaching high concentrations and thereby providing selective conditions for the flourishing of *Acetobacter* species. On decaying roots and other plant materials, succession is commonly initiated by organisms that use simple organic compounds, but when the sugars, organic acids, and other low-molecular-weight substrates are largely gone, these pioneers are replaced by fungi and bacteria capable of utilizing hemicelluloses and cellulose, the first population either being unable to metabolize the polysaccharides of the plant cell wall or faring poorly in the presence of species better adapted to multiply on the polysaccharides. Similarly, the pioneers in animal and plant wounds or the species responsible for lesions or ulcerative growths on higher organisms modify their habitats in a fashion such that the range of successors is typically narrow, as is evident by observations that selected *Staphylococcus* and *Streptococcus* strains commonly find a suitable domicile in human skin ulcers induced by *Leishmania tropica,* specific bacteria regularly are early colonists of skin lesions caused by other pathogens (Marples, 1965), and certain fungi almost invariably follow pioneers in the series of communities developing in wounded plants (Shigo, 1967).

One pioneer species may prevent the successful establishment of another, a phenomenon variously termed *pre-emptive colonization, exclusion,* or *interference.* Pre-emptive colonization is widespread in nature, but it is of especial significance when the organism excluded is a pathogen. An interesting illustration is found in the work of Eichenwald et al. (1965), in which a number of infants less than 24 hours old, none of whom showed the presence of coagulase-positive staphylococci, were colonized by a *Staphylococcus aureus* type carried by a nurse in contact with the babies. None of the infants older than 1 day, most of whom already contained coagulase-positive staphylococci in their noses, was colonized by the nurse's *S. aureus* type. Moreover, artificially induced colonization of newborn infants with a nonpathogenic staphylococcus was notably successful in excluding related organisms. *Staphylococcus* strains present on the skin, in the nose, and even in wounds appear to bar the establishment of other strains, including pathogenic staphylococci, presumably because the niche has already been pre-empted (Marples, 1965).

Pre-emptive colonization is also suggested by the observation that germfree animals initially inoculated with innocuous bacteria are not significantly colonized by pathogens, although infection proceeds rapidly if the animal is exposed to the pathogen alone (Formal et al., 1961). In a similar fashion, prior colonization of plant materials, including wood and crop residues, by certain microorganisms excludes invasion by late arrivals. The exclusion probably results frequently from the assumption by the early colonizer of a niche that might have been occupied by the subsequent species, the first inhabitants performing the activities, using the nutrients, and/or occupying the physical sites in place of the excluded species. However, the production of toxins or a modification of the habitat undoubtedly accounts for interference in occasional circumstances.

BARRIERS TO COLONIZATION

The environment exerts a selective action on those propagules entering it. Local conditions determine which of the viable cells, filaments, or resting bodies will survive, proliferate, or die out. No species is indigenous to all locales into which it is transported, although clearly many species have a nearly global distribution. Diverse bacteria and fungi have remarkably good dispersal mechanisms; yet they are effectively excluded from certain areas because of one or more properties of the potential habitat. Moreover, a species may reproduce its own kind in a given ecosystem, but rarely are communities monospecific; other species encountered prevent the establishment of invaders or oppose the unrestricted replication of native populations. Thus biological or nonbiological factors exist in microbial habitats and operate to exclude new arrivals or maintain their population densities at a low level, interferences reflecting the action of established populations or of factors associated with the environment itself.

The restriction in the numbers of individuals, biomass, or activities of a population imposed by physical, chemical, or biological factors in the ecosystem is referred to as *environmental resistance*. The condition preventing establishment or holding the population size or activity in check is considered to be a *barrier*. The barrier is not necessarily a physical obstacle, although such are in evidence, inasmuch as many chemical substances or viable organisms may also serve as barriers to microbial colonization or population increase. Typical barriers to colonization by individual bacterial, fungal, or viral types are presented in Table 3.1. The table is far from complete, but rather presents selected and general examples of factors implicated in environmental resistance.

Chemical barriers hindering growth or responsible for the loss of viability of introduced propagules have been observed in distinctly different ecosystems. A number of such substances have been identified, and their modes of action in animal or plant tissues, milk, or other habitats or on the external surfaces of higher organisms have been investigated. The evidence for environmental toxicity to species invading some environments is unequivocal, but the effective compound is not yet known. In other instances, particularly those involving human, animal, or plant pathogens, more than one barrier or chemical substance is apparent, but the relative significance of the various factors and occasionally their precise identities remain uncertain.

Mechanical barriers protect or overlie a variety of potential sites for colonization. The external cuticle of fruits excludes an array of microorganisms from the underlying nutrient-rich tissues and may, for example, retard the lactic fermentation of olives. The shell of the avian egg, the gum exuded by wounded fruits, and the suberization associated with damage to potato tissues may likewise hold back po-

Table 3.1

Some Barriers to Microbial Colonization

Environment	Barrier, Postulated or Demonstrated
Skin	Fatty acids
Lungs, stomach	Mucosa
Nose	Lysozyme
Blood	Phagocytes, antibodies
Intestine	Mucosa
Fish tissues	Protamines
Milk	Peroxidase, agglutinins
Roots	Cork layer
Fruits	Cuticle, acids
Trees	Gums, resins, tannins
Plant tissues	Phenolic compounds, glycosides
Virus-infected cells	Interferon

tential invaders (Mossel and Ingram, 1955). The access of micro-organisms is also prevented by the physical obstruction imposed by the outer cork layer of roots or the cork laid down as part of the reaction of plants to wounding. The skin of the normal human and animal teems with bacterial and fungal life, and the underlying tissues are suitable environments for the growth of many skin residents, but the uninjured skin coats these colonizable tissues with a layer impenetrable by most microorganisms. Mucous membranes also present an anatomical barrier. Hordes of bacteria are found on the linings of the intestine and reach the upper respiratory tract, but the tissues below, like the subcutaneous tissues, are typically sterile because of the imposed obstacle. The respiratory mucosa of man is a particularly effective hindrance to *Diplococcus pneumoniae,* for, though 40–70% of humans carry the bacteria, infection and the development of pneumococcal pneumonia is quite rare (MacLeod, 1965). The effectiveness of these mechanical obstructions becomes readily apparent when their integrity is breached so that deeper tissues are exposed; pathogenic staphylococci and other bacteria may develop rapidly following skin laceration, and a perforated appendix is not uncommonly a prelude to peritonitis.

Barriers effective in large part because of their sheer physical presence may act in nonmechanical ways as well. Leaf wax, for example, may play an obstructive role, but it also contains fungistatic phenolic compounds (Martin et al., 1957). Acidity also contributes to environmental resistance. The low pH of tomato products and additional foodstuffs of concern to the canning industry is probably a major factor limiting microbial development. The killing of the vast numbers of microorganisms continually ingested along with foods bearing them is linked with the inhibitory action of the highly acid gastric juice, the gastric fluids apparently barring entry of many of these viable bacteria into the intestine. Acidity, however, many not be the sole factor responsible for the bactericidal action of gastric fluids (Rosebury, 1962).

Specific fatty acids also are components of environmental resistance. Although desiccation is undoubtedly involved in the death of bacteria alighting on the skin, fatty acids contribute to the bactericidal action (Marples, 1956); long-chain unsaturated fatty acids appear to be of greatest significance. Both short- and long-chain fatty acids are found in animal tissues in concentrations that, on the basis of tests in vitro, are toxic to certain bacteria, but the importance of such inhibitors in vivo is uncertain. Long-chain fatty acids accumulate in pneumonia lesions and may aid in the elimination of the local *D. pneumoniae* population, and *Mycobacterium tuberculosis* and *Staphylococcus* cells present in sites of inflammation may be inhibited by the lactic acid formed by the body (Dubos, 1954). Even if the contribution of fatty acids to the development and survival of pathogenic bacteria in vivo remains unsure, the significance of members of this class of compounds to the survival of alien bacteria entering the rumen has recently been verified (Wolin, 1969).

Several enzymes are believed to be important in environmental resistance. Particular attention has been given to two, one of which, lysozyme, acts directly on the microbial cell. The second, peroxidase, catalyzes the formation of products toxic to microorganisms. A reasonable percentage of bacteria are susceptible to the lytic action of lysozyme, and a cloudy suspension containing a suitable bacterium, such as a strain of *Micrococcus,* can be completely cleared by a minute amount of the enzyme. Lysozyme is found in many animal body fluids and tissues, including tears, nasal mucosa, saliva, and lungs. The effectiveness of lysozyme as a barrier to infection has been seriously questioned, particularly because selected pathogenic bacteria are not lysed by the enzyme; however, the possibility exists that otherwise harmless bacteria might indeed be able to invade animal tissues and produce disease were it not for the presence of this lytic agent. Furthermore, although the enzyme itself is without effect on potential

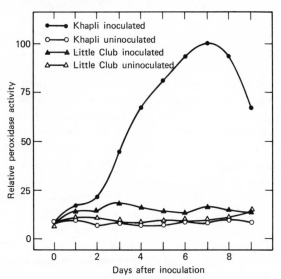

Figure 3.2 Peroxidase activity in inoculated and healthy leaves of resistant (Khapli) and susceptible (Little Club) varieties of wheat (Macko et al., 1968).

intruders, it may aid in the destruction of pathogens in conjunction with factors concerned in immunity.

Far less tenuous, by contrast, is the evidence that peroxidase constitutes a barrier to the colonization of raw milk by bacteria like *Streptococcus* and *Salmonella* spp. Raw milk contains several substances that effectively deter the establishment of some bacteria, substances that are destroyed by heating, with a resulting decline in the potential keeping quality of the milk. One of the deterrents is peroxidase, an enzyme that in the presence of H_2O_2 catalyzes the oxidation of thiocyanate to yield a bactericidal product. Barriers involving peroxidase and thiocyanate may be of general importance, for both are found in many types of animal tissues (Oram and Reiter, 1966). Peroxidases may also be involved in the resistance of plants to microbial colonists, including pathogenic fungi, by catalyzing the formation of inhibitory compounds; consistent with this view are the observations that peroxidase activity increases markedly in wheat plants resistant but not those susceptible to *Puccinia graminis* f. sp. *tritici* (Fig. 3.2) and that the enzyme in the presence of H_2O_2 inhibits hyphal development in culture (Macko et al., 1968).

A variety of proteins as well as low-molecular-weight compounds bar the development of untold microbial species. These include the protamines—proteins associated with nucleic acids of the cell—of fish organs, selectively bactericidal proteins of fresh human serum, and interferon. Interferon is currently attracting much interest. It is a protein, or one of a class of proteins, synthesized in the cells of higher animals as a result of exposure to a virus, or sometimes particular bacteria, rickettsiae, or mycoplasmas. Interferons may be integral parts of the host's defense against disease, functioning to lessen or abolish the capacity of the cell to support the proliferation of another infective virus. The interferons, like some protein barriers but differing from antibodies, inhibit a wide variety of viruses.

Plants contain a number of discrete chemical substances that retard microbial colonization. Interferons or interferon-like substances are synthesized when higher plants are exposed to selected viruses, but owing to their economic importance, the fungal pathogens have been the microorganisms most thoroughly investigated. Oats, for example, have a glucoside, avenacin, toxic to various fungi, but the inhibition is bypassed as the avenacin is detoxified by a fungus like *Ophiobolus graminis* var. *avenae* that colonizes and produces disease in this plant (Turner, 1961). Hydrocyanic acid is produced from the plant glucoside linamarin, and flax root excretions containing cyanide selectively destroy many microorganisms inhabiting the root environment (Timonin, 1941). Phenolic compounds are formed by higher plants, and these substances retard microbial colonization; for example, protocatechuic acid, catechol, chlorogenic acid, polyphenols, and related materials make cultivated plants resistant to infection by species of *Colletotrichum, Helminthosporium, Phymatotrichum, Sclerotinia, Venturia, Verticillium,* and the actinomycete *Streptomyces scabies.* Trees also impose barriers that are surmounted only with considerable difficulty; gums and resins may retard deep microbial penetration into the woody tissue, resin flowing from wounds may hold back *Fomes annosus,* and durable woods likewise contain antimicrobial agents (Shigo, 1967). The surfaces of young green leaves frequently harbor few fungi, but the as yet uncharacterized barrier is destroyed at the onset of senescence, at which time a phase of rapid fungal colonization is initiated. Soils, too, have barriers to fungal spore germination and the growth of certain bacteria, but the elusive mycostatic and bacteriostatic factors have still not been identified.

An established community is among the most effective of barriers to invasion by new arrivals. Pre-emptive colonization, as stated previously, is a means of exclusion of latecomers, and it is, together with the physical and chemical barriers in the locale, one of the components of environmental resistance. Bacteria, fungi, and actinomycetes can overcome the physical and chemical impediments in soil and grow extensively therein if the resident community is eliminated by sterilization, but few become established if the indigenous organisms are still viable and functioning normally (Alexander, 1961). Germ-free animals are notoriously prone to infection, whereas the micro-organisms borne by the normal animal hold in check pathogens resident in the community or those entering from outside, except in those rare circumstances associated with disease (Formal et al., 1961). Pathogenic fungi abound on plants roots, but only occasionally do they induce major deleterious effects, provided that the roots contain their typical microflora; by contrast, aseptically grown plants exposed to the mildest of fungal parasites frequently come down with disease, often at a rapid rate. These various kinds of biological barriers may be modified, breached, or entirely destroyed in several ways. An infamous example of the lowering of a normally effective barrier to colonization is associated with the medical use of antibacterial antibiotics: the antimicrobial agent may be responsible for the elimination of the bacterial infection, but the natural barriers will have been lowered so that opportunistic fungi, formerly maintained at a low population density by the indigenous biota or originating outside the indigenous community, assume a greater degree of dominance and on occasion become clinically significant. The mechanisms of these sorts of environmental resistance will be considered in later chapters.

ENVIRONMENTAL FEEDBACK

Infectious diseases represent a category of population-environment interactions involving a host plus a microorganism with the potential for both colonization and pathogenesis. From the ecological viewpoint the governing feature of this ecosystem is the living animal or plant. The host or *suscept* must be colonizable; that is, it must be receptive to invasion by the particular disease agent. Not all higher organisms are receptive to given pathogenic bacteria, fungi, protozoa, actinomycetes, rickettsiae, or viruses, one animal or plant species providing a totally unsuitable habitat for a microorganism that is capable of invading and doing harm to another higher organism.

Three kinds of barriers underlie the lack of receptiveness: (a) the barriers of the nonreceptive or nonhost species, (b) the factors associated with resistance of the receptive host prior to its first contact with the infective agent, and (c) the obstacles to further microbial development or activity that appear as a consequence of infection. By contrast with aquatic and terrestrial habitats, the many ecosystems localized within the animal or plant respond to microorganisms, and a multitude of processes come into play because of microbial penetration into the tissues or body fluids. In effect, there is an *environmental feedback,* a modification in the habitat resulting from the presence of one or more microbial populations, a change that can affect the size, activity, or survival of the invading population or of one or more segments of the community. This feedback is a dynamic component of the environmental resistance of the suscept and contrasts with the more static devices or barriers that often provide the first line of defense but do not respond directly to the microorganism's presence.

Several *defense mechanisms* of living hosts are well characterized, and their contribution to environmental feedback is clearly defined, but the identities and modes of action of additional barriers erected are unknown. Some of these barriers are relatively nonspecific and act on a wide range of microorganisms coming into intimate contact with living animals and plants, while others are quite specific and retard the development of a small group of closely related bacteria, fungi, or viruses. The defense mechanisms and the factors contributing to environmental feedback vary according to the tissue or organ in which the invader finds itself, several defense mechanisms being mobilized in some such microenvironments for elimination of pathogens, with few barriers available in spatially separate areas. Thus direct inoculation of *Diplococcus pneumoniae* into the bloodstream leads to its successful establishment and often a lethal outcome, while introduction of the bacteria into the anterior chamber of the eye, into the peritoneum, under the skin, and into the lung are decreasingly effective in allowing for colonization (White, 1938). Similarly, fungi applied to various parts of the plant encounter effective, partially effective, or ineffective barriers.

Burnet (1962) has pointed out the potential evolutionary significance of an interesting instance of host defense. An injury to or operation on the knee joint permitting bacteria to remain in the joint cavity commonly leads to an acute bacterial infection because of the paucity of vigorous protective devices, whereas even as drastic a surgical step as the extraction of a number of teeth rarely leads to microbial complications. In a toothed animal species, damage to the teeth and intrusion of bacteria into wounds of the mouth are quite common, and were there no particularly effective barriers in this region, the chances that the species would survive would be small. On the other hand, knee injuries are uncommon, and the absence of an effective mechanism of resistance in that potential microbial habitat is thus of little consequence to the survival of the animal species.

Two variables associated with environmental feedback have already been cited: the appearance of interferon or interferon-like substances and the increase in the peroxidase activity after infection of certain plants with fungi. Other barriers erected by hosts are known, however, some whose ecological and pathological importance is better understood. Animals provoked by a microbial invasion may react by synthesizing antibodies and releasing them to participate in the inactivation of microscopic intruders. The animal typically produces antibodies that are quite specific for the microorganism, often a pathogen, provoking their formation, and the level of these antibodies in the bloodstream rises as the feedback becomes more pronounced. The critical role of antibodies in the defense against infectious disease is particularly evident in individuals lacking antibodies or in whom the ability to produce antibodies is suppressed as a result of therapy. Although antibodies to a number of microorganisms indigenous to man can be found in the body fluids of healthy persons, for example, to species of *Staphylococcus*, *Pseudomonas*, *Hemophilus*, *Actinomyces*, and *Candida*, the antibodies still are assumed to reflect a host response mechanism, presumably aroused by past invasion or subclinical disease caused by the species against which antibodies were formed (Rosebury, 1965).

In contrast to animals, which form chemically complex antibodies that are highly specific for the microorganisms they affect, plants may react to a potential colonist by producing phytoalexins, compounds simpler in structure and apparently less specific for the incitant microorganism than antibodies. Phytoalexins either are generated de novo after the host is exposed to a potential invader, or their concentration increases dramatically as a result of such exposure. Phytoalexin research has focused on compounds made in response to fungal development, and a number of these plant fungitoxins have been characterized and identified. They are generally phenolic compounds and are inhibitory not only to the provoking fungus but to other species and genera as well. The phytoalexin barrier to invasion is effective only when the concentration of the compound attains a level sufficiently high to hold back the microorganism; should the concentration be too low, the fungus may then proceed to cause damage, provided, of course, that it meets no additional barriers and that the plant environment is conducive to its further proliferation (Goodman et al., 1967). Phytoalexins inhibitory to a large number of filamentous fungi gaining access to the surfaces of higher plants have been

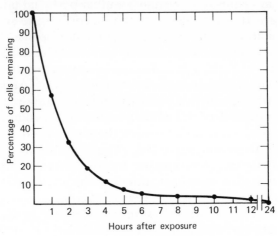

Figure 3.3 Killing of *Staphylococcus aureus* in the lungs of mice exposed to a suspension of these bacteria (Kass et al., 1966).

observed, and these compounds appear to be potent contributory factors to disease resistance and to the prevention of appreciable colonization of plants by pathogenic fungi.

Specialized cells, the phagocytes, are present in animal suscepts, and the increase in abundance of these specialized cells may pose an insurmountable obstacle to bacteria, viruses, rickettsiae, or protozoa reaching sites in the body that otherwise are hospitable and would allow for microbial growth. Phagocytes are often key factors in determining resistance to microbial infection and in ridding the body of harmful microorganisms. A good example of the potency of this barrier is seen in the fact that the bronchial tube–lung structure from the primary bronchi downward is normally sterile because pulmonary phagocytes function to clear the lung of the inhaled bacteria constantly being introduced from the surrounding atmosphere (Kass et al., 1966). Figure 3.3 shows the rapid decline in viable bacterial cells in the lungs of mice exposed to a large population of an aerosolized suspension of *Staphylococcus aureus*. Phagocytic cells commonly migrate from the blood, where they circulate, to the site in the tissue containing the alien microorganisms. The phagocytes may ingest and destroy the potential colonist and thus protect the animal from more serious consequences of infection. On the other hand, phagocytes may not be an insurmountable barrier or an obstacle; the phagocyte may reach an infective bacterium and be unable to ingest it, or the microbial cell may be ingested but may not only be unharmed but it may even proliferate in the cytoplasm of the phagocyte.

Environmental feedback may also be attributable to isolation mechanisms of the host, localized structures or mechanical barriers that restrict the range of microbial development and prevent the colonist from gaining access to an adjacent tissue site where further proliferation is possible. The spread of *Pasteurella pestis* introduced into the skin, for example, may be limited because of the formation of local vesicles or pustules, and fibrin walls formed in abscesses in injured animal tissues likewise may impede microbial proliferation. Isolation mechanisms also are characteristic of living trees, inasmuch as the spread of fungi seems to be obstructed by gums and resins formed by the trees, and nonwoody plants restrict invaders by forming corky layers or by laying down occlusions in the vessels through which microorganisms are not able to grow.

Many of the barriers and mechanisms of environmental resistance have a selective or differential influence, retarding or inhibiting some colonists but not others. For example, the properdin system of blood has no effect on species of *Streptococcus* but is inhibitory to numerous coliform bacilli; serum is toxic to most coagulase-positive but not to coagulase-negative *Staphylococcus*; and phagocytes do not ingest certain encapsulated bacteria but rapidly ingest and destroy closely related nonencapsulated strains. Bactericidal or fungicidal substances in root excretions, vegetables, and milk also act on selected groups. Such differential effects of the factors contributing to environmental resistance have a profound influence on the ecological success of an invader, the succession that may be initiated, the outcome of an infection, or the biochemical changes in the habitat.

References

REVIEWS

Alexander, M. 1961. *Introduction to Soil Microbiology.* Wiley, New York.

Blum, J. L. 1956. The ecology of river algae. *Botan. Rev.,* **22,** 291–341.

Burnet, F. M. 1962. *Natural History of Infectious Disease.* Cambridge University Press, London.

Chapman, V. J. 1957. Marine algal ecology. *Botan. Rev.,* **23,** 320–350.

Dubos, R. J. 1954. *Biochemical Determinants of Microbial Diseases.* Harvard University Press, Cambridge, Mass.

Goodman, R. N., Kiraly, Z., and Zaitlin, M. 1967. *The Biochemistry and Physiology of Infectious Plant Disease.* Van Nostrand, Princeton.

Haynes, F. N. 1964. Lichens. *Viewpoints Biol.,* **3,** 64–115.

Hoffman, H. 1966. Oral microbiology. *Advan. Appl. Microbiol.,* **8,** 195–251.

Marples, M. J. 1965. *The Ecology of the Human Skin.* Thomas, Springfield, Ill.

Rosebury, T. 1962. *Microorganisms Indigenous to Man.* McGraw-Hill, New York.

Shigo, A. L. 1967. Succession of organisms in discoloration and decay of wood. *Intern. Rev. Forestry Res.,* **2,** 237–299.

White, B. 1938. *The Biology of Pneumococcus.* Commonwealth Fund, New York.

OTHER LITERATURE CITED

Booth, W. E. 1941. *Ecology,* **22,** 38–46.

Dubos, R., Schaedler, R. W., and Costello, R. 1963. *Federation Proc.,* **22,** 1322–1329.

Eichenwald, H. F., Shinefield, H. R., Boris, M., and Ribble, J. C. 1965. *Ann. N.Y. Acad. Sci.,* **128,** 365–380.

Fletcher, J. E., and Martin, W. P. 1948. *Ecology,* **29,** 95–100.

Formal, S. B., Dammin, G., Sprinz, H., Kundel, D., Schneider, H., Horowitz, R. E., and Forbes, M. 1961. *J. Bacteriol.*, **82**, 284–287.

Kass, E. H., Green, G. M., and Goldstein, E. 1966. *Bacteriol. Rev.*, **30**, 488–496.

Macko, V., Woodbury, W., and Stahmann, M. A. 1968. *Phytopathology*, **58**, 1250–1254.

MacLeod, C. M. 1965. *In* R. J. Dubos and J. G. Hirsch, Eds., *Bacterial and Mycotic Infections of Man*. Lippincott, Philadelphia. pp. 391–411.

Martin, J. T., Batt, R. F., and Burchill, R. T. 1957. *Nature*, **180**, 796–797.

McCarthy, C., Snyder, M. L., and Parker, R. B. 1965. *Arch. Oral Biol.*, **10**, 61–70.

Mossel, D. A. A., and Ingram, M. 1955. *J. Appl. Bacteriol.*, **18**, 232–268.

Oram, J. D., and Reiter, B. 1966. *Biochem. J.*, **100**, 373–381.

Rosebury, T. 1965. *In* R. J. Dubos and J. G. Hirsch, Eds., *Bacterial and Mycotic Infections of Man*. Lippincott, Philadelphia. pp. 326–355.

Timonin, M. I. 1941. *Soil Sci.*, **52**, 395–413.

Turner, E. M. C. 1961. *J. Exptl. Bot.* **12**, 169–175.

Ueyama, A. 1965. *Mater. Organismen Beih.*, 325–332.

Wolin, M. J. 1969. *Appl. Microbiol.*, **17**, 83–87.

4

Succession and the Climax

Successions are characterized by shifts both in the species composition and in the relative abundance of the resident species of a community. Several groups are particularly numerous at one stage and relatively infrequent or totally absent at another as new dominants come to the fore. The rise and subsequent decline of freshwater and marine algae (Blum, 1956; Hutchinson, 1967; Raymont, 1966), fungi participating in wood decay (Shigo, 1967), and components of the bacterial flora of the gastrointestinal tract have been well characterized.

An excellent example of succession is the sequence of bacteria taking place in the mouse intestine. At the time of birth the mouse is free of cultivable microorganisms, but its various tissues and organs are rapidly invaded immediately after birth. As shown in Fig. 4.1, pioneers like the flavobacteria grow rapidly but then disappear, while pioneers such as the lactobacilli and anaerobic streptococci attain large population sizes and maintain their numbers. Still other groups proliferate slowly or are absent initially but reach high cell densities only to decline in abundance, while members of the genus *Bacteroides* and related forms appear late. Similar types of bacterial succession have been observed in the fecal flora of a number of animals, including man (Smith and Crabb, 1961).

Many kinds of microbial succession have been investigated. For example, in a denuded region initially containing little or no organic matter, as in certain bodies of water or on the surfaces of rocks, algae initiate a succession and generate the organic material that sustains the succeeding waves of heterotrophs. Heterotrophs constitute the initial colonizers of sterile tissues exposed when an animal or plant is wounded, setting the stage for specialized organisms appearing subsequently. The microbe-free surfaces of the plant root as it emerges

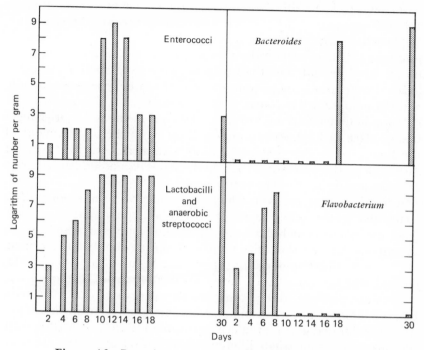

Figure 4.1 Bacteria in the intestine of mice at various ages after birth. Data are expressed as logarithm of bacterial number per gram of large intestine homogenate (Schaedler et al., 1965).

from the seed and the surfaces of the skin and intestinal system of the newborn provide hospitable sites for the development of appropriate bacteria and fungi, but these pioneers are soon replaced by different species more suited to the now inhabited site than were the pioneers. When simple organic materials are introduced into soil, waste water, or sewage treatment facilities, specialized groups appear first and use the added compounds; these are replaced by species that utilize the excretions or parasitize the cells of the pioneers. Hence even a simple organic compound will, when introduced into a habitat containing small numbers of a variety of microbial types, engender a sequence of events involving fluctuations in the abundance of scores of populations.

Successions of this sort involve a progressive development from the pioneer community, containing few species, through a series of stages characterized by increasing numbers of microbial groups, and finally to a phase typified by a reasonably high degree of species diversity.

This trend to increasing diversity with the course of succession probably reflects the few niches in the uncolonized site and the multitude of potential niches in the fully populated habitat. The pioneers are uniquely able to cope with the stresses, limitations, or nutrients of the initial environment, but the waves of succeeding species make the locale biochemically more complex so that more niches are available and the community becomes more heterogeneous. The trend to increasing heterogeneity with time is observed in the series of changes taking place in food undergoing spoilage or in the development of the algal crust in desert soils following a rain; in both instances the initiators of the succession soon find themselves accompanied by a growing multiplicity of heterotrophs.

A vast literature exists in which the sequences of microorganisms concerned in the colonization of particular habitats are described. The environments include animal and plant wounds, exposed tissues, extensively burned skin, the mouth of the newborn, hair, rumen, milk, roots, wood undergoing decay, leaves and leaf litter, rocks, natural and fumigated soil, lakes, streams, estuaries, the open sea, waste water, sewage, dung, compost, and decaying plant materials. Each of these unique environments becomes populated initially by a select and limited group of microorganisms, and each is subsequently colonized by a somewhat different constellation of species. Despite this vast literature, however, the reasons for the sequences and the characteristics determining success in colonization or prominence in the succession are largely unknown.

PROGRESS OF SUCCESSION

Two kinds of succession are readily distinguishable. In one, the sequence is brought about because the resident populations alter their surroundings in a manner such that they are replaced by species better suited to the modified habitat. This is an *autogenic succession*. The late developers have no selective advantage if they arrive early, but conditions conducive to their proliferation are created by the pioneers. The environment is made progressively more suitable for the secondary feeders, and the pioneers lose the selective advantage that permitted them to be initial exploiters of the habitat and hence are usually eliminated. Alternatively, the pioneers may bring about their own demise as they make the habitat increasingly unfavorable to themselves by the removal of nutrients or by the formation of acid or autoinhibitory products. In effect, a succession of the autogenic

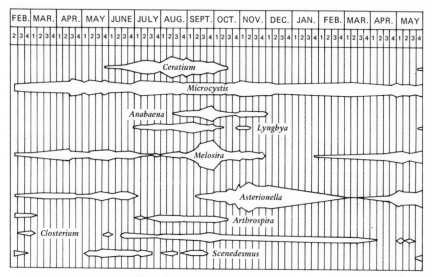

Figure 4.2 Seasonal changes in algal abundance in Lake Mendota. The relative abundance is shown in proportion to the cube root of algal density (Birge and Juday, 1922).

type proceeds and is governed by the activities and interrelationships of the inhabitants of the site and by their interactions with the inanimate environment.

In an *allogenic succession,* on the other hand, one type of community is replaced by another because the habitat is altered by nonmicrobial factors—by changes in ecologically significant physical or chemical properties of the region induced by abiotic agencies or by modifications in the plant or animal host within or upon which the microorganisms reside. Certain allogenic successions are cyclic, such as those involving aquatic algae exposed to seasonal fluctuations in temperature, light intensity, and nutrient concentration. Many are noncyclic and do not recur at that site. A seasonal succession, one governed by both nonmicrobial and microbial factors, is depicted in Fig. 4.2.

A number of well-defined factors determine or contribute to succession. These include (a) the provision by one community of a nutrient that confers an ecological advantage on the species constituting the next stage in the succession; (b) the making available by one population of an element present in insufficient supply to allow for growth of later populations, typically the generation of organic carbon by algal photosynthesis or the fixation of N_2 by bacteria, Cyanophyta, or lichens; (c) the alteration in concentration of an inorganic nutrient; (d) the modification of heterogeneous substrates, such as plant remains or animal tissues, so that constituents favoring the growth of new species are exposed to attack, while constituents metabolized by prior populations are lost as a result of decomposition; (e) an autointoxication effected by the existing community; (f) the elimination of microorganisms by physical means; (g) the appearance of barriers associated with environmental feedback; (h) the selective feeding by animals on microbial populations; and (i) changes in temperature and light intensity in the course of the year. These factors are discussed individually below. Regardless of the factors involved, the available data show that pioneer populations are not necessarily able to maintain themselves merely because they have proliferated and colonized a site. Natural selection determined which of the arrivals were successful as pioneers, and natural selection likewise governs which of the pioneers or later arrivals will maintain their foothold or subsequently assume a dominant status.

The pioneer community is restricted to species capable of growing by use of just those nutrients already in the surroundings. In some areas little or no organic carbon is present, but the availability of sunlight and inorganic nutrients allows for the growth of algae and possibly photosynthetic bacteria. In sterile regions containing one or several organic compounds but few growth factors, nonfastidious heterotrophs will proliferate initially. In still different environments, for example, in sterile food products, animal tissues, or plant materials to be ensiled, growth factors abound, and even a highly fastidious population may assume a major primary role.

Members of the primary community often excrete carbonaceous materials utilizable by other species already present or subsequently invading the site. In the wine fermentation, sugar-metabolizing *Saccharomyces* produces ethanol, which, if oxygen is available, may support an abundant population of one or another *Acetobacter*. In the natural spoilage or controlled fermentation of milk, the lactic acid generated from sugar in the milk sustains a number of different bacteria. Cellulose is decomposed anaerobically to fatty acids by *Clostridium* spp., and the fatty acids formed are excellent energy sources for unique noncellulolytic bacteria responsible for methane evolution. Algae produce during photosynthesis a variety of organic acids and simple carbohydrates, and these are ideal nutrients for various aquatic heterotrophs. The primary feeders, that is, the organisms at the first trophic level, frequently exist side by side with the secondary ones for some time in such a way that a new stage in the succession has been initiated before the prior stage is terminated, the overlapping resulting from the continuing availability of nutrients for the pioneers even as they create carbon and energy sources for their successors.

Elements in addition to carbon are transformed to a state capable of supporting groups unable to assimilate the element in its existing form. Thus proteolytic bacteria and fungi liberate ammonium during the metabolism of proteins, and nitrifying autotrophs use the ammonium as a source of energy. The sulfate generated microbiologically from proteins is an essential nutrient for many species, and it also serves as a terminal electron acceptor for the growth of *Desulfovibrio* in marine and lake muds, estuaries, and soil. The sudden blooming of lake and river algae is sometimes triggered by the appearance of nitrate, but nitrate formation in water is attributable largely to biological processes and involves organisms other than the algae causing the bloom.

Progressive biochemical enrichment of nutritionally impoverished areas parallels the progressive enrichment with additional populations. A widespread characteristic of growing heterotrophs and autotrophs is their ability to excrete certain metabolites, including compounds that are needed by neighboring individuals. The biochemical enrichment acts selectively, allowing for the increase of only particular species, the selectivity depending on, among other things, the molecules liberated and the requirements of those organisms reaching the particular place at the appropriate time. Algae inhabiting aquatic localities, although photosynthetic, occasionally require one or several B vitamins, and various heterotrophic bacteria and fungi will not reproduce unless one or more B vitamins, amino acids, or other growth factors are present. These organisms only appear in habitats where the growth factor needs can be satisfied, and frequently their requirements are met by the excretions of a prior resident. Thus aquatic algae demanding growth factors commonly proliferate in a late stage of succession, while the algae appearing early, when the environment is biochemically simpler, are generally able to develop in simple inorganic media (Margalef, 1958). Similarly, certain fungi unable to synthesize all the vitamins they require are often late to appear in successions.

The mere formation of cell constituents paves the way for new types of inhabitants, namely, predators, parasites, and microorganisms synthesizing enzymes that act on cells or filaments of components of the existing community. The parasite or predator, in order to become established, must not only find a suscept but must compete with other parasites and predators capable of feeding on the existing microflora and microfauna. In communities inhabiting waste water and sewage treatment facilities, heterotrophic bacteria often dominate in the pioneer stage, but their appearance is a prelude to an increase in numbers of protozoa—largely rhizopods, flagellates, and ciliates— feeding on the bacteria. Successions are evident, too, among the protozoa, given groups dominating at one stage but then disappearing as others assume prominence. Fungal hyphae growing at the expense of organic materials in soil frequently are colonized by bacteria and *Streptomyces* excreting extracellular enzymes that digest the fungal cell walls with the release of protoplasmic constituents, and algal blooms in lakes are not infrequently parasitized by fungi and bacteria. Species exhibiting a predatory or parasitic mode of life almost invariably appear late in the succession, either because they are obligate predators or parasites or because, though capable of independent life, they are unable to cope or compete with free-living organisms.

Some aquatic or terrestrial sites contain little organic carbon or minute amounts of nitrogen, and the only species able to participate in the initial colonization of such habitats are those having the capacity to assimilate CO_2 or N_2. The pioneers are thus usually algae or N_2-utilizing microorganisms. In nature, nitrogen deficiency is at times associated with a lack of organic carbon, so here it is the N_2-assimilating Cyanophyta that initiate the succession. As organic carbon or nongaseous nitrogen compounds accumulate, less versatile species make their appearance and gradually assume a more dominant role. Sequences of this sort, initiated by blue-green algae and followed by heterotrophs, are quite common in lakes and reservoirs. For instance, a bloom of the N_2-utilizing *Anabaena* in a reservoir of the White Nile was apparently sufficient to allow for the growth of other organisms that rapidly populated the area, making use of the nitrogen and presumably the carbon initially tied up and then released by the algae (Prowse and Talling, 1958).

The concentration of inorganic nutrients also seems to regulate the appearance and disappearance of populations. Various algae, for example, are not tolerant of high but do well at low concentrations of nutrients, a sensitivity that may explain why some species bloom in lakes after neighbors grow and presumably lower the inorganic nutrient levels (Fogg, 1965). Seasonal overturns in lakes cause nutrients from the deep waters to mix with the nutrient-poor surface zone containing the bulk of the algae, and this fresh nutrient supply induces a bloom as part of an allogenic succession. Changes in concentrations or ratios of nutrients also have been proposed to account for successions involving marine diatoms, for the replacement of one species by another as water flowing downstream gains nutrients, and for the outbreaks of the so-called red tide or discolored waters frequently caused by extensive growth of a species of *Gymnodinium*.

An entirely different kind of succession is initiated when a heterogeneous carbonaceous material, such as a plant residue, is subjected to microbial decay. Plant remains contain readily degradable simple molecules, more resistant hemicelluloses and cellulose, and a highly refractory lignin fraction. Phycomycetes develop rapidly and utilize readily available but not resistant plant constituents; these fungi are thus restricted to a pioneer role and fail to maintain their foothold in the decaying organic matter. Following the initial flare-up of the fast growers, slow-growing fungi able to utilize cellulose and lignin come to the forefront. Although the latter might be capable of using the substrates metabolized by the pioneers, their slow growth rate excludes them from the initial community while their greater enzymatic versatility suits them to a late stage in the progression.

The shift from one group of species to another frequently results from the creation by the first community of conditions unfavorable to its own existence, as by the accumulation of acid or toxic products. Despite the autointoxication of the pioneers, the growth of species unable to colonize initially is not precluded, and they assume dominance as the condition causing the demise of the pioneer community is intensified. Acidity appears to be largely responsible for the successions that take place in certain environments rich in carbohydrates but poor in O_2; in the fermentation of cabbage, cucumbers, milk, and ensiled plant remains, bacteria such as *Lactobacillus* or *Streptococcus* spp. form considerable lactic acid, the accumulation of which ultimately eliminates the pioneers and favors the proliferation of acid-tolerant anaerobes. Biologically induced changes in pH are also believed to account for occasional protozoan successions. Similarly, lactobacilli and possibly corynebacteria seem to generate sufficient lactic acid in the vagina to inhibit all but acid-tolerant organisms, a process of microbial exclusion that may be the basis for the vaginal defense against possible harmful invaders (Rosebury, 1962). An analogous kind of succession is evident in water and soil receiving sulfides or elemental sulfur; thiobacilli proliferate by oxidizing the inorganic sulfur, and the enormous amount of sulfuric acid they produce is the determinative factor in favoring the establishment of a new community.

By contrast with the unquestionable significance of acidity, the

Table 4.1

Population Changes in *Brucella abortus* Cultures [a]

	Nonsmooth Mutant Cells (% of total)		
Days	In Synthetic Medium	In Synthetic Medium + Old Culture Filtrate	In Synthetic Medium + Alanine
0	0[b]	0[b]	0[b]
2	0	0	0
4	0	0	2
6	0	1	19
8	0	2	36
10	—	64	—
11	2	—	83
14	6	—	86
15	—	91	—

[a] From Braun (1952).
[b] The original cultures contained only cells giving rise to smooth colonies on agar media.

evidence for a role of microbial toxins in succession is spotty or equivocal. Toxins have been hypothesized to be important in algal successions, and autotoxic substances are believed by some to explain the termination of certain freshwater blooms. The results of Vance (1965) suggest that inhibitors excreted by *Microcystis aeruginosa* account for its prevalence in ponds and for the abrupt decline of nearby algae. In addition to organic inhibitors, inorganic toxins may be significant; H_2S, for example, is highly toxic and is produced in large amounts from sulfate or sulfur-containing amino acids by bacteria.

In a few of the microbial successions taking place within the animal body, one population may be displaced by virtue of its excretion of products to which it is sensitive but which have little influence on new populations arising or gaining entry to the habitat. Such a disturbance is beautifully illustrated in studies of *Brucella abortus*. Strains of this human pathogen produce an amino acid, D-alanine, that inhibits the metabolite-producing bacterium in vitro but allows for the growth of D-alanine-resistant mutants arising in cultures of the sensitive parent (Table 4.1). The succession in vitro involving a D-alanine-producing, sensitive strain that is overgrown by a resistant population may be associated with an alteration in virulence of the culture. Population changes analogous to those noted in vitro are observed when alanine is administered to guinea pigs infected with one *B. abortus* type, the community in the animals shifting to a more virulent, more alanine-resistant type (Braun, 1952; Mika et al., 1952).

Pioneers may modify the properties of the environment in additional ways so that they are harmed while simultaneously a succeeding species gains a distinct advantage. The first organisms to propagate in sugar-rich solutions lower the osmotic pressure and often are replaced by less sugar-tolerant but more competitive heterotrophs. Obligate aerobes lower the O_2 tension in aerated surroundings and ultimately can create conditions leading to their own demise and replacement by anaerobes; pathogenic *Clostridium* species thus may appear in wounds where their initial colonization was not feasible. Similarly, organic materials introduced into fresh water, sewage treatment plants, or soil promote the development of a large array of aerobic bacteria and fungi that consume the available O_2, and unless the diffusion of O_2 from the atmosphere is sufficiently rapid to meet the biological demand, the aerobes are displaced by facultative and obligate anaerobes. Biologically induced modifications of the physical properties of a habitat may be a contributing factor or essential to succession, as in the dependence of certain algae on others for support and for possible protection against rough seas and high light intensities. A physical change, but in the cell rather than in the environment, occasionally is responsible for seasonal succession, as illustrated in the sinking of populations of aquatic organisms as they metabolize intracellular constituents serving to keep them afloat; the site vacated provides a home for a new arrival.

Environmental feedback sometimes decisively alters the course of a progression. Feedback mechanisms, as they become more pronounced in the animal or plant, tend to exclude species unable to cope with the increasingly effective host response, but microorganisms resistant to the colonization barriers erected persist. Antibodies, phagocytes, and phytoalexins may contribute to this sort of allogenic succession by acting selectively, destroying or retarding the growth of many but not all of the microorganisms living on the host. On the other hand, the feedback itself is known to provoke a response resulting in the displacement of a population sensitive to a host defense mechanism by a resistant mutant. In the animal inoculated with avirulent strains of diverse bacterial pathogens, for example, the avirulent population is eliminated but is supplanted by a virulent population that arose from a mutant, apparently derived from the avirulent culture, that possesses structural or enzymatic features protecting it against the host's defense reactions. Quite commonly the mutation gives rise to cells less readily destroyed by phagocytosis.

Selective grazing by larger animals on algae has been proposed to have a significant impact upon aquatic successions. Herbivorous animals seek out some algae and reject others, the rejected group thereby gaining a marked benefit. Nevertheless, further data are necessary before the importance of grazing in determining the outcome of microbial successions can be unequivocally established.

Cyclic, allogenic successions associated with season of the year may result from a change in the nutrient concentration attributable to mixing of rich water with nutrient-poor waters or from the appearance of herbivores with selective feeding habits. In addition, changes in light intensity, day length, and temperature occur with the move from season to season. Evidence exists that one or more of these factors contribute to the assumption of dominance or decline of individual algal species, as the prevailing light conditions or temperature become more suitable for particular inhabitants of the region (Braarud, 1961).

It must be stressed that many factors operating in concert, not just one, may determine the course of succession. Even when temperature, light intensity, the concentration or type of nutrient, host response, O_2 tension, toxin concentration, predators, or another ecological determinant is clearly correlated with the progressive changes in community composition, no more than a select few of the crucial environmental characteristics may have been examined by the investigator; one or a combination of several unstudied factors may be the underlying reason for the sequence of populations. Causation is not easy to establish, and correlations frequently provide initial avenues for investigation but rarely are the data obtained conclusive. The need remains for intensive study of many microorganisms and different ecosystems to establish the factors that are of major consequence in determining why individual species assume dominance and then recede to the background with the passage of time.

CHARACTERISTICS FOR SUCCESS IN COLONIZATION AND SUCCESSION

A microorganism reaching a potentially habitable location may die immediately after encountering the site, it may remain viable for short or long periods without initiating active metabolism, or it

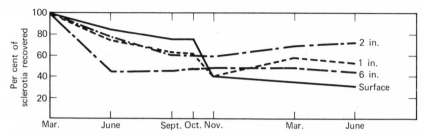

Figure 4.3   Survival of sclerotia of *Sclerotinia trifoliorum* at different depths in soil (Williams and Western, 1965).

may grow and establish itself as a member of the newly evolving community. Many physiological or morphological characteristics might appear to be the bases for success in colonization, but no more than one or possibly several may be responsible for the establishment of a particular organism at a given place and time. It is axiomatic that every existing species is successful by one means or another, but the biological traits of paramount importance depend on the organism, the environment, and the prevailing conditions. The following list, though not complete, probably gives the major attributes contributing to success.

(a) Presence at the colonizable place at the correct time. This implies not only an effective dispersal mechanism but also one allowing the microorganism to invade its new habitat when conditions are suitable for proliferation. The susceptibility of animate and inanimate environments to microbial intrusion is frequently modified because of changes associated with season, physical forces, aging, and so on, and the invader must be at the site when its growth needs are satisfied. Some species are poorly dispersed, but they are able to reach a colonizable region because their filaments extend from one microsite to another.

(b) Possession of a survival capability permitting prolonged viability in deleterious circumstances. When favorable conditions return, the endospore, cyst, conidium, chlamydospore, sclerotium, or other resting stage gives rise to the metabolically active organism, which then may have a distinct advantage by reason of its intrinsic aggressiveness or because it persisted sufficiently long that cells of neighboring species died out. Species of *Bacillus*, for example, assume prominence because they but not their neighbors endure long droughts. The persistence of viable sclerotia is depicted in Fig. 4.3.

(c) Ability to obtain *all* nutrients from the ecosystem. The required nutrients may be present even in the uncolonized habitat, or they may have been synthesized by prior populations. The necessary nutrients include sources of energy, carbon, inorganic elements, an electron acceptor, and possibly one or more growth factors. That the absence of a single essential nutrient precludes establishment has been demonstrated most dramatically with mutants derived from animal or plant pathogens; although the parent culture is invasive and multiplies in the host, mutants requiring distinct growth factors fail to develop and produce no disease since the appropriate habitats in the plant or animal do not contain enough of the growth factor in question.

Some species owe their prevalence to the possession of enzymes catalyzing the degradation of natural products or novel molecules that few indigenous microorganisms use; thus the ability to grow with lignin as carbon source is common to Basidiomycetes but is rare among soil heterotrophs, and only rare bacteria seem able to multiply by using various synthetic organic compounds as carbon sources. Likewise, N_2-fixing blue-green algae obtain all requisite nutrient elements from habitats containing neither organic carbon nor nongaseous forms of nitrogen, a blessing accorded to few microorganisms, and just autotrophic bacteria proliferate in the dark in solutions containing inorganic ingredients alone.

(d) Capacity to tolerate all the ecologically significant abiotic factors of the environment. The pH, O_2 level, temperature, osmotic and hydrostatic pressure, oxidation-reduction potential, light intensity, and moisture level must be such as to allow for survival and, at occasional intervals at least, vegetative growth. The intensity or level of these factors may vary at irregular or regular intervals, as in the seasonal or diurnal fluctuations in temperature and light intensity of aquatic ecosystems and the frequent moisture changes in soil, and the successful species must be able to endure all these fluctuations. In habitats subject to wide fluctuations, the colonizing organism and indeed all indigenous species must have a sufficiently broad tolerance range to the environmental factor subject to variation to withstand all but the rare cataclysmic occurrence.

(e) Possession of mechanisms to overcome or cope with environmental resistance attributable to inanimate components of the habitat or to viable hosts. All barriers must be breached by the successful invader so that it can obtain what it requires for replication. Skin,

mucous membranes, plant surfaces, tissue and serum toxins, phagocytosis, and inhibitors added deliberately to foods and manufactured
products must be bypassed or overcome. Even the physical dimensions of the microbial cell or filament may be the basis for exclusion
or, alternatively, the ability to colonize, since small pore size in a
nonfluid environment, like soil, may be an insurmountable barrier to
a population of protozoa with large cells but pose no problem to
small-celled species. Intriguing pathogens eliminate structural barriers by producing enzymes that digest the structural component
restricting their colonization and spread, as in the action of *Clostridium* spp. implicated in gas gangrene. Among the enzymes catalyzing the destruction of animal structural elements and presumably
facilitating bacterial spread are lecithinase, hyaluronidase, and collagenase, which act on and destroy cellular lipids, connective tissue,
and muscle collagen. Other penetration-enhancing factors are known
to increase microbial virulence—for example, the enzyme making
mammalian cells more readily penetrated by *Toxoplasma gondii*
(Lycke et al., 1968)—but the identity of the structural barrier in the
host environment is often uncertain.

The environmental resistance resulting from inhibitors in tissues
or body fluids is abolished by populations synthesizing enzymes that
catalyze the destruction of the antimicrobial principles. Production of
detoxifying enzymes seems to account for the establishment of coliform bacilli in serum (Colebrook et al., 1960) and of the pathogen
Ophiobolus graminis var. *avenae* on oats (Turner, 1961), despite the
presence initially of toxins active against these organisms. Surface
components of gram-negative bacteria may protect the cells against
the bactericidal antibody-complement system of blood, and bacterial
polysaccharides may protect the possessors against nonspecific antibacterial agents in serum.

To cope with the phagocyte defense of higher animals, a number
of mechanisms have evolved among the bacteria. It is not surprising
that avirulent cells of pathogens are commonly far more readily
phagocytized than cells of closely related, virulent strains (Fig. 4.4).
The resistance to elimination by phagocytosis is frequently determined by the existence of a polysaccharide capsule surrounding the
cell, the polymer apparently acting largely mechanically to retard ingestion. Among the organisms possessing polysaccharide shields to retard phagocytosis are *Diplococcus pneumoniae* and *Hemophilus in-
fluenzae*. Somewhat different structural devices to protect against
phagocytosis are evident in the hyaluronic acid capsules of select

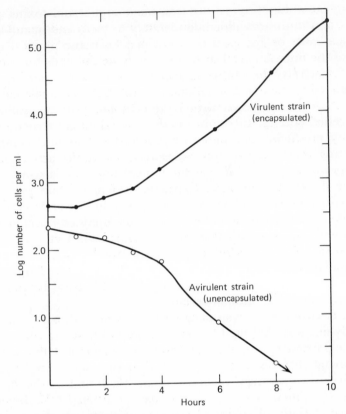

Figure 4.4 Clearance of *Diplococcus pneumoniae* from the blood of rabbits injected with bacterial strains of varying virulence (Wright, 1927).

Streptococcus strains and the host-derived fibrin layer that is laid down on the surfaces of cells of coagulase-producing *Staphylococcus* (MacLeod and Bernheimer, 1965).

(f) Ability to overcome or cope with environmental resistance attributable to microorganisms already present in the habitat. Competitiveness and resistance to predators, parasites, or microbiologically synthesized toxins are characteristics that may contribute individually or collectively to ecological success. The excretion of a toxin also reduces stresses set up by the microflora.

(g) Capacity to grow rapidly. The ability of a species to grow and make use of environmental resources at a greater rate than neighboring organisms constitutes an undeniable ecological advantage. Greater efficiency in converting organic substrates to cell material may likewise be advantageous. Rapid multiplication probably accounts for the quick colonization by *Pseudomonas* in many ecosystems, the early appearance of fungi like *Pythium* or *Trichoderma* in soils treated with sufficient fungicide to kill most microorganisms, and the dominance of *Clostridium* spp. in some foods undergoing spoilage. A rapid rate of development on plant and animal tissues may also be beneficial in allowing the invader to establish itself before environmental feedback becomes pronounced; thus *Erwinia aroideae* proliferates so rapidly on plant tissues that it is entirely at home before the host's defense reactions are fully mobilized, whereas extensive development of *Pseudomonas syringae,* an organism with a similar potential in terms of colonization and pathogenesis, does not occur, for the host's response becomes effective before the slower-growing bacterium gains a foothold (Goodman et al., 1967).

(h) Ability of the resting bodies of the species to germinate promptly and for the metabolically active cells emerging from them to proliferate readily when inhospitable conditions terminate and circumstances are conducive for vegetative life. This trait may account in part for the appearance of mucoraceous Phycomycetes as pioneers in soil.

(i) Production of specialized structures or surface components permitting the possessor to colonize surfaces from which other microorganisms are washed off. The early prominence of specific algal or fungal genera on stones and rocks in streams, rivers, or trickling filter beds seems to result from the specialized structures they have to make them fit for these habitats.

Each species making up the pioneer community, the participants in the various stages of succession, and those found in the final organism assemblage have individual and peculiar advantages in their characteristic milieux. The advantage may be directly or indirectly linked to discrete nutritional, physiological, or morphological traits. A small number of these traits are known; many are not. The lack of a character contributing to success of one species, such as high growth rate or resistance to toxins, may not be a major shortcoming for a second, since the latter may have a feature that possesses appreciable ecological value, like the synthesis of a rare enzyme or the capacity to produce a persistent structure. The microbial world is composed of thousands of genera and species, and the number of different overt and covert traits contributing to or underlying ecological success is undoubtedly enormous. The absence or low level of one trait is compensated for by the presence or increased intensity of another. The heterogeneity in the microbial world is indeed a reflection of the variety and combinations of characteristics providing existing species with their unique reasons for being.

THE CLIMAX COMMUNITY

Once an opportunity for colonization is created, wave after wave of populations appear and then recede. The replacement usually proceeds quite rapidly, although slow progressions from one stage to another do occur in especially harsh or nutrient-deficient ecosystems. In previously sterile but nutrient-rich ecosystems as deep wounds, extensively burned skin, canned food products that have been opened, or crop residues mixed into soil, succession is notably rapid. The initially sterile intestinal tract and the oral cavity of the newborn are extensively colonized within the first day or two, and the communities that arise are enriched with new groups and progressively modified with time (Hoffman, 1966; Schaedler et al., 1965). The turnover of algal populations in various rivers is quite rapid, too. Because microbial successions usually proceed rapidly, the initial and intermediary stages may not be observed. Occasionally, although the pioneer community is quickly overgrown, the subsequent displacement of species requires considerable periods of time; for example, the initial, rapid colonization of fumigated soil by fungi is followed by a very gradual but orderly replacement of fungal species, and considerable time elapses before the fungal community of treated soil is the same as in untreated land.

The reciprocal interaction between the microbial and nonmicrobial components of the ecosystem ultimately leads to a form of stabilization in which living and nonliving exist in harmony and in equilibrium with their environment. This final microbial assemblage in the area is known as the *climax community*. At the climax stage the species composition is maintained reasonably constant with the passage of time. The stability in the composition of the community does not result from a static set of circumstances but rather reflects a highly dynamic situation in which cells are constantly dying only to be replaced by new cells and new filaments. The climax is essentially a self-replicating entity that reproduces itself with remarkable fidelity. Microbial climaxes are typical of many aquatic ecosystems, soils, the activated sludge and trickling filter systems of sewage disposal, the bovine rumen, root surfaces, and presumably also the human skin, mouth, and intestinal tract and the leaf surfaces of perennial plants. The physical and chemical properties of the environment prevailing after colonization and succession have proceeded and biological factors imposed by the host or microbial coinhabitants of the site regulate the composition of this final community, both qualitatively and quantitatively. Given the same initial physical and chemical site characteristics or identical hosts, the same general types of successional sequences will be initiated and fostered; hence similar but spatially separate habitats tend to support remarkably similar climax communities, whereas the climaxes arising in environments with different properties usually are unalike. The abundance of some species may fluctuate to a modest extent, but the fluctuations are within reasonably narrow limits, the death rate of individuals in the various indigenous populations largely equaling the rate of formation of fresh cell material. Despite the potential of microorganisms for rapid growth, the population densities of the species making up the climax are, as a rule, reasonably fixed.

Nevertheless, the climax may be modified from time to time as the ecosystem is exposed to drastic exogenous forces. Soil organisms may be subjected to a prolonged drought, a stream may suddenly receive a large amount of pollutants, or animals may undergo a severe physiological stress, all of which will affect the composition of the indigenous community. The modification may be occasional or it may be recurrent, but the change in the climax is almost invariably transitory. The microbial equilibrium tends to be restored as the habitat returns to its original state. On the other hand, environments become irreversibly altered, too, as when the host ages, the nutrient level in a lake increases, or virgin land comes into cultivation; in such circumstances the climax is deflected and shifts to a different steady state, one in equilibrium with the new ecosystem.

The climax may contain either a few or a wealth of species, as indicated briefly in Chapter 1. Most of the habitats that so far have been studied extensively support stable communities with an abundance of species, if account is taken of all taxonomic groupings, although a single species of algae, fungi, or bacteria may stand out. Exceptions to the species diversity of climaxes are found in extreme environments, extreme in the sense of being near the biological tolerance limits of temperature, pH, organic or inorganic toxins, and so on. As cases in point, the flagellate *Bodo minimus* may dominate in H_2S-rich polluted water because few protozoa tolerate the sulfide levels (Fjerdingstad, 1954), halophilic bacteria are especially prominent on cured fish, a mere handful of species make up the community on foods containing extremely high sugar levels (Mossel and Ingram, 1955), while the biota of the healthy human vagina is restricted by the acidity and has only a few bacterial species (Rosebury, 1962).

Climax communities frequently occupy a remarkably small area. A stream or bay may show different communities in adjacent areas, and the biota inhabiting the surface of actively growing plant roots varies from place to place. A leaf may have a local lesion populated by pathogenic bacteria or fungi near healthy regions supporting solely nonpathogens, and an infected wound exhibits one group of bacteria whereas healthy skin just centimeters away supports a totally different community.

The climax contains many niches, and the species occupying each niche is uniquely fit, at least among those organisms having access to the locale, for the function associated with the niche in that specific ecosystem. Nevertheless, inasmuch as there are numerous niches or potential functions, particularly where there is an intermeshing of food chains, many physiologically different groups of organisms coexist indefinitely. The deliberate addition of an alien to a climax rarely leads to its successful or permanent establishment: most potential invaders are ubiquitous and have previously been introduced into the environment by natural forces, and the fact that they are not already established in the climax community suggests that their possible niches are filled by more fit organisms. This topic will be considered subsequently in greater detail.

To maintain the climax, energy must be supplied either frequently or continuously. The energy sources for heterotrophic life in nature include the organic carbon generated by microbial photosynthesis in oceans and lakes, root excretions or plant remains incorporated into the soil, host-derived products found on leaves and on the surfaces of animals, foods swallowed by vertebrates, or organic matter introduced into streams or sewage disposal systems. In certain ecosystems the microflora and microfauna developing at the expense of these carbon sources metabolize their organic nutrients at a rate such that the energy inflow—that is, the influx of organic carbon usable by heterotrophs—equals the rate of energy lost from the system. In climaxes of this sort, heterotrophs metabolize and degrade a quantity of carbon equal to that entering their environment, the net effect being that the carbon level remains constant. This balance between inflow and outflow is well exemplified by the microbial community of soil, a community acting in a way that the amount of CO_2 carbon lost yearly is essentially equivalent to the quantity of organic carbon added. The same is not true of bogs, marshes, septic tanks, or other anaerobic environments which receive regular increments of plant remains or carbonaceous wastes; in these anaerobic, closed environments, the organic matter level rises with time.

The available supply of many elements is constantly regenerated by the climax communities of water and soil. These nutrient elements are assimilated during active growth, and they are subsequently released upon the death and decomposition of the cells, only to be reused by succeeding cells and filaments. One or more nutrients may be lost from the ecosystem, but these must be replaced through microbial or other means for the climax to be self-perpetuating. Owing to the depletion of an essential nutrient element, some successions may not even terminate in a climax but rather lead to their own demise; thus, in the succession accompanying the decomposition of organic materials added not continuously but just once to a specific site, the available supply of one or more nutrient elements—frequently carbon—is exhausted by the metabolic activities of the series of heterotrophs that appear and disappear, and the biota thus destroys the basis for its own sustenance.

The climax concept, though quite useful in microbial ecology, is not applicable to all habitats. Many communities are ephemeral and are readily upset or completely eliminated. Moreover, cyclic changes in abiotic properties of the environment may be comparatively rapid by comparison with the rate of species replacement in succession, and a new cycle of physical and chemical changes may be initiated before a climax is reached; this seems to occur in some aquatic ecosystems. In several instances, as in bacteremias and other animal infections, there is no succession but rather the first community that appears—which may even be composed of a single population—maintains itself until the environment is drastically altered, as, for example, the result of the recovery of the diseased host and the elimination of the pathogen or, conversely, the death of the host. In wounds, on the other hand, a succession is initiated and continues to progress, but the sequence of population changes terminates abruptly as the defense mechanisms of the host result in elimination of the pathogen and accompanying invaders. Nevertheless, a wide variety of habitats populated by microorganisms support climax communities, and for these a delineation of the characteristics and properties of the climax microflora and microfauna is quite relevant.

References

REVIEWS

Blum, J. L. 1956. The ecology of river algae. *Bot. Rev.,* **22,** 291–341.

Fogg, G. E. 1965. *Algal Cultures and Phytoplankton Ecology.* University of Wisconsin Press, Madison.

Goodman, R. N., Kiraly, Z., and Zaitlin, M. 1967. *The Biochemistry and Physiology of Infectious Plant Disease.* Van Nostrand, Princeton.

Hoffman, H. 1966. Oral microbiology. *Advan. Appl. Microbiol.,* **8,** 195–251.

Hutchinson, G. E. 1967. *A Treatise on Limnology,* Vol. 2. Wiley, New York.

Margalef, R. 1958. Temporal succession and spatial heterogeneity in phytoplankton. *In* A. A. Buzzati-Traverso, Ed., *Perspectives in Marine Biology.* University of California Press, Berkeley. pp. 323–347.

Raymont, J. E. G. 1966. The production of marine plankton. *Advan. Ecol. Res.,* **3,** 117–205.

Rosebury, T. 1962. *Microorganisms Indigenous to Man.* McGraw-Hill, New York.

Selleck, G. W. 1960. The climax concept. *Bot. Rev.,* **26,** 534–545.

Shigo, A. L. 1967. Succession of organisms in discoloration and decay of wood. *Intern. Rev. Forestry Res.,* **2,** 237–299.

OTHER LITERATURE CITED

Birge, E. A., and Juday, C. 1922. *Wisc. Geol. Natl. History Surv. Bull.* 64 (Sci. Ser. 13).

Braarud, T. 1961. *In* M. Sears, Ed., *Oceanography.* American Association for the Advancement of Science, Washington, D.C. pp. 271–298.

Braun, W. 1952. *Am. Naturalist,* **86,** 355–371.

Colebrook, L., Lowbury, E. J. L., and Hurst, L. 1960. *J. Hyg.,* **58,** 357–366.

Fjerdingstad, E. 1954. *Hydrobiologia,* **6,** 328–330.

Lycke, E., Norrby, R., and Remington, J. 1968. *J. Bacteriol.*, **96**, 785–788.

MacLeod, C. M., and Bernheimer, A. W. 1965. *In* R. J. Dubos and J. G. Hirsch, Eds., *Bacterial and Mycotic Infections of Man.* Lippincott, Philadelphia. pp. 146–169.

Mika, L. A., Goodlow, R. J., and Braun, W. 1952. *Bacteriol. Proc.*, 77–78.

Mossel, D. A. A., and Ingram, M. 1955. *J. Appl. Bacteriol.*, **18**, 232–268.

Prowse, G. A., and Talling, J. F. 1958. *Limnol. Oceanogr.*, **3**, 222–238.

Schaedler, R. W., Dubos, R., and Costello, R. 1965. *J. Exptl. Med.*, **122**, 59–66.

Smith, H. W., and Crabb, W. E. 1961. *J. Pathol. Bacteriol.*, **82**, 53–66.

Turner, E. M. C. 1961. *J. Exptl. Bot.*, **12**, 169–175.

Vance, B. D. 1965. *J. Phycol.*, **1**, 81–86.

Williams, G. H., and Western, J. H. 1965. *Ann. Appl. Biol.*, **56**, 261–268.

Wright, H. D. 1927. *J. Pathol. Bacteriol.*, **30**, 185–252.

5

Nutrition

The habitat is the source of all nutrients required by viable micro-organisms, and the component members of a community are entirely dependent on their surroundings for the various inorganic and organic substances necessary for their continued proliferation. The requirements may be few in number, as with the wholly nonfastidious algae or bacteria requiring a single energy source and several inorganic ions, or they may be particularly abundant. But whether it is a nonexacting inhabitant, a species needing a few organic molecules, or the highly fastidious parasite restricted to one kind of host, the organism relies on its native milieu for the substances essential for its life.

Because of this dependency, the absence of a species from an individual locality may be ascribed to the lack of one or more nutrients it requires, or the population size may be regulated by the level or turnover of an essential factor whose concentration or rate of formation is too low to maintain maximum cell numbers. Hence the concentration, rate of formation, or absence of a nutrient may well determine the composition of a community. The factor in question may be light, an organic energy source, some other macronutrient, or an inorganic or organic micronutrient, but the lack of any one precludes development of the dependent species. Conversely, the growth of an organism in an environment is of itself evidence that the essential nutrient, or a physiologically equivalent substance, is present, although analytical techniques may not reveal its presence because, though constantly generated, it is utilized as quickly as it appears. It is convenient also to assume that organisms requiring specific metabolites in vitro need them in vivo, so that the growth of a population in nature can be taken as evidence that the compound or ion known to be essential, on the basis of laboratory studies, is found in the natural habitat (Lucas, 1955).

Not every essential nutrient is of ecological significance. A needed substance present at all times in concentrations sufficiently high to satisfy the demand of all indigenous and invading populations is of little ecological consequence, unless the level is so high that toxicity results. Organic compounds or inorganic ions are sometimes present initially in amounts high enough to meet the demand of the pioneers, but growth leads to a diminution in the supply so that late arrivals may fare poorly by virtue of the inadequate levels remaining.

Chemical analysis of an environment prior to invasion tells more about the pioneers that cannot multiply than those that do; organisms requiring substances not initially present are excluded, but many that should be able to grow, at least on the basis of their nutritional patterns, fail to become established for reasons not related to nutritional habits. Nevertheless, fastidious species appear in heterogeneous communities occupying localities not initially containing metabolites critical for the exacting organisms, because the early colonists synthesize the metabolites needed by the more demanding groups. Analysis of a habitat supporting a climax likewise provides limited information on the nutrition of the indigenous forms, because nondemanding organisms multiply where growth factors are available, ignoring the complex compounds all around, and because, conversely, a fastidious population may use the entire quantity of a factor it requires, leaving none for the analyst to find. For these reasons assessments of nutritional patterns in vitro frequently provide more useful information than chemical analyses of the ecosystem.

If a species is excluded from an area because a nutrient it needs is absent or if its population density in nature is governed by the quantity of a particular substance, that substance is considered to be a *limiting nutrient*. Similarly, if the activity or biomass of a community is governed by one nutrient, it too is said to be limiting. Various inorganic ions and organic compounds are known to restrict the distribution and abundance of individual species or the biochemical transformations catalyzed by complex microbial assemblages. The restriction is eased as the concentration of the deficient factor is raised.

NUTRIENT REQUIREMENTS

Among the substances that must be obtained from the surroundings for vegetative development to proceed are the following: (a) a supply of energy in the form of sunlight, H_2, ammonium, nitrite, an inorganic sulfur compound, ferric iron, or an organic substrate to support the synthesis of protoplasmic constituents; (b) a carbon source, either organic or CO_2; (c) a suitable terminal electron acceptor—O_2 for aerobes, nitra : for denitrifying bacteria, and CO_2, sulfate, or a simple organic molecule for anaerobes; (d) a source of nitrogen, phosphorus, and sulfur; and (e) several other nutrient elements, usually in an inorganic state. In addition to serving as a terminal electron acceptor for growth, O_2 is required by certain species in one or more of the reactions by which the carbonaceous substrate is metabolized, as in the initial phase of hydrocarbon degradation by bacteria.

No difficulty is encountered in establishing an obligate dependence on potassium, magnesium, and iron, for appreciable amounts are necessary. It is likewise simple to demonstrate that significant quantities of nitrogen, sulfur, and phosphorus are needed. Some populations grow as well with organic as with inorganic forms of the latter three elements, although the organic molecule is probably decomposed first to ammonium, H_2S, or orthophosphate prior to assimilation of the nutrient element by the majority of organisms using the organic complexes. On the other hand, groups specifically relying on an organic form of nitrogen or sulfur are not difficult to isolate. By contrast with the foregoing, requirements for micronutrients are less readily established because of the minute amounts involved and the frequent presence of the element in question as a contaminant in culture medium ingredients, but the currently available data indicate that ions of elements other than the six listed above may be either essential or stimulatory (Table 5.1). Needs for a few of the elements listed in Table 5.1 are rare, and only a narrow spectrum of microorganisms may show the requirement, as is the case for silicate and chloride ions. A need for cobalt, zinc, and copper, though difficult to establish, has been recorded for some species, and the elements may prove to be essential for a significant number of genera.

The concentration of many elements that suffices for growth depends on the species (as with calcium and iron) and cultural conditions (molybdenum). Furthermore, an organism may be able to use only a limited array of compounds or a single form of the element; thus not all species possess the ability to use nitrate, organic sulfur, or tetravalent manganese. The presence or amount of particular elements affects pigmentation, a feature striking to the eye but also one sometimes of ecological significance. The precise influence on pigmentation differs according to the organism, but such effects have been noted with copper, boron, molybdenum, cobalt, and manganese. It is also known that occasionally one element can be partially or

Table 5.1

Elements Essential for Selected Microbial Groups

Element	Typical Organisms Requiring the Element
Sodium	Halophilic and marine bacteria, marine fungi
Calcium	Algae, bacteria, protozoa
Silicon	Diatoms, possibly some chrysomonads
Manganese	Algae, protozoa, bacteria
Molybdenum	Blue-green algae, certain bacteria
Boron	Some algae and bacteria
Vanadium	A few algae
Chlorine	Certain marine bacteria

totally replaced by another. Examples of replacements are potassium for sodium, strontium for calcium, vanadium for molybdenum, and bromine for chlorine in selected aerobes and anaerobes.

The microbial cell is composed of a multitude of organic compounds that together are responsible for structure and physiological activity. These include amino acids, B vitamins, purines, and pyrimidines, among others. Most simple organic substances that exist free or are bound in polymers are synthesized by the cell, but certain organisms cannot produce one or more of these essential metabolites. Hence, for vegetative growth to proceed, an external source must be provided. These exogenously supplied substances are the *growth factors*, the term referring to organic compounds that in low concentrations promote growth. The carbon, energy, and nitrogen sources must be available in large amounts, and the micronutrients considered above are inorganic, so that they are not deemed to be growth factors.

A broad range of competence exists in the ability to synthesize growth factors, and hence there is wide variation in the reliance of microorganisms on their habitat for a supply of these compounds. Unique bacteria and many algae proliferating readily in the absence of any organic molecules and those bacteria, fungi, and actinomycetes needing but a single carbon compound, one used for both carbon and energy, have no such wants. Less versatile populations need one or two, while occasional bacteria and protozoa require five or more. Among the growth factors shown to be essential for one or another species are the B vitamins, diverse amino acids, purines and pyrimidines, nucleosides and nucleotides, sterols, saturated and unsaturated fatty acids, heme compounds, coenzymes such as coenzyme A and NAD, compounds of the vitamin K group, choline, lecithin, hydroxamic acids, di- or polyamines, and N-acetylglucosamine. Many fastidious protozoa, fungi, and algae multiply in highly complex media but not in those containing defined chemicals, and others have yet to be grown in axenic culture or in the absence of a host, so that the list of nutritional requisites remains far from complete.

An exacting species might, at first glance, seem to be handicapped in nature when compared with its nondemanding neighbor. If, however, the environment contains the substances needed by the more fastidious organism, it is not necessarily at a disadvantage. And if the exacting organism has physiological attributes that make it more fit than a population with a simple nutrition, with the biochemical gains possibly being paid for by nutritional losses, then it may well become dominant in the community and even exclude the nutritionally more versatile coinhabitant of the site.

Strains and varieties of a single species or species of the same genus may have dissimilar growth factor requirements and use entirely different carbon and nitrogen sources. Such dissimilarities might, at least in part, underlie the patterns of geographical distribution, host selection, prey preferences, or choices of microhabitats among members of individual taxonomic categories.

Specific growth factors are essential, and replication does not proceed without them. Others are beneficial or stimulatory to a particular organism, but development is feasible in their absence. Stimulation by a substance that is not essential is commonly attributed to a slow rate of synthesis of the metabolite by the cell, a pace too slow to allow for maximum multiplication rates, and the externally supplied factor then supplements the sluggish intracellular production.

The optimum concentrations of the sources of energy, carbon, inorganic nutrients, and growth factors often vary among the several potential invaders or inhabitants of an environment, so that a given level might be ideal for some, excessive for others, and suboptimal for a third group. Every species also has a maximum level for each nutrient, above which it will not grow, and some, possibly most, organisms have a minimum concentration of specified essential factors below which vegetative development does not occur. The ecological role of nutrient concentration has been thoroughly explored with the freshwater algae, particularly in regard to the components of communities appearing in *oligotrophic* and *eutrophic* lakes. An oligotrophic body of water is poor in inorganic nutrients, while one which is eutrophic is rich in inorganic and often organic substances and hence commonly supports a sizable algal bloom.

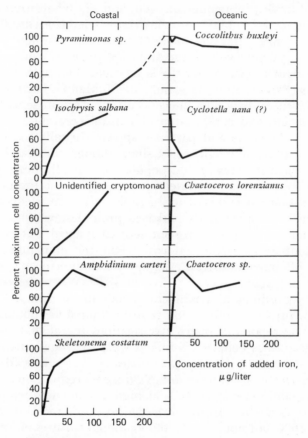

Figure 5.1 Relative iron requirements of coastal and oceanic algae (Ryther and Kramer, 1961).

The significance of nutrient concentrations in determining distribution patterns is widely evident, and populations capable of development at low food levels are favored in nutrient-deficient conditions while tolerant microbial groups are dominant in habitats rich in dissolved inorganic or organic metabolites. This is nicely illustrated in a study of iron requirements of aquatic algae (Fig. 5.1). The algae developing in coastal waters have a high iron demand, one apparently satisfied by water draining from land masses. Waters of the open sea, by contrast, contain little dissolved iron, so that the algae demanding little of this element have a distinct advantage in such ecosystems. In addition, light intensities and concentrations of organic substrates, inorganic nitrogen, and phosphate optimum for some free-living microbial groups are quite deleterious to others. For example, dinoflagellates and many Cyanophyta grow well in waters with little nitrogen, while some Chlorophyta withstand or require higher nutrient levels, so that the latter algae dominate in nutrient-rich waters (Barker, 1935; Provasoli, 1958). Many other illustrations could be given to show how the concentration of nutrients regulates the distribution and composition of natural communities. On the other hand, predators and parasites appear partially to escape the problem of dilute nutrient levels, since, though the search may be prolonged when the prey or host population is small, these organisms invariably feed on a concentrated food supply in the form of a prey or host cell that has gathered together the sparse nutrients and provided them in a conveniently packaged protoplasmic container.

Requirements for growth are not necessarily fixed, but rather they may be modified by factors external to the cell or filament. Microorganisms often synthesize an essential metabolite in one but not a second set of circumstances, at least in vitro. When the conditions preclude biosynthesis of a requisite metabolite within the confines of the cell, further proliferation requires that it be obtained from without. The reports of many investigations reveal that this newly acquired fastidious behavior is linked with a changed environment, as witnessed by the following: (a) isolates of *Pasteurella, Mycobacterium, Escherichia, Sclerotinia,* and *Neurospora* require one or several B vitamins, amino acids, or CO_2 at high but not low temperature; (b) a strain of *Bacillus stearothermophilus* needs nicotinic acid and leucine at 36°C but not at 45°C; (c) occasional cultures of *Salmonella* assume a demand for vitamins as the water activity decreases; (d) anaerobically cultivated *Staphylococcus, Cytophaga, Saccharomyces,* and *Mucor* strains must be supplied with uracil, CO_2, sterols, unsaturated fatty acids, or vitamins although the cultures show no equivalent dependency in air; (e) a *Streptococcus faecalis* isolate must have lipoic acid and thiamine for aerobic but not for anaerobic development; and (f) bacteria and blue-green algae require much molybdenum when the nitrogen source is N_2 or nitrate but little or none if the organisms are supplied with other nitrogen sources.

As conditions are altered, not only do new absolute requirements arise but the extent of an existing need also may be modified, as in the increased need by *Ochromonas malhamensis* for thiamine and vitamin B_{12} with a temperature increase (Hutner et al., 1957) and the greater manganese demand of some algae for growth in the light than in the dark (Wiessner, 1962). Environmentally induced changes in microbial nutrition are not restricted to free-living forms; for example, the intracellular protozoan parasite *Leishmania donovani* requires a substance found in the living cells of its host for growth at 37°C, although it will replicate in vitro but at lower temperatures (Trager, 1960). This phenomenon will be considered further, but from another viewpoint, in the subsequent discussion of tolerance range.

PATTERNS OF MICROBIAL NUTRITION

Considerable differences exist in the terms used to categorize the modes of microbial nutrition. Bacteriologists, mycologists, phycologists, and protozoologists each have their own sets of expressions, many of which overlap or conflict. The terms used here, therefore, cannot possibly satisfy contending microbiologists with acute semantic sensitivities, but the definitions should, at the very least, allow the reader to understand what groups are under consideration.

Autotrophs obtain their energy either from light or by the oxidation of inorganic compounds or ions, and acquire carbon for cell synthesis largely or entirely from CO_2. Chlorophyll-containing organisms that derive their energy by photochemical processes constitute the *photoautotrophs*, and a small number of bacterial genera, collectively called *chemoautotrophs,* obtain the necessary energy for cell synthesis by the oxidation of H_2 or inorganic nitrogen, sulfur, or iron. A few photosynthetic algae require one or several growth factors, although CO_2 still serves as the chief carbon source; these are referred to as *photoauxotrophs, auxotrophy* designating a requirement for growth factors. *Heterotrophs* rely on organic compounds for energy and for most or all of the carbon incorporated into cellular constituents.

Three categories of heterotrophs can be delineated: *phagotrophs, saprobes,* and *parasites.* The phagotrophs, organisms sometimes designated as holozoic, ingest solid food and are typified by protozoa preying on living cells as well as by heterotrophs ingesting inanimate particulate matter. Phagotrophs commonly have organelles that aid in food capture. Saprobes utilize either soluble organic nutrients or insoluble carbonaceous materials which are solubilized before entering the cell, and they do not generally possess specialized food-capturing organelles. Organisms feeding on only dissolved foods, organic or inorganic, are sometimes called osmotrophs, a term that includes both autotrophs and saprobes. Parasites develop at the expense of living animals, plants, or microorganisms, but, in contrast to predators, without ingesting the cell of the associate.

To species restricted to one type of nutrition, the adjectives obligate, strict, or compulsory are applied. Species with two or more modes of nutrition are termed facultative, but organisms exhibiting two nutritional patterns in vitro may, in particular natural environments, be restricted to only one. Thus strains of *Thiobacillus* multiply either as heterotrophs or as chemoautotrophs in the laboratory, but only as autotrophs in sulfide deposits penetrated by O_2; facultatively saprobic protozoa are probably usually restricted to phagotrophy in polluted waters; certain facultatively heterotrophic algae live largely or solely as photoautotrophs in bodies of water; and the vegetative growth of diverse facultatively saprobic fungi occurs only when they parasitize higher plants. The difficulties in ascertaining the nutritional pattern employed in a given environment is illustrated by some algae, which, on the basis of investigations in vitro, unquestionably have the capability of multiplying heterotrophically in the dark as well as living autotrophically in the light. Algae of this sort are found at depths in the ocean or soil where no light penetrates, depths where the resident community is dominated by bacteria, fungi, and actinomycetes, and the questions arise whether they make use of their heterotrophic potential or whether the viable cells merely represent inactive individuals that have settled downward into the dark zone. Authorities still disagree, and no unequivocal statements are possible in regard to the extent of heterotrophy among these organisms in nature.

One cannot be but amazed at the scores of nutritional habits exhibited in the microbial world. Chrysophyta, for example, are known to photosynthesize; yet they are likewise omnivorous phagotrophs. A protozoan like *Didinium nasutum,* despite its predatory characteristics, may suffer dietary deficiencies if the prey has itself not partaken of an adequate diet. The malaria parasite, *Plasmodium,* may behave in its intracellular habitat partly as a phagotroph and partly as a saprobe. Among the most interesting groups from the nutritional viewpoint are the obligate parasites, organisms that have not yet been cultivated in inanimate media. Viruses, rickettsiae, representatives of selected fungal genera, some trypanosomes, *Mycobacterium leprae,* and members of other groups are restricted in this way to viable substrates that constitute the compulsory host-provided habitat.

The degree of nutrient or *substrate specificity* exhibited by microorganisms varies appreciably. Many species or strains are restricted to one or a few carbon or energy sources both in the laboratory and in nature. Others metabolize a range of energy-yielding substrates in pure culture tests, but in nature they cannot compete effectively with their more vigorous neighbors for a high proportion of these substances and hence are limited to the handful for which they have a peculiar advantage, either by virtue of having a rare enzyme system, a faster growth rate on these compounds, or an advantage linked to the population's physical proximity to the substrates. The truly versatile species have an array of soluble or particulate nutrients, prey organisms, or hosts available for their nourishment in nature. Narrow substrate specificities in vivo, in vitro, or both characterize obligate parasites, obligate autotrophs, *Acetobacter* species limited to habitats where ethanol is being formed, and selected lignin-decomposing Basidiomycetes. Parasites of animals, plants, and microorganisms invariably live on a narrow range of hosts, and this *host specificity* is of considerable importance in pathology. Protozoa have been described as feeding on a single prey genus—that is, they are *monophagous,* ecologically at least; for example, *Didinium nasutum, Actinobolina radians,* and *Perispira ovum* are notoriously exacting in this manner (Noland and Gojdics, 1967). Limited specificities are reflected in distributions governed by the availability and geographical spread of the particular nutrient source, whether the nutrient in question is sunlight, a soluble inorganic or organic compound, animal or plant tissue components, or microbial prey.

A perennial temptation exists to extrapolate from laboratory studies of nutrition to nature. Such extrapolations are fraught with danger, however. True, the distribution of a substrate-specific strain or species will be patterned after that of its substrate. It is also true that the presence of a population of actively dividing cells is evidence that its required growth factors and micronutrients are found in the locality and are being assimilated. Nevertheless, a majority of heterotrophs have the capacity to metabolize a large number of carbon sources, and the mere fact that a substrate coexists with a species known, from tests in vitro, to degrade it is not sufficient evidence to conclude that the organism of interest is indeed responsible for the decomposition of the food source. A different autochthonous population may have a selective advantage and be far more vigorous in utilizing the nutrient in question under the prevailing conditions. This is well exemplified by the cellulose-decomposing fungi, which, though using simple organic molecules as carbon sources in axenic culture, often fail to compete effectively in nature with fast-growing bacteria for the sugars and organic acids in plant remains; cellulose breakdown is their ecological role. Laboratory trials disclose the potential for activity, but only tests in nature reveal which of the latent biochemical traits are expressed.

NUTRIENTS IN THE ECOSYSTEM

The number of energy sources available to saprobes is at least equal to and quite likely somewhat greater than the number of biologically formed organic compounds. It is generally believed that all naturally occurring organic compounds can be degraded by one or another microbial species, and no cogent reasons are evident to argue against this contention. The catalog of utilizable carbonaceous substances is longer than the list of biochemicals, because numerous synthetic compounds, a variety of which enter water and soil or are applied to animals and plants, sustain microscopic life. The types of substrates metabolized by parasites and predators in the tissues or cells on which they feed are, however, unknown. Parenthetically, it is noteworthy that not a few biologically generated organic molecules are resistant to decomposition in certain environments into which they enter, although the same substances are avidly attacked elsewhere.

A multitude of simple and a significant collection of complex substrates are accessible to an immense group of organisms in nature. Quite a few compounds are utilized by an array of species in tests conducted in axenic cultures, but the prevailing conditions or stresses imposed by metabolically active populations in the habitat dictate that only one or a few of the indigenous species actually will exploit the energy source; chitin, for example, is a carbon source for fungi and bacteria, but *Streptomyces* often dominates in soil communities, as shown in Fig. 5.2. Still different kinds of substrates, like lignin and diverse synthetic aromatic hydrocarbons, are used by a meager range of species, either in nature or in vitro (Alexander, 1961).

The supply of one or more essential nutrients is frequently discontinuous, and food may arrive at very infrequent intervals or in irregular bursts. Heterotrophs relying on components of leaf litter and photoautotrophs limited by the presence of nutrients in water receive materials for their sustenance seasonally. In ecosystems getting nutrients intermittently, the energy-containing substances or other nutrients tend to be soon dissipated, with a resulting dramatic modification in community composition. Ecosystems of several sorts have a reasonably continuous inflow of nutrients, as in the gastrointestinal system, that region of the soil receiving exudates from roots, inhabited

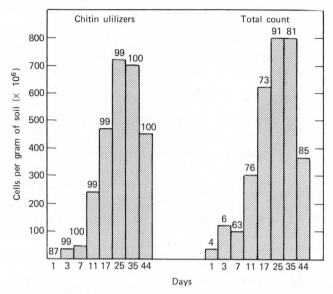

Figure 5.2 Effect of chitin addition to soil on the total number of microorganisms and abundance of chitin utilizers. The figures within the drawing refer to the percentage of actinomycetes, largely *Streptomyces* (Veldkamp, 1955).

tissues of higher organisms, and municipal sewage treatment plants, and the communities of these environments do not generally exhibit the abrupt fluctuations noted in localities showered with utilizable nutrients irregularly.

The biotic changes induced by organic substrates have been studied most thoroughly in three kinds of habitats: soil, rumen, and water. The types and complexities of community modification resulting from the introduction of carbonaceous matter are well illustrated in soils receiving plant remains. The tissues of green plants contain organic acids, sugars, and polysaccharides like cellulose and hemicelluloses, as well as variable amounts of lignin. Upon the colonization of such a heterogeneous substrate, wave after wave of organisms arise, assume dominance, and then recede, a sequence of population displacements attributable in large part to the utilization initially of the highly available low-molecular-weight compounds, then the less readily degradable polysaccharides, and finally the resistant lignin. The multiplicity of microbial cell constituents synthesized by these populations in turn supports parasites, predatory protozoa, and lytic actinomycetes and bacteria. The end result is a succession involving a diversity of heterotrophic bacteria, filamentous fungi, actinomycetes, Basidiomycetes, and protozoa. The bovine rumen too receives a nearly endless torrent of organic materials, but because the initial substrates and the surroundings are different, the identity of the saprobes and the course of the succession is wholly dissimilar from that in soil.

Increasing urbanization, industrialization, and use of pesticides in agriculture have provided new classes of chemicals and increasing rates of nutrient influx into microbial habitats. Rivers, lakes, estuaries, and soils are being charged with such abundant quantities and varieties of synthetic and natural organic compounds, as well as with inorganic elements which normally limit algal productivity, that investigations of the ecological effects of pollution and of the microbiological modification of pollutants have rapidly assumed critical importance. From the microbiological standpoint, the majority of these water and soil pollutants are little more than sources of energy or of one of the elements required for growth. However, the capacity of microorganisms to use the pollutant, or their inability to do so at significant rates, has a profound impact on man and his surroundings. Synthetic organic compounds reaching water or soil, if biodegradable, are attacked mainly by heterotrophic bacteria or fungi, and depending on environmental conditions and the particular chemical, the substance is converted either rapidly or slowly to the inorganic state.

Although the quantity of synthetic chemicals at any one point in soil or water is rarely great, the same is far from true for the natural products or biological wastes discharged directly or indirectly through man's action in urban, industrialized, or rural localities. Typically, massive amounts are discharged in short periods of time or into limited volumes of water. This leads to a rapid, extensive proliferation of aerobic heterotrophs capable of using constituents of the polluting material, with a consequent precipitous fall in the O_2 concentration. The resulting shift from aerobiosis to anaerobiosis not only sets the stage for the advent of a new community in the aquatic habitat but also is a prelude to putrefaction and the generation of foul odors. Fish, of course, will not tolerate O_2-free water and are killed. Protozoa suited to such a milieu enjoy a harvest of plenty at the expense of the hordes of bacteria responsible for the decomposition. With time, the organic matter is completely degraded and O_2 becomes plentiful again, a process slow under stagnant conditions but more rapid if the tainted water is flowing away from the source of pollutant discharge.

The tale of enhanced microbial growth does not terminate when all the carbonaceous matter is destroyed, inasmuch as the ultimate products of the aerobic destruction of organic nitrogen and phosphorus are nitrate and phosphate, ideal anions for the photoautotrophic algae that already have a good supply of carbon as CO_2 and of energy as sunlight. For the heterotrophic biota of water, the limiting nutrient usually is organic carbon, but it is frequently the nitrate or phosphate concentration that determines the mass of algae that can be supported. The nitrate and phosphate in rivers and lakes may have an inorganic origin as well, detergents providing waterways with millions of tons of phosphates and agricultural land serving as the likely source of large amounts of nitrate nitrogen.

Habitats constantly receiving nutrients often maintain a steady-state concentration of the nutrient substances, the level representing a balance between supply or release and withdrawal rates. The steady-state level is likely to be high where the substance in question is not limiting, but the concentration will be quite low where the particular nutrient element or growth factor governs the community biomass. Quantitative measurements have indeed provided evidence that the concentration of readily utilizable organic substrates in soil and water exists at a low, steady-state level, except following the introduction of fresh supplies, as does the quantity of certain growth factors and inorganic nutrients in water. A reasonably steady-state balance between nutrient replenishment and utilization might also be expected in the intestine and in activated sludge and trickling filter systems for the biological treatment of wastes. Clearly, chemical analysis showing the existence of a steady state provides limited information of relevance to ecology whereas data on the simultaneous formation and consumption of nutrients would be enormously helpful. However, only a modest start has been made in measuring such nutrient turnover rates of organic metabolites in undisturbed ecosystems (Hobbie et al., 1968).

Fluctuations in nutrient concentrations with time are quite common in microbial ecosystems too. Where the supply of an essential perennially exceeds the demand for it, a change in the quantity of this substance is of little ecological significance. However, frequently it is the limiting nutrient that is subject to appreciable increases and decreases, with immediate and profound influences on the size and composition of the community (Fig. 5.3). Such a rise and fall is obvious in environments with discontinuous introductions of carbonaceous foods, the store of the organic material added falling steadily from the time of the first addition.

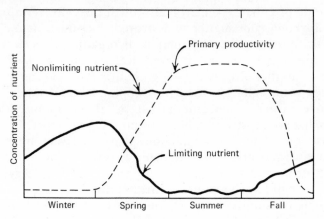

Figure 5.3 Relationship of limiting nutrient concentration in lake water to primary productivity (modified from Sawyer, 1968). Primary productivity here refers to the rate of energy storage by algal photosynthesis.

A great deal of attention has been given to the relation between nutrient supply and density of *phytoplankton,* the floating plant life in bodies of water. In occasional ponds, as one case in point, phosphate seems to be the ion limiting algal growth for part of the year, but nitrate replaces phosphate as the factor regulating the mass of photosynthetic organisms in the spring and summer (Potash, 1956). Summer nitrogen deficiencies are encountered with regularity in waters. A seasonal cycle in which magnesium becomes the limiting element at one time of year has also been described (Goldman, 1961), and similar deficiencies and modifications in supply undoubtedly hold for additional elements essential for the development of photoautotrophs or photoauxotrophs. Seasonal exhaustion and replenishment is widespread in marine or fresh waters, where the supply of many algal nutrients is high in the winter, but the level of at least one nutrient falls almost to zero as the algae multiply in the spring, only to be regenerated and reused in the autumn as the existing cells are decomposed or as vertical turbulence introduces a fresh supply. Factors other than the availability of critical elements may limit the phytoplankton density also during portions of the year, low temperature and light intensity taking on great significance in the winter months.

The mere presence of an essential factor in the ecosystem does not mean that it will be assimilated, because selected nutrients are susceptible to complexing or combination in ways that make them no longer readily utilizable. Unavailability rather than absence of an essential for life can limit the growth of both heterotrophs and autotrophs, and nutrient inaccessibility is probably of far greater consequence than is generally realized. Deficiencies resulting from the unavailability rather than the absence of organic carbon sources, inorganic factors required in large amounts, and micronutrients have been described in soil, the open sea, and lakes, and similar observations undoubtedly will be made in the future with the saprobes and parasites residing in other environments. Among the processes rendering a potential nutrient unavailable are its sorption by clay, its encrustation by an organic substance highly resistant to attack, its precipitation by salts, or its conversion to an oxidation state not readily assimilable; thus cations and both simple and complex carbonaceous substrates are strongly retained by clay, polysaccharides normally metabolized with ease are made inaccessible when coated with lignin, needed elements are precipitated by the $CaCO_3$ of marl lakes, and the unavailability of iron and manganese is characteristic of alkaline habitats because the cations are converted spontaneously to the poorly assimilable ferric or manganic state.

Any discussion of nutrients in natural ecosystems must take into account the formation, utilization, and concentration of growth factors. Auxotrophs are ubiquitous, and though their proliferation denotes that their needs are satisfied, the level of growth factors in nature and their significance in governing the composition of the biota have been the subject of only modest inquiry, except in water and soil. Analyses of natural bodies of water reveal that vitamin B_{12}, thiamine, nicotinic acid, biotin, auxin-like compounds, and an assortment of amino acids are present in one area or another, and studies of soils show the presence of varying quantities of the same substances in addition to riboflavin, pyridoxine, inositol, and p-aminobenzoic acid. Plant and animal tissues and body fluids are, as expected, likewise rich in compounds capable of satisfying such microbial requirements.

It is reasonable to hypothesize that the nutrition of the vegetative state of an organism mirrors the chemistry of the environment where it flourishes. In this regard, information on auxotrophs is particularly valuable, for the occurrence of vegetative cells of fastidious populations, but not species with a simple nutrition, reflects the sorts of nutrients supplied in vivo, and the geography of auxotrophs provides a guide to the distribution of the compounds or ions on which they rely. In a habitat devoid of B vitamins and amino acids essential for fastidious species, natural selection acts against auxotrophs, and only *prototrophs*—species needing no growth factors—are able to exploit the water, soil, tissue, or foodstuff. Prototrophs are likewise found in a rich environment inasmuch as they are not necessarily selected against in a nutritionally complex milieu.

Numerous organisms show a degree of nutritional uniqueness in that they require for proliferation in laboratory media, in addition to the widespread growth factors, a decoction obtained from their environment; for example, rumen fluid or constituents of serum, soil, or plants. In time, the compounds in these decoctions will be characterized, but nevertheless the view that the organism depends on the chemistry of its environment is reinforced. Evidence for a relationship between nutrition and the presence in nature of characterized chemical substances is seen in the following: (a) rumen-inhabiting strains of *Bacteroides* need volatile fatty acids for growth (Koser, 1968), acids that are always present in the bovine rumen; (b) strains of *Corynebacterium* from the human skin require oleic acid (Pollock et al., 1949), and unsaturated fatty acids like oleic acid are located on the skin; (c) *Dictyostelium*, a fungus obtaining its nourishment by feeding on bacteria, is dependent on a protein derived from bacterial cells (Sussman, 1955); (d) *Plasmodium lophurae*, the intracellular protozoan causing avian malaria, has an obligate requirement for a coenzyme present in the red blood cells where it lives (Trager, 1965); and (e) certain marine fungi and bacteria have an absolute reliance on sodium, a cation abundant in the sea. Probably nowhere is the relationship between nutrition and environmental chemistry better illustrated than among the obligate parasites and predators, which, because of their reliance on molecules synthesized by another organism, cannot multiply, so far as is known, in any habitat or feed on any nutrient other than that provided by the cells of the host or prey. Even an inhabitant of the mucous membrane like *Borrelia vincentii* exhibits a highly complex set of requirements, presumably reflecting its ready access to fluids derived from the host (Hutner, 1962).

NUTRIENT EFFECTS ON DISTRIBUTION AND ACTIVITY

Unequivocal evidence demonstrates how energy sources and key elements affect the distribution and activity of microorganisms in nature. Few environments or taxonomic groups have been examined as yet, and thus generalizations are still risky. Nevertheless, the experimental results reveal the kinds of approaches that have

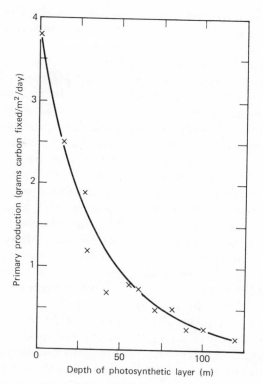

Figure 5.4   The effect of depth of water on primary productivity (Steeman Nielsen and Jensen, 1957).

been successfully employed and point the way to new avenues for exploration. Without question, prime interest has been in the aquatic algae, and these therefore serve as the topic of discussion.

The quantity of an energy source, an essential element—provided that it is in an assimilable form—or growth factors may affect the total mass of autotrophic or heterotrophic cells, the species composition of the community, or both. The oceans provide an almost ideal environment for algae, for they contain sufficient moisture, CO_2, micronutrients, a favorable temperature, and a reasonable supply of nitrogen and phosphorus at considerable depths below the surface, yet 95% of the water of the seas is essentially in total darkness. Algae are apparently unable to multiply to a significant extent in these otherwise rich environments. For them, light is frequently a limiting factor. The region where photosynthetic development occurs extends down to about 120 meters in clear sea water (Fig. 5.4), but the zone is considerably restricted when growth occurs and the algae and associated organisms absorb and scatter the light penetrating the water. Below this shallow layer of the oceans where photosynthesis proceeds, light is permanently in short supply (Ryther, 1960). The abundance of heterotrophic cells and filaments or the activity of heterotrophic communities is also commonly held in check by an inadequate quantity of biologically useful energy, but the energy source in these instances is not light but rather readily usable organic compounds.

The introduction of a limiting nutrient has an explosive and widespread effect in surface waters. When neither light nor CO_2 is in short supply, frequently it is either phosphate or nitrogen that governs the extent of algal growth, as indicated above; these two nutrients stand out because they are needed in reasonably large amounts per unit of cell material produced and because the supply of other essentials usually exceeds the immediate demand. Many studies of the open sea, lakes, or rivers have demonstrated that nitrogen or phosphorus scarcity typifies a high proportion of aquatic habitats, and these two nutrients thereby control the extent of algal development and hence the fertility of the area. As expected of a limiting element, the concentration falls to an almost undetectable point as the photoautotrophs multiply, as during the widespread spring outbursts of the phytoplankton, and increments of these nutrients flowing in from an external source exert a profound influence on the community.

The specific locality and sometimes the season of year determine whether phosphorus, nitrogen, or a third factor governs algal productivity. Phosphorus is, without a doubt, the nutrient of concern in many lakes, ponds, rivers, and areas of the sea, as shown by the effect of supplemental phosphate on the indigenous algal community, the low phosphate levels in relation to potential demand, and the marked decline in phosphate content when massive growths appear. Considering that many of the communities are dominated by bluegreens, a group unique among the algae on account of their capacity to satisfy the need for nitrogen by assimilating N_2, it is quite likely that such communities are also, as a rule, limited by the phosphorus content, since N_2 is never lacking, provided, of course, that the N_2 is being assimilated. In natural waters phosphorus exists not only as phosphate but also in organic complexes, and a reasonable number of algae can make use of organic phosphorus compounds; nevertheless, inasmuch as the bacteria exhibit far greater activity in degrading organic phosphorus, it is probable that algae assimilate much of the phosphorus only following decomposition of the organic molecules to phosphate by heterotrophic populations.

Nitrogen too governs the degree of photoautotrophic development in numerous waters. Evidence for this limitation has been obtained, as in the case of phosphorus, by bioassay techniques, by measuring the influence of supplemental nitrogen, and by demonstrating the almost complete removal of the element, especially the nitrate nitrogen, from surface waters coincident with algal multiplication. Chief interest is centered on nitrate, although use is made of ammonium and simple organic nitrogen compounds by some algae, because the metabolism of heterotrophic residents of oxygenated waters converts the simple organic nitrogenous materials to ammonium, which in turn is rapidly oxidized by chemoautotrophic bacteria to nitrate.

Considerable ecological importance probably can be assigned to the appreciable variation in phosphorus requirements of algal species. A concentration optimum for one may be excessive for a second and suboptimal for a third. *Dinobryon divergens*, for example, grows well at low and *Scenedesmus quadricauda* at high phosphorus concentrations (Rodhe, 1948). The optimum level of nitrogen similarly varies from species to species. Low phosphorus requirements exhibited in axenic cultures are occasionally in accord with observations in vivo, but some species, like *Asterionella formosa*, are able to take up phosphate from lake water containing the anion in concentrations not allowing for assimilation in culture media (Rodhe, 1948; Mackereth, 1953), suggesting the existence in nature of substances facilitating the utilization of low inorganic nutrient levels.

Eutrophication, the enrichment of oligotrophic waters by the influx of nutrients promoting algal proliferation, has recently achieved the unfortunate status of a major economic and esthetic problem. Enrichment of lakes with phosphorus and nitrogen is a natural process, but its rate is usually measured in geological time. Man's activities, however, have immensely speeded up the normally slow fertilization achieved by natural phenomena, and algal blooms now follow man wherever he congregates or establishes centers for industry or intensive agriculture. The resulting algal *bloom* is composed of numbers of cells so immense that the surface water takes on a distinct green, blue-green, brown, or red coloration. The bloom is usually dominated by a single species, for example, a representative of *Microcystis, Anabaena, Asterionella,* or *Ceratium*, and eutrophic lakes are known to contain as much as 177 kg of floating organisms per hectare (Prescott, 1960).

The sources of these nutrients include (a) urban centers which, even when employing well-designed waste treatment facilities, liberate considerable nitrate and phosphate into the surrounding waters because natural microbial processes of decomposition as well as conventional waste treatment procedures convert the organic matter to CO_2, phosphate, nitrate, and other inorganic ions; (b) industrial operations that not uncommonly release manufacturing wastes into neighboring waterways; (c) runoff from agricultural and nonfarmed land; and (d) animal manure. Wastes from cattle fecal matter spread on frozen soil are carried laterally into a watershed, and those from domesticated ducks are immediately accessible to waterways while the wastes from chicken farms are often deliberately introduced into streams. The vast amounts of detergents used in the home constitute excellent material for the initiation of photoautotrophic development because of their high phosphate content. The materials need not be inorganic, for heterotrophs are quite capable of decomposing carbonaceous effluents and converting them to precisely the correct state to foster the masses of algae causing the unsightly surface scums, the green coating on beaches, the clogging of filters at water treatment facilities, and the foul odors generated as the blooms themselves are subjected to heterotrophic attack.

Nevertheless, the algal community or the reproduction of individual populations is frequently independent of the available nitrogen and phosphorus content of waters receiving sufficient light. In selected habitats clear evidence is at hand to show that this independence results from a deficiency of nutrient elements other than nitrogen and phosphorus. Silicon is a particularly interesting nutrient, as there is no demonstrable need for it by most microorganisms, yet diatoms and some species of Chrysophyta and Xanthophyta have an absolute silicon requirement. Hence the distribution of those algae requiring this uncommon nutrient will be conditioned by the presence of available silicate. Usually the concentration of silicate in water is quite ample, but occasional deficiencies have been reported (Table 5.2). It has also been observed that enrichments of aquatic organisms contain diatoms if the enrichment medium is fortified with silicon, but commonly none appears in silicon-free solutions. Moreover, diatoms with thin walls, a sign of silicon deficiency, have been noted in natural waters.

Deficiencies of other elements sometimes restrict the growth or photosynthetic activity of aquatic algae. These include potassium, sulfur, iron, and molybdenum (Table 5.2). These findings are based on studies involving biological assays of water samples, chemical measurements revealing the coincidence in time of nutrient disappearance with cessation of multiplication, and the effect of supplemental

Table 5.2

Inorganic Nutrients Limiting Development of Aquatic Algae

Limiting Element	Location	Reference
Silicon	Cayuga Lake, N.Y.	Hamilton (1969)
Silicon	Sargasso Sea	Ryther and Guillard (1959)
Silicon	English lakes	Hughes and Lund (1962)
Potassium	Castle Lake, Calif.	Goldman (1960a)
Sulfur	Castle Lake, Calif.	Goldman (1960a)
Sulfur	Lake Victoria	Fish (1955)
Iron	Sargasso Sea	Menzel and Ryther (1961)
Iron	A marl lake	Schelske et al. (1962)
Molybdenum	Castle Lake, Calif.	Goldman (1960b)

salts on the rate of photosynthesis. The absence of still other inorganic nutrients may retard the proliferation of aquatic photoautotrophs in as yet uncharacterized ecosystems.

NUTRIENT REGENERATION

Inorganic nutrients in short supply are assimilated and enter into unavailable protoplasmic complexes in terrestrial and aquatic habitats. In soil these unavailable complexes remain in place because of the virtual absence of vertical and horizontal disturbance, whereas the algae, other microorganisms, and higher plants using nutrients in the surface layers of oceans, lakes, reservoirs, and ponds tend to sink so that not only are nutrients in bodies of water converted to unavailable states but the elements are lost from the site of photoautotrophic proliferation to subsurface zones where the light intensity is too weak to support photosynthesis. Still, reproduction by a significant portion of the indigenous populations does not terminate upon the cessation of fresh inputs of limiting nutrients. Life continues because *nutrient regeneration* takes place, and the elements assimilated are made available once again to succeeding generations.

Replenishment of the supply of available forms of an element may result, to a lesser or greater extent, from the microbial decomposition of the cells that previously removed the nutrient from circulation. Nitrogen, phosphorus, and sulfur are regenerated in this way in soil and water. Carbon is also circulated in aerobic, heterotrophic communities, but the storehouse of this element is constantly being depleted as much of the carbon is irretrievably lost from the ecosystem in the form of microbiologically generated CO_2. In water the elements rendered unavailable spatially by the sinking of cellular materials are returned to circulation in the surface layers only when the element is reintroduced into the region from which it was lost. This may be accomplished by upwelling or other types of vertical circulation that carry nutrient-rich waters from below to the zone that has been depleted of its nutrients. Surface turbulence, such as that induced by wind action, also brings about a mixing of the topmost water with the richer, underlying solution.

The quantity of inorganic nitrogen, phosphorus, sulfur, or silicate recirculated in these ways may be small, but it is sufficient to maintain the continued existence of a variety of species. The rate of recycling is on occasion surprisingly rapid, as in some aquatic environments, or it may be excruciatingly slow, as in the transformations of certain elements in soil. The depletion-regeneration cycle is often reasonably continuous throughout the year, but in waters of the temperate zone, to cite one example, a distinct seasonal influence on nutrient turnover is in evidence as the cooling of the surface waters affects both the vertical mixing and microbial metabolism.

GROWTH FACTORS AND MICROBIAL ECOLOGY

Bacteria, fungi, protozoa, and algae with an obligate requirement for one or more growth factors are ubiquitous. Every major type of environment contains nutritionally dependent species, and though many animal and plant pathogens are acknowledged to be exacting, the same is true of heterotrophs inhabiting soil, sea and fresh water, sewage, waste water, animal surfaces, food products, the rumen, and the area of soil immediately adjacent to plant roots, the *rhizosphere*. Not a few aquatic photoautotrophs are also auxotrophic. The ubiquity and abundance of organisms relying on the presence of growth substances suggest they have a distinct selective advantage. The availability and turnover of the requisite growth factors, therefore, are of profound ecological importance.

The nutritional patterns of the bacterial community of soil and root surroundings have been well defined. A large percentage of the native bacteria of soil, the rhizosphere, and the root surface exhibit an absolute requirement for amino acids and B vitamins. A sizable number develop without these compounds, but their generation times

Table 5.3

Percentage Incidence of Vitamin-Requiring Aerobic Bacteria in Different Habitats [a]

Growth Factor	Per Cent of Bacteria Requiring Growth Factor			
	Soil	Rhizosphere	Root	Marine Muds
Thiamine	44.9	15.2	17	28.7
Biotin	18.7	6.1	7	28.9
Pantothenic acid	3.7	3.0	3	0.86
Folic acid	1.8	3.0	4	—
Nicotinic acid	5.6	6.1	5	5.5
Riboflavin	1.8	2.0	4	0.06
Pyridoxine	1.8	1.0	5	—
Vitamin B_{12}	19.6	2.0	1	8.8
Terregens factor	1.8	<1	1	—
p-Aminobenzoic acid	<0.9	<1	<1	—
Choline	<0.9	<1	<1	—
Inositol	<0.9	<1	<1	—

[a] Based on Burkholder (1963) and Rouatt (1967).

are decreased when specific growth factors are present. Only a small proportion of the bacteria from these habitats develop best in media devoid of growth factors. Selected data illustrating vitamin needs are given in Table 5.3. Thiamine stands out because of the frequent necessity for it, while other vitamins are essential for many bacteria derived from one environment but not another. The relative rarity in these communities of bacteria requiring p-aminobenzoic acid, choline, and inositol suggests that microorganisms demanding these for replication probably either have no selective advantage or find no ready source of supply. In comparison with the bacteria, a much higher percentage of free-living terrestrial fungi multiply in the absence of growth factors, but most of the fungi are markedly stimulated by vitamins, amino acids, or uncharacterized growth factors (Atkinson and Robinson, 1955).

Large numbers of the aerobic heterotrophic bacteria of marine environments also cannot grow well, if at all, in the absence of growth factors. A notably high percentage of the indigenous bacteria of marine sediments and of surface and deep oceanic waters require biotin, thiamine, and vitamin B_{12}, whereas reliance on other B vitamins is reasonably uncommon. The proportion needing no vitamins or amino acids is small in this environment also, and many marine bacteria must have available more than one of these compounds (Burkholder, 1963; Skerman, 1963). Vitamin B_{12}, thiamine, and biotin requirements are attributes not only of marine bacteria but also of a significant number of aquatic algae; for example, a sizable percentage of the Chlorophyta, Euglenophyta, Cryptophyta, Pyrrophyta, Chrysophyta, and Bacillariophyta, but few of the Cyanophyta, whose vitamin requirements have been elucidated require vitamin B_{12}, thiamine, or biotin (Provasoli, 1961). The abundance of aquatic species needing just these substances is unquestionably not fortuitous. Phagotrophic flagellates like *Peranema*, however, do require additional growth factors.

Vitamins, amino acids, or both are likewise essential for many bacteria indigenous to activated sludge and raw sewage. Moreover, various heterotrophs in these ecosystems too fail to develop in the synthetic media employed to date, suggesting a highly fastidious nutrition (Dias and Bhat, 1964). Rumen bacteria are exacting also, some needing vitamins, branched chain fatty acids, vitamin K, hemin, and as yet uncharacterized constituents of rumen fluid (Koser, 1968).

While most blue-greens and many other algae have no vitamin requirements and hence are not regulated by the formation of these metabolites in nature, algal photoauxotrophy has provoked considerable research. Although some algae need amino acids, prime attention has been given to the widespread vitamin B_{12}, thiamine, and biotin dependency in species thriving in the light in solutions containing no extra organic carbon sources. The stereotyped needs among algal auxotrophs—B_{12}, thiamine, and biotin—are so common and so notable that no alternative exists but to assign ecological significance to the nutritional pattern, provided, of course, that it is valid to assume that a requirement for a growth factor in vitro is an adequate basis for proposing that the compound is produced in the environment where the species naturally flourishes.

When a particular growth factor falls below the level that supports maximum growth, or the highest replication rate, of native species and thus becomes limiting, the compound can govern the species composition of the community. Auxotrophs multiply poorly in such circumstances, if at all, and organisms not dependent on the limiting substance proliferate with little or no competition from dependent populations. On the other hand, when the supply is plentiful, auxotrophs are no longer at a necessary disadvantage, and they may become prominent in the community. This is well illustrated by the vitamin B_{12}-requiring diatom *Skeletonema costatum*, which is abundant in surface waters of Long Island Sound and selected inshore waters containing adequate concentrations of the vitamin but is unable to establish itself in aquatic habitats where little B_{12} is formed (Curl, 1962; Vishniac and Riley, 1961). Occasional freshwater and marine phytoplankton communities, moreover, carry out photosynthesis slowly by reason of the inadequate quantities of available vitamin B_{12}, and certain successions among marine algae seem to be correlated with, if not explained by, seasonal fluctuations in the concentration of this vitamin (Fogg, 1965). It has also been suggested that yeasts and fungi are favored in the colonization of fruit products, which usually are poor in B vitamins, because such colonists do not depend on the environment for these compounds, whereas no such nutritional self-reliance is necessary for the colonists of animal food products rich in microbial growth factors (Mossel and Ingram, 1955).

The sources of growth factors for auxotrophs are plentiful and diverse. The habitat itself may contain all that is required for the pioneers or for the members of the climax community, as is likely in many plant and animal tissues, body fluids, food products, and sewage. Chemical analyses of tissues, blood, saliva, and foods demonstrate that they indeed contain compounds of the appropriate sorts. Moreover, the fact that pioneers in these habitats are auxotrophic itself indicates that all the necessities for microbial life are present. Plant roots excrete an assortment of amino acids, an activity that may at least partially account for the abundance of amino acid auxotrophs in the rhizosphere. Inasmuch as the capacity to synthesize and excrete B vitamins and amino acids is characteristic of populations in pioneer and preclimax communities, it is reasonable to believe that such organisms provide metabolites of these kinds to their neighbors or successors. Excretion of water-soluble vitamins and amino acids is well documented for soil, rhizosphere, and root surface bacteria (Lochhead, 1959; Rouatt, 1967), and excretions of bacteria and possibly more complex organisms are implicated in the formation of the vitamin B_{12} needed by aquatic algae and the growth factors required by marine and freshwater heterotrophs (Burkholder, 1963; Provasoli, 1963). Compounds not excreted may be made available to succeeding populations as cells autolyze or are subjected to attack. The release of a multitude of metabolites through excretion, autolysis, or decomposition of microbial cells and filaments by neighbors is surely widespread.

HOST NUTRITION AND COLONIZATION

Pathogenicity requires that a parasite reach its potential host, be capable of invading it, and then multiply sufficiently to bring about an adverse effect. Although the unsuitability of animal or plant tissues as environments for microbial growth frequently results from defense mechanisms that prevent or restrict replication of a potential colonist, unsuitability also may be attributable to the lack of a full complement of nutrients required by the propagule. A deficiency of any essential at the site of introduction or localization of an individual would prevent its multiplication, so that disease would not ensue.

Direct lines of evidence support the contention that hosts do indeed supply specific nutrients, the lack of which renders harmless the virulent pathogen. To study nutrients in tissue ecosystems associated with colonization and disease, mutants of virulent pathogens have been obtained that are unable to synthesize specific vitamins, amino acids, purines, and pyrimidines and hence must be provided with an external source. When the ability of these auxotrophic mutants to reproduce and cause disease is tested, many are found to be as virulent as the parent cultures. Some of the mutants, however, retain little or none of the virulence or fail to become established in the host normally serving as a suitable habitat for the original strains (Table 5.4). The addition to the host of the factor needed by the auxotrophic mutant, with the means of application depending on the animal or plant employed, frequently leads to the onset of typical disease symptoms. Furthermore, when the apparently avirulent cultures of the bacterial species listed in Table 5.4 revert to give rise to populations now independent of the compound needed by the mutants, virulence is restored. This provides striking evidence that the availability of particular nutrients governs whether a microorganism will be established in an environment.

Clearly the substance in question is present in the host, but the concentration may be too low or the compound may be unavailable because, for instance, it is incorporated into a protein or nucleic acid molecule. Results of this sort should not be taken to mean that all differences between nonpathogens and pathogens have a nutritional basis but rather that a key component of disease resistance may be the absence from the host environment of metabolites necessary for the invader. This nutritional component in the host-parasite interaction is simply a single case of the more general principle that

Table 5.4

Changes in Virulence of Auxotrophic Mutants Derived from Pathogenic
Bacteria and Fungi

Microorganism	Auxotrophic Trait Leading to Loss of Virulence	Host	Reference
Agrobacterium tumefaciens	Methionine	Pinto bean	Lippincott and Lippincott (1966)
Bacillus anthracis	Adenine	Mouse	Ivanovics et al. (1968)
Erwinia aroideae	Arginine	Radish	Garber et al. (1956)
Klebsiella pneumoniae	Adenine	Mouse	Garber et al. (1952)
Pseudomonas tabaci	Amino acids	Tobacco plant	Garber and Heggestad (1958)
Salmonella typhosa	Xanthine	Mouse	Formal et al. (1954)
Venturia inaequalis	Choline	Apple	Keitt et al. (1959)

microorganisms develop only in environments where their food wants
are wholly satisfied.

Other experimental approaches have also provided evidence that
inadequate nutrition of the host or deficiencies in specific substances
affect the host-parasite balance in such a way that the parasite fares
poorly and the apparent resistance of the host increases. For example,
rats and monkeys fed on a diet deficient in p-aminobenzoate are
resistant to infections by species of *Plasmodium burghei*, but susceptibility to malaria caused by these protozoa is pronounced if the diet
is supplemented with the vitamin (Hawking, 1953; Hawking, 1954).
As expected, the malaria parasites require p-aminobenzoate. Similarly, pantothenate deficiency is associated with retarded development of *Plasmodium gallinaceum* in chickens (Brackett et al., 1946).
Rhodosticta quercina, a myo-inositol-requiring fungus causing cankers
on the bark of plum trees, affords a particularly interesting example.
The bark of resistant varieties of trees is poorer in myo-inositol than
susceptible varieties, but resistance of the former to attack by *R.
quercina* can be overcome readily by additions of inositol (Lukezic
and DeVay, 1964). In man the extent to which findings with individual growth factors are relevant to disease is still open to dispute,
although malnutrition is known to make humans less susceptible to
a few communicable diseases. Observations that the starvation during

an acute famine in India was linked with a retarded multiplication of the human malarial parasite and that clinical cases of malaria arose from subclinical infections when the food supply was improved constitute an interesting analogy to findings that malaria parasites of animals are less virulent in malnourished than in well-fed hosts (Geiman, 1958).

Conversely, a general dietary inadequacy or a deficiency for specific nutrients frequently increases the severity of animal or human disease. Similar effects have been noted in plants receiving inadequate quantities of individual inorganic nutrients. Nutritional impoverishment has now been shown to have an influence, presumably by increasing host susceptibility, on the enhanced development of diseases caused by various bacteria, viruses, fungi, rickettsiae, and protozoa, and protein, biotin, riboflavin, thiamine, and folic acid deficiencies have so far been implicated. As yet, sad to say, it is not possible to establish generalizations in view of the incomplete status of the research and the large number of hosts, parasites, and resistance mechanisms involved, but it is abundantly clear nevertheless that the nutritional state of the host affects antibody formation, as yet inadequately characterized antimicrobial properties of body fluids, and the production, activity, and efficiency of phagocytes, as well as totally unresolved mechanisms by which the host destroys its parasites (Clark, 1950; Geiman, 1958; Sprunt and Flanigan, 1960).

References

REVIEWS

Alexander, M. 1961. *Introduction to Soil Microbiology*. Wiley, New York.

Clark, P. F. 1950. The influence of nutrition in experimental infection. *Ann. Rev. Microbiol.*, 4, 343–358.

Fogg, G. E. 1965. *Algal Cultures and Phytoplankton Ecology*. University of Wisconsin Press, Madison.

Geiman, Q. M. 1958. Nutritional effects of parasitic infections and disease. *Vitamins Hormones*, 16, 1–33.

Hutner, S. H. 1962. Nutrition of protists. *In* W. H. Johnson and W. C. Steere, Eds., *This is Life*. Holt, Rinehart and Winston, New York. pp. 109–137.

Koser, S. A. 1968. *Vitamin Requirements of Bacteria and Yeasts*. Thomas, Springfield, Ill.

Lochhead, A. G. 1959. Rhizosphere microorganisms in relation to root-disease fungi. *In* C. S. Holton, Ed., *Plant Pathology: Problems and Progress*. University of Wisconsin Press, Madison. pp. 327–338.

Lucas, C. E. 1955. External metabolites in the sea. *Deep-Sea Res.* (Suppl.), 3, 139–148.

Noland, L. E., and Gojdics, M. 1967. Ecology of free-living protozoa. *In* T.-T. Chen, Ed., *Research in Protozoology*, Vol. 2. Pergamon Press, New York. pp. 215–266.

O'Kelley, J. C. 1968. Mineral nutrition of algae. *Ann. Rev. Plant Physiol.* 19, 89–112.

Provasoli, L. 1958. Nutrition and ecology of protozoa and algae. *Ann. Rev. Microbiol.*, 12, 279–308.

Provasoli, L. 1961. Micronutrients and heterotrophy as possible factors in bloom production in natural waters. *In Algae and Metropolitan Wastes*. U.S. Public Health Service Publ. SEC-TR-W61-3, Cincinnati. pp. 48–56.

Provasoli, L. 1963. Organic regulation of phytoplankton fertility. *In* M. N. Hill, Ed., *The Sea*, Vol. 2. Interscience Publishers, New York. pp. 165–219.

123

Ryther, J. H. 1960. Organic production by plankton algae, and its environmental control. *In* C. A. Tryon and R. T. Hartman, Eds., *The Ecology of Algae.* Pymatuning Laboratory of Field Biology, University of Pittsburgh. pp. 72–83.

Sprunt, D. H., and Flanigan, C. 1960. The effect of nutrition on the production of disease by bacteria, rickettsiae and viruses. *Advan. Vet. Sci.,* 6, 79–110.

OTHER LITERATURE CITED

Atkinson, R. G., and Robinson, J. B. 1955. *Can. J. Bot.,* 33, 281–288.

Barker, H. A. 1935. *Arch. Mikrobiol.,* 6, 157–181.

Brackett, S., Waletsky, E., and Baker, M. 1946. *J. Parasitol.,* 32, 453–462.

Burkholder, P. M. 1963. *In* C. H. Oppenheimer, Ed., *Symposium on Marine Microbiology.* Thomas, Springfield, Ill. pp. 133–150.

Curl, H. 1962. *Limnol. Oceanog.* 7, 422–424.

Dias, F. F., and Bhat, J. V. 1964. *Appl. Microbiol.,* 12, 412–417.

Fish, G. R. 1955. *E. African Agr. J.,* 21, 152–158.

Formal, S. B., Baron, L. S., and Spilman, W. 1954. *J. Bacteriol.,* 68, 117–121.

Garber, E. D., Hackett, A. J., and Franklin, R. 1952. *Proc. Natl. Acad. Sci. U.S.,* 38, 693–697.

Garber, E. D., and Heggestad, H. E. 1958. *Phytopathology,* 48, 535–537.

Garber, E. D., Schaeffer, S. G., and Goldman, M. 1956. *J. Gen. Microbiol.,* 14, 261–267.

Goldman, C. R. 1960a. *Ecol. Monogr.,* 30, 207–230.

Goldman, C. R. 1960b. *Science,* 132, 1016–1017.

Goldman, C. R. 1961. *Verh. Intern. Ver. Limnol.,* 14, 120–124.

Hamilton, D. H. 1969. *Limnol. Oceanog.,* 14, 579–590.

Hawking, F. 1953. *Brit. Med. J.,* 1201–1202.

Hawking, F. 1954. *Brit. Med. J.,* 425–429.

Hobbie, J. E., Crawford, C. C., and Webb, K. L. 1968. *Science,* 159, 1463–1464.

Hughes, J. C., and Lund, J. W. G. 1962. *Arch. Mikrobiol.,* 42, 117–129.

Hutner, S. H., Baker, H., Aaronson, S., Nathan, H. A., Rodriguez, E., Lockwood, S., Sanders, M., and Petersen, R. A. 1957. *J. Protozool.,* 4, 259–269.

Ivanovics, G., Marjai, E., and Dobozy, A. 1968. *J. Gen. Microbiol.,* 53, 147–162.

Keitt, G. W., Boone, D. M., and Shay, J. R. 1959. *In* C. S. Holton, Ed., *Plant Pathology: Problems and Progress.* University of Wisconsin Press, Madison. pp. 157–167.

Lippincott, B. B., and Lippincott, J. A. 1966. *J. Bacteriol.,* 92, 937–945.

Lukezic, F. L., and DeVay, J. E. 1964. *Phytopathology,* 54, 697–700.

Mackereth, F. J. 1953. *J. Exptl. Bot.,* 4, 296–313.

Menzel, D. W., and Ryther, J. H. 1961. *Deep-Sea Res.*, 7, 276–281.

Mossel, D. A. A., and Ingram, M. 1955. *J. Appl. Bacteriol.*, 18, 232–268.

Pollock, M. R., Wainwright, S. D., and Manson, E. E. D. 1949. *J. Pathol. Bacteriol.*, 61, 274–276.

Potash, M. 1956. *Ecology*, 37, 631–639.

Prescott, G. W. 1960. *In* C. A. Tryon and R. T. Hartman, Eds., *The Ecology of Algae.* Pymatuning Laboratory of Field Biology, University of Pittsburgh. pp. 22–37.

Rodhe, W. 1948. *Symbolae Bot. Upsal.*, 10, 1–149.

Rouatt, J. W. 1967. *In* T. R. G. Gray and D. Parkinson, Eds., *The Ecology of Soil Bacteria.* Liverpool University Press, Liverpool. pp. 360–370.

Ryther, J. H., and Guillard, R. R. L. 1959. *Deep-Sea Res.*, 6, 65–69.

Ryther, J. H., and Kramer, D. D. 1961. *Ecology*, 42, 444–446.

Sawyer, C. N. 1968. *J. Water Pollution Control Federation*, 40, 363–370.

Schelske, C. L., Hooper, F. F., and Haertl, E. J. 1962. *Ecology*, 43, 646–653.

Skerman, T. M. 1963. *In* C. H. Oppenheimer, Ed., *Symposium on Marine Microbiology.* Thomas, Springfield, Ill. pp. 685–698.

Steeman Nielsen, E., and Jensen, E. A. 1957. *Galathea Rept.*, 1, 49–136.

Sussman, M. 1955. *In* S. H. Hutner and A. Lwoff, Eds., *Biochemistry and Physiology of Protozoa*, Vol. 2. Academic Press, New York. pp. 201–223.

Trager, W. 1960. *In* J. Brachet and A. E. Mirsky, Eds., *The Cell*, Vol. 4. Academic Press, New York. pp. 151–213.

Trager, W. 1965. *In Progress in Protozoology.* Excerpta Medica, Amsterdam. pp. 97–98.

Veldkamp, H. 1955. *Meded. Landbouw. Wageningen*, 55, 127–174.

Vishniac, H. S., and Riley, G. A. 1961. *Limnol. Oceanog.*, 6, 36–41.

Wiessner, W. 1962. *In* R. A. Lewin, Ed., *Physiology and Biochemistry of Algae.* Academic Press, New York. pp. 267–286.

6

Tolerance Range

Each species grows, reproduces, and survives within a definite
range of external conditions, which represent its *tolerance range* or
ecological amplitude for critical environmental factors. Thus, for
individuals in a population to proliferate, produce daughter cells,
generate reproductive structures, or merely survive when not growing,
the key environmental factors must not exceed the limits of tolerance
of the population. These abiotic factors include temperature, moisture or available water level, light intensity, pH, osmotic pressure,
oxidation-reduction potential (E_h), sometimes hydrostatic pressure,
and toxins arising from inanimate constituents of the habitat. Biologically produced toxins will be considered in a later chapter. The
ranges of intensity or levels of these factors that a population can
withstand define its tolerance.

Unique microorganisms have become established in polar seas or
snowfields, while others are indigenous to hot springs or locales that
are heated as bulk quantities of organic materials undergo decomposition. Some bacteria require several hundred atmospheres of pressure to grow, whereas rare fungi and bacteria develop only where the
salt level is extremely high. Each has an ecological amplitude attuned
to the environments where it resides, a tolerance range that prevents
it from gaining a foothold in innumerable ecosystems.

The amplitude of the ecologically significant abiotic factors permitting life of one organism or another is wide. A large number of
bacteria multiply in media almost entirely free of salts, whereas
halophilic bacteria and the alga *Dunaliella* are metabolically active
in extremely salty brines. Heterotrophs and autotrophs of a variety of
sorts are well suited to regions at 1 atmosphere pressure, but baro-
philic bacteria residing near the bottom of the sea tolerate 1400
atmospheres. The fungus *Acontium velatum* has been found in solu-

126

tions with a pH of less than 1.0, and *Plectonema nostocorum* possibly withstands alkaline conditions as extreme as pH 13. Anaerobic bacteria grow readily in locales having a low E_h, while at the other extreme are the aerobes that require highly oxidized surroundings. Occasional bacteria and fungi seem capable of reproducing appreciably below 0°C, provided that liquid water is present, a not uncommon occurrence where the solution contains significant quantities of salts or sugar (Vallentyne, 1963). Considerable controversy exists on the upper temperature limit for microbial life, but recent evidence indicates that cells resembling flexibacteria actually replicate in hot springs at temperatures up to 91°C, whereas the blue-green alga *Synechococcus* is active in such springs at 73–75°C. The maximum for fungi is in the vicinity of 56–60°C, while that for animals, including protozoa, is approximately 51°C (Brock, 1967). The range for other factors is quite narrow, however; the concentration limits of ions of certain heavy metals and possibly sulfide are so narrow that probably localities exceeding these narrow limits either are sterile or support no active metabolism.

The range for a given organism covers only a small portion of these biologically tolerable limits. Terms such as obligate halophile, thermophile, barophile, osmophile, psychrophile, xerophile, and anaerobe have been coined specifically to designate microorganisms living in restricted habitats and possessing narrow ecological amplitudes. The expressions refer to organisms requiring high salt, temperature, pressure, or sugar levels, low temperature, little moisture, and the absence of oxygen, respectively. Broader ranges, of course, commonly typify populations that are facultative for these traits.

Although the activities and very existence of a population are associated with a multitude of abiotic factors, a maximum and a minimum can usually be established, in the laboratory at least, for each of these factors. These are the physiological limits beyond which the population is unable to maintain itself or to perform a vital function. In some instances no minimum exists, because growth is possible to a point where the factor is essentially no longer present, as with toxins or salinity. For each factor an optimum level or range can be established, in addition to an upper limit. As the intensity of the factor exceeds or falls short of the optimum, the response of the organism diminishes, as measured in terms of growth, reproduction, sporulation, or the carrying out of a given metabolic process. Tolerance ranges of one bacterial strain to temperature, water activity, and pH are depicted in Fig. 6.1. Comparable data are available to show how many other organisms respond in vitro not only to

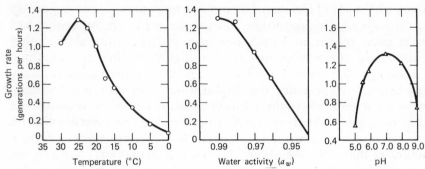

Figure 6.1 The effect of temperature, water activity, and pH on a *Microbacterium* species (Brownlie, 1966). Water activity is the ratio of the vapor pressure of the solution to the vapor pressure of the solvent, low water activity indicating high osmotic pressure.

temperature, water activity, and pH but also to toxic cations, organic inhibitors, E_h, etc.

Strains, varieties, or races of the same species commonly show differences in their tolerance and in their optima for most, if not all, of the major environmental factors. The differences presumably arise because the populations have been subjected to a natural selection for those phenotypes best suited for the specific environmental factors. These populations of identical species that differ because of an adjustment to local circumstances represent *ecotypes* of the species. The dissimilar tolerance ranges and optima of distinct ecotypes reflect the variations in tolerance to the factor in question as dictated by the genetic makeup of the cells; the environment merely selects for those variants most suited to the locale.

Species similar or identical in many of their biochemical and morphological characteristics exhibit widely different tolerance ranges. A few species of autotrophic sulfur-oxidizing bacteria of the genus *Thiobacillus*, for example, are restricted solely to acidic environments, while others are unable to grow in these regions under any conditions. Many such examples are known among the bacteria, fungi, algae, protozoa, and actinomycetes with respect to pH as well as other abiotic factors.

The minimum, maximum, and optimum levels or intensities of a factor are not fixed for a given population, but depend on environmental circumstances. Caution must be exercised, therefore, in extrapolating from experiments wherein only a single factor is varied. In effect, an organism may develop under conditions that appear on

Table 6.1

Factors Affecting Microbial Tolerance Ranges

Tolerance Range Modified	Factor Altering Range	Organism	Reference
Temperature	Nutrients	Ochromonas	Hutner et al. (1957)
	CO_2	Neurospora	Charles (1962)
	Hydrostatic pressure	Desulfovibrio	ZoBell (1958)
	NaCl	Brevibacterium	Mulder et al. (1966)
	pH	Mastigocladus	Gessner (1955)
	Toxins	Bacillus	Ljunger (1968)
Humidity	Temperature	Uncinula	Delp (1954)
Available water	Temperature	Pseudomonas	Wodzinski and Frazier (1960)
pH	NaCl	Brevibacterium	Mulder et al. (1966)
pH	Pyridoxine	Neurospora	Stokes et al. (1943)
O_2	Thiamine, nicotinic acid	Mucor	Bartnicki-Garcia and Nickerson (1961)

first sight to be outside its tolerance range. This anomalous behavior is designated *compensation*, inasmuch as one factor compensates for a second. Nevertheless, even under the best of circumstances, the extension of the range is modest. How the limiting influence of several abiotic determinants regulating microbial growth is modified by compensation is illustrated in Table 6.1. The results of many investigators show that the temperature range for microbial growth can be extended by specific nutrients. In some instances a mesophile provided with a compound it does not require at one temperature will grow at a higher temperature in the presence of the nutrient; in other instances a thermophile, like *Bacillus stearothermophilus*, will multiply at lower temperatures if provided with growth factors than when it does not have these substances (Campbell, 1954). Carbon dioxide allows a strain of *Neurospora* developing at 20°C but not at 30°C to proliferate at the higher temperature, and thiamine and nicotinic acid permit an isolate of the fungus *Mucor rouxii* to grow without O_2, a feat not possible in the absence of these vitamins. The explanation presumably often lies in the inability of the organisms to synthesize the nutrient in question at a point somewhat outside their presumed initial tolerance ranges, although they can do so within the range, so that replication is permitted if an exogenous source of the metabolite is supplied.

The limiting effect of extremes of other abiotic factors may also be overcome, as indicated in Table 6.1. The response to humidity or available water is modified by temperature, and the maximum temperature for an organism may be increased by raising the hydrostatic pressure or changing the pH. It is likely that the tolerance range of one or another species to each of the ecologically important abiotic factors will be broadened by most of the other factors. Yet the range may not always be extended; for example, although *Brevibacterium linens* grows at more acid and alkaline extremes in the presence than in the absence of NaCl, *Arthrobacter globiformis* loses its tolerance to both high and low hydrogen ion concentrations in solutions rich in salt (Mulder et al., 1966). Similarly, the minimum temperature for *B. stearothermophilus* is raised by toxic agents (Ljunger, 1968).

The tolerance range for survival of a species or its ecotypes is frequently far greater than that for multiplication. Often this is attributable to the formation of specialized resting structures that are appreciably less susceptible than actively metabolizing cells to extremes of temperature, drought, acidity, or toxins. Bacterial endospores, protozoan cysts, fungal chlamydospores or sclerotia, and actinomycete conidia all thus aid in perpetuating the species in particular locales. With survival, as with growth, compensation can occur; thus the ability of bacteria to withstand heat is modified by pH and is increased as the cells become drier, and the survival of sporelings of red algae in water of reduced salinity is altered by temperature (Boney, 1966). Studies of bacteria have provided a multitude of illustrations demonstrating how resistance to several kinds of stress is affected by modification in the environment where the stress is being imposed.

The tolerance range should never be considered, however, as anything more than a permissive range, one defining limits, admittedly modifiable through compensation, outside which multiplication or survival is impossible. Within the boundaries, development is feasible but it still may not occur, for any one of a variety of reasons. The habitat in vivo is far different from that defined by the glass walls of a culture vessel, and either abiotic or biological factors or stresses may restrict an organism in nature to an amplitude considerably narrower than that typifying it in axenic culture in liquid or agar media.

Organisms utilizing nutrients or substrates whose availability is regulated by abiotic factors present a special situation in regard to ecological amplitude. The availability of many soluble or insoluble nutrients, both organic and inorganic, changes as the level or intensity of these factors is modified; for example, pH regulates the extent of retention of numerous soluble nutrients on colloidal surfaces, whereas the prevailing E_h determines whether or not several inorganic nutrient elements are in a form readily assimilated. Predators and parasites living on microorganisms represent special cases, too, in that, although restricted to a set of ranges defining the conditions allowing their reproduction, they are further limited by the ecological amplitude of their prey or host. The actual tolerance range of saprobes, parasites, or predators in nature is shortened, therefore, according to the manner by which each individual abiotic factor determines the availability of its inanimate or viable food supply. Similarly, the actual ecological amplitude of plant and animal pathogens is a reflection of the interaction between the micro- and the macroorganisms' respective ranges, and inasmuch as parasites often exhibit a narrower tolerance than their hosts, the latter are free of the former and show no symptoms of disease produced by pathogenic agents in environments suitable for the plant or animal alone.

LIMITING FACTORS

A variable controlling the distribution or density of a population is said to be a *limiting factor*. Among the limiting abiotic factors are pH, temperature, osmotic or hydrostatic pressure, humidity, light intensity, and salinity. Only the aforementioned variables are considered in the present chapter, though clearly the crucial determinant of the occurrence or abundance of a species may be nutritional, environmental feedback, or stresses imposed by established inhabitants of the site. When the population's tolerance is incompatible with the factor, because the intensity exceeds either the maximum or the minimum acceptable level, the organisms fail to develop and frequently disappear. The limiting factors in dissimilar surroundings are frequently different for the same species, as when excess heat operates in one region and acid in another. In an individual habitat, moreover, the critical limiting factor varies according to the species, so that rarely can a single variable be singled out as the sole ecological determinant for all potential residents of a particular locality.

Nevertheless, autecological investigations have clearly demonstrated that the absence of selected species in natural habitats is directly attributable to a single feature of the habitat. Results of typical studies depicting environmental regulators of prevalence or occurrence in vivo are summarized in Table 6.2. Because of the un-

Table 6.2

Factors Limiting Distribution of Selected Microorganisms

Limiting Factor	Environment	Organism	Reference
Acidity	Food	*Clostridium botulinum*	Mossel and Ingram (1955)
	Vagina	*Mycoplasma*	Morton (1965)
	Soil	*Streptomyces scabies*	Alexander (1961)
	Soil	*Azotobacter* spp.	Alexander (1961)
Alkalinity	Rock pools	Red algae	Atkins (1922)
Temperature, high	Rabbits	*Diplococcus pneumoniae*	White (1938)
Temperature >54°C	Hot springs	*Oscillatoria terebriformis*	Castenholz (1968)
<54°C	Hot springs	*Synechococcus lividus*	Castenholz (1968)
>43°C	Hot springs	Protozoa	Copeland (1936)
>19°C	River	Certain diatoms	Blum (1956)
O_2	Sewage digestors	Aerobic protozoa	Noland and Gojdics (1967)
Salinity, high	Great Salt Lake	Diatoms	Patrick (1936)
Light intensity	Atoll waters	*Tydemania expeditionis*	Gilmartin (1966)

favorable pH, temperature, O_2 levels, salinity, or light intensity, the organisms listed do not become established, although they do successfully invade and colonize the same kinds of habitats when the intensity of the factor indicated does not exceed their tolerances. The practical implications are considerable, inasmuch as the absence of many disease agents, such as *Diplococcus pneumoniae* or *Streptomyces scabies,* from particular regions or hosts can be ascribed to a well-defined abiotic factor controlling their invasion or survival. Conversely, a group of environmental properties rather than a single one undoubtedly is responsible for the inability of a high proportion of invaders to gain a foothold in areas they reach.

The lack of virulence of *D. pneumoniae* strains for rabbits provides an interesting example of the operation of temperature as a limiting factor. Isolates unable to grow at 41°C cause no disease in rabbits, whereas thermotolerant pneumococci are pathogenic. The reason is that following intravenous introduction of the bacteria the body temperature of the animal rises to 41°C during fever, and a strain incapable of multiplying at the elevated temperature of the infected rabbit is eliminated and no disease ensues (White, 1938). Susceptibility to freezing and thawing probably restricts the distribution of a variety of microorganisms, for they would be excluded from regions alternately above and below the freezing point of water; in this light it is interesting that algae from Antarctica do not lose their viability when exposed to a freezing-thawing cycle, whereas temperate-zone strains of the genera *Phormidium* and *Oscillatoria* may not survive a single freezing (Holm-Hansen, 1967).

Pronounced environmental stresses are evident in intertidal regions—that is, the areas along a coastline between high- and low-tide marks—and a large proportion of the marine flora coming into contact with these habitats are incapable of enduring the extreme conditions. Intertidal algae, on the other hand, are uniquely capable of withstanding the stresses and establish themselves where other algae cannot. They retain their viability despite the desiccation occurring at low tide, the large fluctuations in salt concentration as the ocean water evaporates or as rains dilute the salt water, the diurnal and seasonal changes in temperature, the mechanical force of wave action, the high light intensities, and the freezing and thawing in the winter season of temperate latitudes. Tolerances to this assortment of entirely different types of environmental shock are not widespread among marine algae, and hence they do not colonize those habitats for which the intertidal algae are so uniquely suited. The salinity range of many algae is also insufficiently wide to permit their multiplication in both fresh and ocean water, but salt tolerance underlies the fitness of those species adhering to and growing on the sides of ships that travel from rivers to the seas and back again (Biebl, 1962). A basis for habitat selection by certain algae is their tolerance to light of various intensities, the species of intertidal zones being chosen from among those not destroyed by high light intensities; aquatic algae inhibited by excessive sunlight apparently often occur at depths where the light intensity is low (Biebl, 1962; Gilmartin, 1966). One might also expect that protozoa and algae prone to inactivation by desiccation and producing no drought-resistant structures will not be indigenous to habitats subject to frequent drying cycles, as at the soil surface or on rocks, and that sulfide-intolerant cells will not be located in waters in which sulfide is generated in copious quantities.

The intensity of abiotic factors in many habitats is reasonably constant, but of particular value in assessing the impact of these variables in nature are environments wherein a pronounced gradient exists. Salinity gradients are evident in estuaries and in artificial ponds erected for salt production, and distinct effects on distribution based on algal tolerance to salt are readily discernible there. Thermal

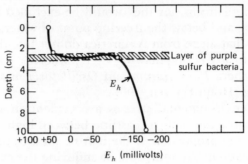

Figure 6.2 Distribution of purple sulfur bacteria in salt water lakes as a function of depth (Gorlenko, 1968).

gradients are not difficult to find in natural ecosystems, and those of hot springs show interesting patterns of algal distribution. This is well illustrated by *Oscillatoria terebriformis,* which has an upper limit of about 54°C for growth. This blue-green species is native to hot springs of Oregon at sites where the temperature ranges from 35° to 54°C, and it migrates to keep itself within this temperature range. During the fall and winter the thermal gradient is modified, but the population of *O. terebriformis* responds by altering its location so that it is found throughout the year in the zone having the appropriate temperatures (Castenholz, 1968). Gradients in E_h are also widely evident in nature, and the location of many populations is quite obviously correlated with the potential that, on the basis of laboratory tests, allows for their development. An example of the relation between gradient in E_h and vertical distribution of purple sulfur bacteria is shown in Fig. 6.2. These bacteria are anaerobes that require a low E_h and use H_2S in their metabolism. Often the organisms are found near the water's surface; in the zone in which they multiply, however, no O_2 is present and the E_h is invariably quite low.

The breadth of the tolerance range of an organism is described by the prefixes *steno* and *eury*. The former denotes a narrow and the latter a wide range. Thus strains with a limited tolerance to temperature, salt concentrations, or moisture levels are termed stenothermal, stenohaline, and stenohydric; those developing over a wide span of temperatures, salt concentrations, or moisture levels are described as eurythermal, euryhaline, and euryhydric. Species restricted to a narrow range of a particular environmental condition are said to be *stenoecious* for that factor, while those with a large amplitude are designated *euryoecious*. A species possessing an obligate dependency on some factor, such as the obligate psychrophile or halophile, is often, though not necessarily, more stenoecious than the facultative organism; for example, the strict thermophile or anaerobe is restricted to localities that have high temperatures or are O_2-free, whereas their facultative counterparts multiply at moderate as well as at high temperatures or, alternatively, in the absence or presence of O_2. Euryoecious populations characterize environments where the intensity of the abiotic factor fluctuates, though stenoecious individuals may persist for some time even in the face of the restrictive condition and become abundant when the habitat once more is favorable to them. Alternatively, the stenoecious type may have an efficient dispersal mechanism and, though readily eliminated at certain periods, repopulate the briefly favorable habitat from an adjoining site where it maintained itself. By contrast, environments in which the level of the factor is reasonably constant may contain both stenoecious and euryoecious groups. Notable examples of euryoecious behavior are the intertidal algae tolerant of extreme temperature fluctuations; the diatom *Pinnularia appendiculata,* which is found in both cold mountain lakes and bodies of water with temperatures as high as 70°C; polar algae and lichens growing on rock surfaces where the temperature changes appreciably in short time periods; and the euryhaline diatom *Melosira nummuloides,* which has been observed in salt waters with a wide range of salinities. Stenoecious behavior has been recorded for the many stenohaline marine diatoms, the so-called snow algae growing only in narrow temperature ranges, and species of *Clostridium* and *Desulfovibrio* resident in habitats with a narrow and consistently low E_h range (Marre, 1962; Wood, 1965).

Species developing in ecosystems that differ in the intensity of a particular abiotic factor, whether the intensity in a given region is relatively fixed or fluctuating, are euryoecious for that variable. Ubiquity is not the exclusive property of euryoecious groups, for many stenoecious microorganisms are cosmopolitan too, apparently because the conditions they require are widespread. Nevertheless, species tolerant of extremes of one factor may have a surprisingly limited distribution, a result of the fact that cells or filaments euryoecious for one critical ecological determinant are often stenoecious for a second; thus facultatively anaerobic bacteria are frequently quite pH-sensitive and eurythermal algae may be stenohaline.

It should be obvious that a stenoecious organism will, with regard to the factor for which it has a narrow tolerance, reflect its environment; this is well exemplified by the stenothermal, stenohaline *Thalassiosira antarctica*, an alga bearing a species epithet suggesting its geographical distribution. Euryoecious species, by contrast, do not reflect the character of their habitat in regard to the variable for which they have a broad range.

When one ecologically significant abiotic factor is at the optimal level, a number of organisms show an unexpected degree of tolerance to large fluctuations in a second factor. The span of moisture levels allowing for germination of fungal spores is frequently widest at the optimum temperature for germination, and the range narrows at temperatures distant from those most favorable; certain bacteria and fungi are least affected by high osmotic pressures at their optimum temperature or optimum pH; and germination of the spores of selected fungi is possible over a wider temperature amplitude when the organisms are at a favorable acidity. Still, microorganisms frequently do not thrive in natural habitats when a factor of interest is at its optimal point for them, and communities are typically composed of organisms living and actively metabolizing at intensities of particular factors significantly above or below their optima. Indeed, data on optima in vitro have little known ecological significance, although it can be argued that a population may compete most vigorously with its neighbors at conditions close to those most ideal. Many physiological or environmental properties contribute to determining the optimum in vivo, and only one, but likely a major one, is the most suitable intensity in vitro.

Levels of abiotic factors vary both in space and in time. Marked daily temperature fluctuations typify polar and desert areas, and seasonal changes are features of surface waters and soil. Regular modifications in O_2 status, toxin concentration, acidity, or salinity characterize numerous habitats. Only species withstanding the usual chemical and physical modifications to which the habitat is subject will make up the climax communities, and as a rule stenoecious strains fail to endure such adversities and typically are not native to regions subject to alterations of these sorts. Ecological resilience is a necessary attribute in a changing microcosm, and a factor present, even ephemerally, at a level exceeding a species' resistance can be the reason for the organism's exclusion from an environment. Tolerance is influenced, furthermore, not merely by the existence of the extreme condition but also by its duration, and the capacity to endure short periods of exposure to extremes of temperature, pH, E_h, toxin levels, and so on is typical of organisms not able to remain viable through prolonged stresses. The exposure time must then be considered in defining an organism's ability to withstand critical environmental factors and in attempting to correlate distribution with abiotic determinants of community composition.

Among organisms undergoing morphological changes, individual phases in the life cycle often have an especially narrow tolerance to one or more abiotic factors. The occurrence and distribution of such species may be governed entirely by the width of this range, even though the particularly sensitive phase exists for only a short span of time. Morphogenetic sequences occur in the life cycles of members of diverse taxonomic groups. Age, too, is associated with changes in tolerance range, as illustrated by observations that the protozoan *Ochromonas danica* becomes more sensitive to inhibitory chemicals as the population ages (Aaronson and Bensky, 1967) and that *Aspergillus glaucus* spores derived from a young culture germinate at available water levels that do not permit ready germination of spores from old cultures (Scott, 1957). The tolerance of marine algae, moreover, may vary according to the season (Biebl, 1962).

It must be emphasized, to avoid oversimplification, that the environment is made up of a multitude of factors acting collectively and in unison, and the influences of these many variables tend to be interlaced and interdependent. Furthermore, inasmuch as a large number and great variety of species with like tolerance ranges can or do coexist, the interactions among these populations will determine ultimately which of those with the appropriate tolerances will be abundant, which will have lesser significance, and which will be eliminated or fail to become established. Determinations of ecological amplitudes allow for the delineation of ecosystems where a species may grow, not those in which it does indeed proliferate. Given that an organism has found its way to a particular locality, nutrients of the right kinds are present, environmental feedback is not of consequence, and the tolerance ranges of the species are not exceeded, there may still be neither multiplication nor survival of the invading propagule: individuals already well established eliminate the newcomer through mechanisms to be considered in later chapters. The distribution of species having similar or identical ecological amplitudes is governed to a great degree by the interactions between them, and tolerance therefore becomes of little significance or is irrelevant. When conditions approach the endurance limits of the populations, the ecological amplitudes nevertheless assume overriding importance.

Observations in vitro must also be interpreted with a view that the tolerance in axenic culture is often appreciably different from the breadth of the range noted in the mixtures of populations constituting the majority of natural communities. An organism may fare poorly in nature at its optimum pH, for example, because scores of neighbors develop well at the same reaction, so that there is keen competition for the available nutrient supply; if the species of interest reproduces over a wide range, it may develop more extensively in vivo at a pH suboptimal for its growth in axenic culture owing to the absence of significant competition at the extremes of pH. This probably is the explanation for the abundance in acid soils of fungi having pH optima near neutrality. Similarly, many blue-green algae of hot springs develop in the laboratory at moderate temperatures, but they are excluded from the cooler waters by nonthermophilic algae better able to exploit prevailing conditions.

EXTREME ENVIRONMENTS

The capacity of a microorganism to withstand extreme levels of some abiotic factor may be the basis for its ecological prominence. As a rule, as the intensity of a factor approaches the limits permitting life, the number of different genera, species, or varieties in the community declines. This is admirably illustrated in Copeland's (1936) study of the blue-green algae of Yellowstone National Park, in which it was shown that the abundance of different types was progressively smaller as the temperature and pH of the habitat approached the upper or lower extremes (Fig. 6.3). Conversely, the greatest number of species probably will be found in the moderate portion of the gradient of an abiotic factor such as temperature or pH. Near the limits of biological tolerance few species are evident, and the community may indeed be monospecific or dominated by a single type of bacterium, fungus, alga, or protozoan. In this relatively homogeneous community, where species diversity is small, the number of individuals or the density of filaments can nevertheless be appreciable, as in the masses of algae in hot springs, the green coatings on solid objects in highly acid streams, or the vivid colors resulting from microbial growth on snow surfaces or in the solar evaporation ponds of salt works.

Scores of extreme environments are known on the surface of the globe, but just a few have been examined, and in only rare instances have the studies been extensive. Harsh habitats that have been scrutinized to some extent are the highly saline Great Salt Lake and Dead Sea, evaporation ponds of salt works, alkaline lakes, rivers heavily polluted with H_2SO_4-rich seepage from mine waters, hot springs, H_2S-laden waters, partially dried foods or food products, soil during prolonged droughts, composts undergoing decomposition at high temperature, and snowfields of polar, glacial, and alpine regions.

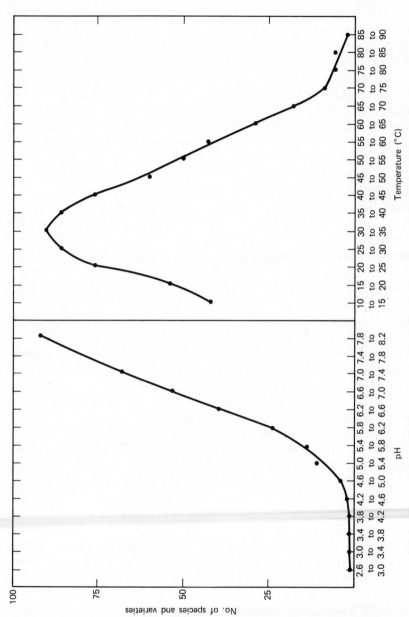

Figure 6.3 Distribution of blue-green algae as related to environmental reaction and temperature.

The dominant taxonomic groups and the prominent genera and species within those groups depend on the identity of the harsh factor and the particular environment. The dominants are selected from among those propagules that have successfully reached the site and possess tolerance limits that enable them to multiply under the stress conditions. All other potential invaders are rejected because they cannot withstand the stress. Not uncommonly the dominants in the community are obligate in their reliance on high levels of the variable in question; that is, they are obligate thermophiles, halophiles, or the like.

The dominants of a select few of these communities have been characterized. Bacteria are usually found in almost all harsh circumstances permitting life, but rarely are they identified taxonomically. Thus highly saline and grossly polluted waters, hot springs, and the heterotrophic communities bringing about decomposition of organic materials at elevated temperatures are rich in bacteria, largely of unknown taxonomic affinities. *Halobacterium, Micrococcus,* and *Sarcina* are common in brines and heavily salted foods, *Bacillus* is abundant in desert soils, and strains of *Thiobacillus* inhabit acid mine waters. Fungi are present in arctic regions and are among the dominants in fabrics, manufactured products, and incompletely dried foods. Unique algae are prominent in hot springs, saline and alkaline lakes, snowfields, and acid mine waters. Among the protozoa, dominance by a single genus or species has been recorded in Great Salt Lake, localities in the antarctic, and heavily polluted waters, whereas actinomycete genera are dominants in the microbial communities of soils during drought and of organic materials decaying at high temperature. Neither fungi nor protozoa appear in significant or detectable numbers at temperatures in excess of about 50°C. Typical inhabitants of extreme environments are listed in Table 6.3. In the artificial communities established in the laboratory by employing solutions highly toxic to most living things, the microflora is largely or actually monospecific, too; instances have been described in which a *Trichosporon* appeared in media containing 1 N H_2SO_4 and bacteria developed in solutions containing very high concentrations of LiCl. The housewife also regularly encounters essentially a monospecific community in jars of jelly or syrup.

Because harsh environments are frequently ephemeral and widely dispersed on the earth's surface, habitats similar in their physical and chemical properties often support somewhat different communities. Dispersal mechanisms may not be sufficiently effective or continuous to permit the introduction of the same microbial types in separate localities. This is evidently true of many of the extreme environ-

Table 6.3

Dominants in Extreme Environments

Environment	Typical Dominant	Reference
Hot springs	Synechococcus	Brock (1967)
Acid hot springs	Cyanidium caldarium	Round (1965)
Heating dung piles	Bacillus, Streptomyces	Waksman et al. (1939)
Snow surfaces	Various algae	Tiffany (1951)
Antarctic grassland	Corythion dubium	Heal et al. (1967)
Brines, salty products	Halophilic bacteria	Larsen (1967)
Great Salt Lake	Uroleptus	Pack (1919)
Solar salt pond	Dunaliella	Gibor (1956)
Alkaline lake	Arthrospira platensis	Jenkin (1936)
Roots of desert plants	Bacillus	Elwan and Mahmoud (1960)
Desert soil	Bacillus	Henis and Eren (1963)
Soil during drought	Streptomyces	Meiklejohn (1957)

ments occupying a minute area, as in the case of preserved foods and manufactured materials. The lack of adequate dissemination applies likewise to residents of environments readily accessible to aerially transported particles; for instance, although hot springs in Yellowstone National Park may contain *Synechococcus* or other blue-green algae growing at 73–75°C (Brock, 1967), comparable areas in hot springs elsewhere are devoid of these algae, presumably because cells tolerating the temperature extremes are not introduced. The algae of distant springs may withstand temperatures no higher than about 60°C, and the dominant is not *Synechococcus* but rather a species like *Mastigocladus laminosus* (Schwabe, 1960).

The biochemical versatility and metabolic activities of the biotas inhabiting such extreme circumstances have received little attention, but because of the low species diversity, the number of different activities is likely to be far fewer than in communities of less harsh localities. Elements assimilated by the climax community are likely to be released slowly by reason of the absence of an array of populations capable of attacking protoplasmic constituents. Many functions typical of other communities are undoubtedly slow or missing. Animals or microbial predators and parasites feeding on members of aquatic microfloras may be unable to grow, so that a check on the biomass is absent. It is not surprising, therefore, that the protoplasmic material may remain intact for long periods of time, though, should the cells and filaments be washed to a nonextreme region or the harsh conditions become upset, a spurt of decomposition is initiated.

References

REVIEWS

Alexander, M. 1961. *Introduction to Soil Microbiology.* Wiley, New York.

Biebl, R. 1962. Seaweeds. *In* R. A. Lewin, Ed., *Physiology and Biochemistry of Algae.* Academic Press, New York. pp. 799–815.

Blum, J. L. 1956. The ecology of river algae. *Bot. Rev.,* **22,** 291–341.

Boney, A. D. 1966. *A Biology of Marine Algae.* Hutchinson, London.

Brock, T. D. 1969. Microbial growth under extreme conditions. *In* P. M. Meadow and S. T. Pirt, Eds., *Microbial Growth.* Cambridge University Press, London. pp. 15–41.

Noland, L. E., and Gojdics, M. 1967. Ecology of free-living protozoa. *In* T.-T. Chen, Ed., *Research in Protozoology,* Vol. 2. Pergamon Press, New York. pp. 215–266.

Round, F. E. 1965. *The Biology of the Algae.* St. Martin's Press, New York.

Tiffany, L. H. 1951. Ecology of freshwater algae. *In* G. M. Smith, Ed., *Manual of Phycology.* Chronica Botanica, Waltham, Mass. pp. 293–311.

Vallentyne, J. R. 1963. Environmental biophysics and microbial ubiquity. *Ann. N.Y. Acad. Sci.,* **108,** 342–352.

Wood, E. J. F. 1965. *Marine Microbial Ecology.* Reinhold, New York.

OTHER LITERATURE CITED

Aaronson, S., and Bensky, B. 1967. *J. Protozool.,* **14,** 76–78.

Atkins, W. R. G. 1922. *J. Marine Biol. Assoc. U. K.,* **12,** 789–791.

Bartnicki-Garcia, S., and Nickerson, W. J. 1961. *J. Bacteriol.,* **82,** 142–148.

Brock, T. D. 1967. *Science,* **158,** 1012–1019.

Brownlie, L. E. 1966. *J. Appl. Bacteriol.,* **29,** 447–454.

142

Campbell, L. L. 1954. *J. Bacteriol.*, **68**, 505–507.

Castenholz, R. W. 1968. *J. Phycol.*, 4, 132–139.

Charles, H. P. 1962. *Nature,* **195**, 359–360.

Copeland, J. J. 1936. *Ann. N.Y. Acad. Sci.*, **36**, 1–223.

Delp, C. J. 1954. *Phytopathology*, 44, 615–626.

Elwan, S. H., and Mahmoud, S. A. Z. 1960. *Arch. Mikrobiol.*, **36**, 360–364.

Gessner, F. 1955. *Hydrobotanik*, Band. I. Deutscher Verlag Wissenschaften, Berlin.

Gibor, A. 1956. *Biol. Bull.*, **111**, 223–229.

Gilmartin, M. 1966. *J. Phycol.*, **2**, 100–105.

Gorlenko, V. M. 1968. *Mikrobiologiya*, **37**, 26–30.

Heal, O. W., Bailey, A. D., and Latter, P. M. 1967. *Phil. Trans. Royal Soc. London*, **B252**, 191–197.

Henis, Y., and Eren, J. 1963. *Can. J. Microbiol.*, **9**, 902–904.

Holm-Hansen, O. 1967. *In Environmental Requirements of Blue-Green Algae.* Pacific Northwest Water Laboratory, Corvallis, Ore. pp. 87–96.

Hutner, S. H., Baker, S., Aaronson, S., Nathan, H. A., Rodriguez, E., Lockwood, S., Sanders, M., and Petersen, R. A. 1957. *J. Protozool.*, 4, 259–269.

Jenkin, P. M. 1936. *Ann. Mag. Nat. Hist., Ser. 10*, **18**, 133–181.

Larsen, H. 1967. *Advan. Microbial Physiol.*, **1**, 97–132.

Ljunger, C. 1968. *Physiol. Plantarum*, 21, 26–34.

Marre, E. 1962. *In* R. A. Lewin, Ed., *Physiology and Biochemistry of Algae.* Academic Press, New York. pp. 541–550.

Meiklejohn, J. 1957. *J. Soil Sci.*, **8**, 240–247.

Morton, H. E. 1965. *In* R. J. Dubos and J. G. Hirsch, Eds., *Bacterial and Mycotic Infections of Man.* Lippincott, Philadelphia. pp. 786–809.

Mossel, D. A. A., and Ingram, M. 1955. *J. Appl. Bacteriol.*, **18**, 232–268.

Mulder, E. G., Adamse, A. D., Antheunisse, J., Deinema, M. H., Woldendorp, J. W., and Zevenhuizen, L. P. T. M. 1966. *J. Appl. Bacteriol.*, **29**, 44–71.

Pack, D. A. 1919. *Biol. Bull.*, **36**, 273–282.

Patrick, R. 1936. *Bull. Torrey Bot. Club*, **63**, 157–166.

Schwabe, G. H. 1960. *Schweiz. Z. Hydrol.*, **22**, 759–792.

Scott, W. J. 1957. *Advan. Food Res.*, **7**, 83–127.

Stokes, J. L., Foster, J. W., and Woodward, C. R. 1943. *Arch. Biochem.*, **2**, 235–245.

Waksman, S. A., Umbreit, W. W., and Cordon, T. C. 1939. *Soil Sci.*, **47**, 37–61.

White, B. 1938. *The Biology of Pneumococcus.* Commonwealth Fund, New York.

Wodzinski, R. J., and Frazier, W. C. 1960. *J. Bacteriol.*, **79**, 572–578.

ZoBell, C. E. 1958. *Producers Monthly*, **22**(7), 12–29.

7

Geography and
Microenvironment
of Microorganisms

Countless genera and species occur worldwide in their distinctive habitats, and although the quantitative composition of the communities containing these ubiquitous microorganisms may vary, physically and geographically isolated regions still contain the same general taxonomic groups. Such cosmopolitan distributions are evident for a multitude of heterotrophs and autotrophs, and additional lists of the natives of well-studied ecosystems consequently fail to provide much new information. Conversely, a restricted rather than a universal distribution is the rule among specific freshwater and marine algae, saprobic and plant- and animal-parasitic fungi, predatory and parasitic protozoa, and bacteria, each exhibiting distinctive geographic preferences. This is illustrated for marine algae in Fig. 7.1.

Biogeography is concerned with the distribution of organisms and the bases for their distributions. Though microscopic forms of life possess unique patterns of occurrence extending over large areas as well as in locales measured in terms of 1 mm or less, use of the term biogeography in a microbiological sense will be restricted here to the study of patterns characteristic of larger regions. The mere existence of a discipline of biogeography connotes either spatial barriers preventing the introduction of propagules of some population into a new area or differences, quite subtle or unmistakable, between habitats usually described as being similar. It can be noted in Fig. 7.1 that algae may exhibit similar or overlapping distributions; these photoautotrophs are *sympatric*. *Allopatric* organisms, by contrast, occupy separate and distinct geographical areas.

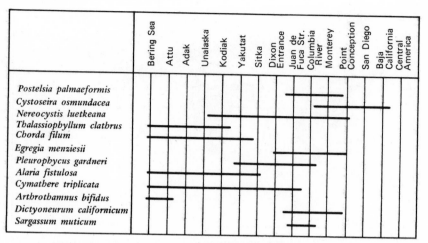

Figure 7.1   Distribution of certain species of marine algae along the west coast of North America (Scagel, 1963).

GEOGRAPHIC DISTRIBUTIONS

A notable example of a unique biogeography among terrestrial organisms is provided by *Beijerinckia*, a genus of free-living, N_2-assimilating heterotrophs. These bacteria are largely confined to tropical soils, and their occurrence has been recorded in tropical parts of Asia, Africa, South America, and Australia. Occasionally small numbers are found in soils of the temperate zone, but these reports have been so infrequent that the bacterium can essentially be considered to have a tropical milieu (Becking, 1961*a*).

Distinct geographical patterns are usual among the marine algae. For example, *Thalassiosira antarctica* is characteristic of antarctic waters, *T. hyalina* is limited to arctic regions, and *Planktoniella sol* is typically a tropical and subtropical diatom (Smayda, 1958). The locations of these algae have been mapped accurately, and each is doubtlessly confined to a limited region of the seas. Fucoid algae are prominent in oceanic waters of the temperate zone, and red algae stand out near the shores of tropical seas. The many maps prepared to depict distributions of the marine phytoplankton reveal quite clearly the predilection of individual species of *Ceratium, Rhizosolenia, Thalassiothrix,* and *Biddulphia* for northern or southern oceanic waters, or

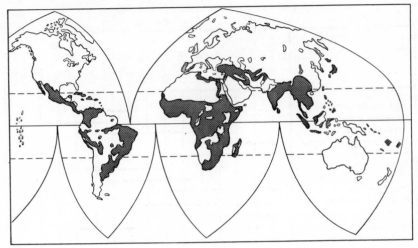

Figure 7.2 World distribution of leprosy (Stamp, 1964). By permission of Cornell University Press.

for either the open sea or habitats closer to the shore (Glover, 1961). Some algae are restricted to habitats always bearing snow, as near the tops of tall mountains, where their presence is easily noted because of the colors they impart to the snow.

Areas inhabited by particular microorganisms may not be too extensive, but distinct patterns of occurrence are evident nonetheless. Patchiness of algal blooms in coastal waters and lakes is nearly universal, the patches varying in size from several centimeters to more than 300 km in length or breadth. The algal patches are on occasion monospecific, but a high proportion are heterogeneous in species composition. Although a variety of diatoms and flagellates are found in distinct and sometimes sizable patches, especial attention has been given to the uneven distribution patterns of those aquatic algae producing the so-called red tides, an interest arising because of the toxins they elaborate (Bainbridge, 1957).

Parasitic microorganisms have been known for centuries to be subject to geographical restrictions (Fig. 7.2), but in only rare instances are the reasons established. Four categories of parasite biogeographies are distinguishable: (a) the specific host is quite localized, and hence parasites feeding on it exhibit a narrow geographic range; (b) the characteristic host is widely distributed, but its parasite extends over only a portion of the host's geographic range; (c) the parasite is not specific for any single host species and is found in a wider range of localities than any one animal or plant host, as in the case of the plant pathogen *Armillaria mellea;* and (d) the parasite grows saprobically in nature and hence is not restricted by the distribution of the plants or animals on or in which it occasionally lives. If a microorganism is nutritionally and ecologically highly specialized and develops on just a single host species, the distribution of the host governs the occurrence of its parasite. When the host plant or animal exists solely in distinct climatic areas, clearly the parasite must too; for example, the Ascomycete *Cyttaria* attacks the antarctic leech and thus is indigenous exclusively to sites in the southern hemisphere containing the leech (Diehl, 1937). Similarly, *Oospora citri* is parasitic on *Citrus* and *Thecaphora solani* lives solely on the Andean potato, so that the former fungus is typically subtropical and tropical while the latter is an inhabitant exclusively of the Andes. In like fashion, microorganisms feeding on selected algae, sponges, and fish are restricted to regions containing the aquatic hosts (Pirozynski, 1968).

Often, parasites attacking cosmopolitan animal or plant species cannot, for one reason or another, become established in environments that are perfectly suitable for the host. When this occurs with a pathogen of man, livestock, or crop plants, the end result is that the disease it causes is geographically confined. The restriction of many diseases to one continent, a group of adjoining countries, or fixed climatic areas is attributable to the environmental boundaries within which the pathogen is forced to live. Among the illustrations of the range of a parasite being narrower than that of its potential host are the following: species of *Plasmodium, Mycobacterium leprae,* and *Treponema pertenue* causing human malaria, leprosy, and yaws in the tropics and subtropics; *Trypanosoma gambiense,* indigenous to tropical Africa and responsible for African sleeping sickness; and the yellow fever virus and *Trypanosoma cruzi,* the causal agent of Chagas' disease, which are localized in Central and South America. Great attention has been given in recent years to the fungi causing human respiratory infections, and some of these, like *Blastomyces dermatitidis, Coccidioides immitis,* and *Histoplasma capsulatum,* exhibit striking territorial patterns. *B. dermatitidis,* the cause of North American blastomycosis, is largely confined to the eastern portion of the United States and Canada, although the disease has been noted in small, circumscribed areas of Africa and possibly Latin America too. *C. immitis,* the agent of coccidioidomycosis, is a Western Hemisphere resident exclusively and is endemic to the southwestern portion of the United States, Mexico, and Central and South America. Histo-

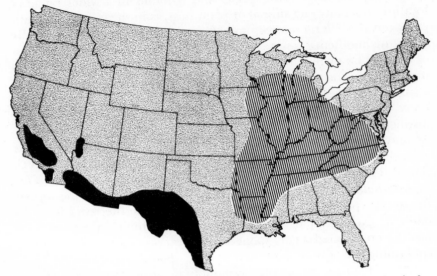

Figure 7.3   Regions of the United States recognized as endemic for coccidioidomycosis (black) and North American blastomycosis (lined area) (Furcolow, 1959).

plasmosis, caused by *H. capsulatum*, is another American disease and is found in the central Mississippi and Ohio River valleys. Figure 7.3 depicts those sections of the United States in which coccidioidomycosis and North American blastomycosis are endemic and illustrates the geographical spread of the two responsible fungi. Species of *Trichophyton* and *Microsporum*, fungi pathogenic for man, also exhibit a high incidence in distinctive localities of Europe, Japan, Southeast Asia, and the South Pacific.

Selected plant pathogens, though limited in distribution by the need for a host, also have significantly narrower geographical ranges than the species they parasitize. *Phymatotrichum omnivorum*, for example, is capable of attacking hundreds of plant species, but still disease caused by the fungus is absent from the United States north of a fixed point, despite the fact that susceptible hosts develop quite well in the more northerly latitudes. Other fungal pathogens developing in a more limited zone than the plants on which they proliferate are *Puccinia glumarum*, *Spongospora subterranea*, and *Helminthosporium maydis*, causing diseases of wheat, potato, and corn, respectively (Bisby, 1943; Diehl, 1937; Menzies, 1963).

DETERMINANTS OF GEOGRAPHIC PATTERNS

The bulk of the literature on microbial geography is concerned with tabulation of the incidence of individual genera and species rather than with establishing the reasons underlying the restricted zones or the lack of ubiquity. Nevertheless a small body of data exists that either helps to account for or points to ecological parameters that can be correlated with the unique biogeographies of a choice few microbial groups. From the available evidence, sparse though it is, it seems likely that microbial geography is determined by the supply of novel nutrients, the availability of suitable vectors, temperature, aridity, rainfall, salinity, soil type, and clay mineralogy, as well as yet undefined factors. One of these ecological factors occasionally appears to be of prime significance in itself, but frequently a combination acting together probably regulates the extent of spread of a particular organism.

The range of a species is limited to regions containing the particular substrates or food sources it can use and on which it has a selective advantage over other potential colonizers. This is well illustrated with the parasites that either are obligate or are restricted in nature to a parasitic way of life. The coprophilic, or dung-loving, fungi and the marine Ascomycetes and Fungi Imperfecti developing on submerged wood are similarly found only where their specialized food requirements are satisfied. *H. capsulatum*, a human pathogen residing in soil, has a fascinating nutritional dependency, for, though it is neither infective for nor actively carried by chickens, its occurrence is closely associated with soils above which chickens live. *H. capsulatum* also inhabits soil beneath the roosts of starlings and pigeons, presumably because the wastes of these animals provide nutrients selectively favoring proliferation of the fungus (Ajello, 1967). Differences in the distribution of algae may be related to the concentration of essential inorganic nutrients in the environment, as suggested by the finding that algae of iron-poor oceanic water tend to require far lower concentrations of iron than photoautotrophs proliferating in iron-rich coastal waters (Ryther and Kramer, 1961). Caution must be exercised in evaluating such information, however, because nonnutritional determinants generally operate simultaneously, as witnessed by the fact that bird droppings are plentiful in ecosystems free of *H. capsulatum* and the observation that inorganic nutrients may be in proper concentrations for a particular photoautotroph that still fails to make an appearance.

Populations that rely entirely on a vector for their dissemination are confined to zones containing the vector. Once the microbial population can no longer be sustained by its food source, whether a viable host or an inanimate material, the continued existence of the species requires transmission by an appropriate vector to a new nutrient supply. On rare occasions the microorganism may be accidentally introduced into a territory alien to it; yet it fails to become established permanently because it has no vector to convey it from the food source it has exhausted in the atypical environment to a site where renewed growth is feasible. Malaria provides an excellent example, since infected humans often move into countries where malaria is not endemic, but because of the absence of the insect vector, in this case the Anopheles mosquito, the *Plasmodium* responsible for the disease does not gain a foothold. Hence the territorial domain of the animate vector, whose occurrence is frequently attributable to climatic variables, regulates the geography of vector-dependent protozoa, viruses, fungi, and bacteria.

Temperature is a critical determinant of the extent of spread of diverse species, and excessive heat or cold serves as an effective means of regulating the geographic range of several algae and fungi. Humid or warm and arid land areas and tropical or subtropical seas support heterotrophs and algae unknown to other localities. Such populations do not require the specialized physiological or structural features that would be necessary for survival through the cold winters of temperate and polar latitudes. Free-living fungi of the genus *Allomyces* (Emerson, 1941) and individual species of the algal genera *Skeletonema* and *Planktoniella* (Hulburt and Guillard, 1968; Smayda, 1958) are found exclusively in the tropics or in warm temperate localities. Similarly, the parasites *Balansia* (Diehl, 1937) and *Phymatotrichum omnivorum* (Menzies, 1963) are present only under warm conditions; yet the plants attacked by these fungi will grow in soils of cooler regions, a difference in temperature sensitivities of parasite and host that accounts for the absence of disease in plants developing in the cooler localities. Occasional protozoa are also apparently favored by warmth, while others are typically residents of cool habitats.

The diatom *Skeletonema tropicum* and the root-rot fungus *Phymatotrichum omnivorum* have been well studied in regard to the regulating effects of temperature on biogeography. The former, an inhabitant of the Atlantic Ocean, cannot become established at latitudes greater than 30° because of its inability to multiply below 13°C (Hulburt and Guillard, 1968). The latter does not occur in soils of North America exposed to low temperatures, inasmuch as neither its hyphae nor sclerotia can withstand the rigors of harsh winters (Menzies, 1963). Conversely, as stated before, the alga *Thalassiosira antarctica* is characteristically a member of the antarctic phytoplankton, *T hyalina* is an arctic form, and *Spongospora subterranea* attacks potatoes in cool but not in warm soils (Bisby, 1943; Smayda, 1958). On the basis of their behavior, these three are stenothermal in nature and either do not have the ability to grow or are eliminated by biological interactions in warm environments. Temperature tolerances also appear to govern the territorial span of species of *Fucus* and *Laminaria* in marine ecosystems (Hedgpeth, 1957).

Soils that are perennially warm and dry have a typical microbiota. Special Basidiomycetes and other fungi are characteristic in surroundings of this kind. *Coccidioides immitis* likewise is well suited to semiarid land, and its preference for the New World is further narrowed by its occurrence in environments with little rainfall, mild winters, and hot summers (Ajello, 1967). At the opposite extreme, *Treponema pertenue*, the yaws spirochete, and species of plant-invading *Phytophthora* are virtually confined to those portions of the tropics where the rainfall and humidity are high, but it is difficult to separate the climatic influence on the human or plant host from that acting on the microorganism (Marples, 1965; Tucker, 1933).

Soil type has a significant bearing on the distribution of various heterotrophs, but the precise properties of the soil underlying the territorial distinctiveness of the particular organism are still not fully resolved. *Beijerinckia*, for example, not only is restricted primarily to acid soils of the tropics but is noted in a higher percentage of lateritic than nonlateritic soils within the tropics. Comparisons of the physiological properties of *Beijerinckia* with the closely related *Azotobacter*, a N_2-fixing bacterium of both tropical and temperate soils, indicate that a favorable effect of high temperature or inability to survive the cold does not account for the tropical proclivity of the former. However, strains of *Beijerinckia* are less sensitive to high iron and aluminum concentrations and are able to multiply at lower phosphate levels than the isolates of *Azotobacter* tested, data coinciding with the abundance of iron and aluminum and the paucity of phosphorus in laterites, which are characteristically tropical soils (Becking, 1961*a*, 1961*b*). Such results may in part account for the preference of *Beijerinckia* for laterites, but the observation that members of the genus are also numerous in acid, nonlateritic soils of the tropics remains unexplained. As stated above, factors in addition to the prevalence of animal wastes must be of significance in governing the incidence of histoplasmosis, for though *Histoplasma capsulatum* is linked to sites

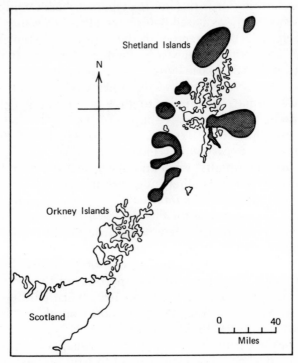

Figure 7.4 The distribution of *Rhizosolenia* in the vicinity of the Shetland Islands (Bainbridge, 1957).

containing bird and bat droppings, the disease exhibits a distinct zonal pattern whereas bird droppings are nearly universal; the prevalence of *H. capsulatum* has, however, been additionally correlated with the occurrence of red-yellow podzolic soils and with the presence of certain types of clay minerals in the soil (Stotzky and Post, 1967; Zeidberg, 1954). A distinct correlation has also been noted between the types of clay minerals in soil and the rate of spread of a wilt disease of bananas caused by *Fusarium oxysporum* f. *cubense*, but the microbiological significance of such relationships are difficult to interpret because the clay mineral, as well as possibly affecting the pathogen directly, may alter the susceptibility of the host plant or modify the saprobic community in a manner that affects the survival or vigor of the pathogen (Stotzky and Martin, 1963).

The reasons for the large or small aggregations of cells in the phytoplankton patches in the sea, as illustrated in Fig. 7.4, may be many. Winds blowing over a mass of water have been deemed responsible for the collection of dispersed algae into well-demarcated, dense assemblages of cells. The upward movement of deep, nutrient-rich waters into a nutrient-poor, lighted zone near the surface or the entry of nutrients into the open sea from rivers and streams probably frequently triggers a locally concentrated phytoplankton growth in an otherwise sparsely populated area. Inasmuch as marine animals grazing on algae are unevenly distributed, their activity or their absence may also contribute to the existence of regionally dense algal communities (Bainbridge, 1957).

The examples presented illustrate several of the territorial peculiarities of microorganisms and demonstrate unequivocally the existence of a microbial geography. Other than showing that unique distributions exist, the available information is far from adequate, and it is possible to do little more than speculate on the precise reasons determining the fascinating zonal patterns.

THE MICROENVIRONMENT

Sharp or gentle gradients extending over a small area exist in nature for many of the physical and chemical factors important to microbial growth and survival. Sites at points remarkably close to one another may be vastly different in their nutrient level and type, moisture status, O_2 content, light intensity, temperature, pH, osmotic pressure, accessibility of solid surfaces, or the kinds and amounts of toxins. Such vertical and horizontal stratifications of environmental features give rise to a marked local heterogeneity commonly expressed biologically in a mosaic of microhabitats or *microenvironments*, each potentially containing a somewhat dissimilar set of populations and a different community. In effect, organisms residing in what seems to be the same general environment may actually be living under quite dissimilar conditions.

The microenvironment is in itself a special microcosm, indeed a distinct ecosystem, possessing a characteristic community made up of populations coexisting and interacting with one another. It is a separate entity and has properties that cannot be appreciated or understood by gross ecological measurements. The mean physical, chemical, and biological composition of the macrohabitat is often far removed from that of the many microenvironments it contains, and hence information gleaned from studies of the general area may be misleading in regard to the properties and components of the miniature ecosystems. The spatial discontinuities that create the minute habitats occasionally are measured in terms of microns or millimeters, but sometimes significant spatial effects are not readily apparent except over greater distances.

Microenvironments may be discontinuous not only in space but also in time. For reasons not under microbial control, the microcosm can be drastically altered so that the indigenous species are faced with an entirely new set of circumstances, some of which they may be wholly unable to cope with. For example, host responses occur at localized sites wherein colonization is progressing, and the established colonists thus become exposed to a microhabitat appreciably modified from the one they originally encountered. Successions occupying an area of small magnitude may likewise bring about changes in the microhabitat with time.

Because of the dissimilarities in the physical and chemical determinants in a heterogeneous environment and the consequent establishment of somewhat distinct communities in the spatially separate microhabitats, one population may be subjected to diverse stresses in the adjoining microenvironments it inhabits. These stresses afford variant cells in the population a selective advantage, and the variants will thereby assume dominance and represent the species in the separate microcosms. In this way physiologically or morphologically distinct ecotypes develop as the forces of natural selection favor those cells most fitted for life in each of the multitude of small, semi-isolated units. The concept of a microecology thus has a special prominence in microbiology, microorganisms displaying not only a territorial distribution over large areas, as discussed above, but a geography on a microcosmic scale. Information obtained in vitro and in vivo must be assessed, therefore, with the understanding that the surroundings of a microbial community encompass a minute area, and that environmental factors of consequence in the environment at large may have little bearing on interspecific interactions or the interplay between populations and abiotic components of their microhabitat.

The zonation of aquatic autotrophs nicely exemplifies a clear and striking pattern of microenvironments. Whether observations are made vertically or horizontally, nonuniform distributions of algae and photosynthetic bacteria in small areas are nearly universal at the shoreline, in deep waters, in flowing streams, or in estuaries. The development of no single aquatic population seems to be so favored that it is able to dominate its neighbors at all points in the physically and chemically heterogeneous environment. One species will have the advantage at the first location, but a second, then a third, and finally a fourth will assume the forefront at a different place along the gradient of light, nutrient concentration, salinity, or temperature. Gradients are at times extremely abrupt, as in the change of light intensity between adjacent lighted and shaded areas or in adjoining exposed and submerged surfaces of rock protruding from the water in a flowing stream, and here the zonation is marked. Indistinct boundaries between dominant microbial types, on the other hand, are typical where gradients are gentle. Some algae displaying marked vertical and horizontal stratification are characteristically found at low light intensities and inhabit deep waters, while others are seemingly favored by intense light. One species dominates at a specific depth near the surface, a second congregates immediately beneath the first, and still a third is localized solely and prevails farther down in the water (Blum, 1956; Raymont, 1966). Likewise, along the coast and notably where the shore is rocky or is intermittently submerged, preferences for selected microenvironments are evident among the brown, red, green, and blue-green algae.

Congregation within narrow boundaries is widely evident among other aquatic microorganisms, too. Photosynthetic sulfur bacteria, for example, are localized in the sunlit, sulfide-rich, anaerobic zones where they obtain the light and sulfur they demand while avoiding the O_2 that inhibits their cell division, whereas algae often occupy the overlying areas where O_2 is plentiful. Protozoa and iron bacteria also show a well-defined vertical localization in lakes, a spatial distribution reflecting a preference for microhabitats where nutrients are obtainable and where abiotic factors are conducive to their multiplication (Webb, 1961). Similarly, zonality is known among the bacteria residing in the sea, and nonuniformity in their distribution is readily detectable by sampling at various depths in the open ocean.

Soils are noted for their spatial heterogeneities and the resulting multiplicity of microenvironments. Soil scientists have unequivocal evidence of small, localized regions differing in pH, O_2 tension, CO_2 supply, E_h, moisture level, and kinds and amounts of nutrients. As expected, moreover, microscopic observations reveal that many of the resident bacteria are situated at different sites in the soil matrix, developing in small colonies or as independent cells (Aristovskaya, 1963). The hyphae of fungi and actinomycetes also have a preference for specific surroundings, particularly those where organic matter is in good supply, and few or no filaments are detectable little more than a few microns away from the colonized place. Algae sometimes thrive at the soil surface, but such growths tend to be patchy in distribution and usually occupy an area no greater than a few square centi-

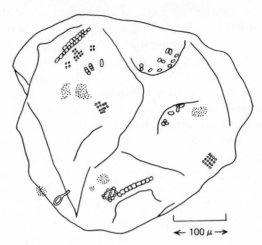

← 100 μ →

Figure 7.5 The distribution of microorganisms on the surface of a marine sand grain (Meadows and Anderson, 1968).

meters. A pronounced microenvironmental distribution is seen among the microorganisms attached to marine sand grains, where the development frequently is in the form of small colonies separated by large areas of exposed surfaces entirely free of life; furthermore, the kinds of indigenous organisms—be they bacteria, blue-green algae, or diatoms—change at points a short distance apart (Fig. 7.5).

Adsorption of cells to solid surfaces and their proliferation thereon happens almost wherever the appropriate surfaces and microorganisms coexist. Bacteria are retained by colloidal constituents of soil, namely, the clay minerals and humus. In polluted waters bacteria, protozoa, and fungi commonly adhere to and grow on solid surfaces; a prominent inhabitant of the attached slime of polluted waters is *Sphaerotilus natans*. Even the surfaces of terrestrial fungi and aquatic algae become coated with a variety of heterotrophs, the cells of these filament-dominated microhabitats possibly utilizing metabolic products excreted by the filaments, living on cell wall constituents, or having no metabolic association whatsoever with the surface on which they congregate. Solid surfaces have many attributes to distinguish them from the ambient fluid as a microhabitat. One of these microenvironmental characteristics has been the subject of careful scrutiny owing to its readily determinable influence on metabolism; this is the property of the negatively charged surface of a colloidal particle, like a clay, to attract hydrogen ions, with the result that the clay microenvironment has a lower pH than the surrounding solution (Mc-

Laren, 1963). A cell adhering to a negatively charged surface of this sort, therefore, exists in a more acid milieu than indicated by pH determinations of the ambient fluid.

The microenvironmental importance of O_2 has been the subject of considerable scrutiny. Where the demand for O_2 is high, because of the availability of readily degradable substrates, and free and rapid diffusion of the gas is retarded, anaerobic microsites may be generated immediately adjacent to localities where the O_2 supply is abundant. Under these circumstances obligate anaerobes multiply in close proximity to aerobes, and physiological processes functioning only in the absence of O_2 take place side by side with metabolic sequences dependent on free O_2. Thus the strictly anaerobic *Clostridium* seems to coexist with strictly aerobic fungi, and the O_2-dependent autotrophic oxidation of ammonium proceeds in what is apparently the same location as the bacterial reduction of nitrate to N_2, the latter reaction sequence being inhibited by O_2. The apparent discrepancy between the behavior in vitro and in vivo is attributable to the heterogeneity of the environment and the coexistence of small oxygenated pockets with anaerobic regions.

Many kinds of microenvironments have been studied either from a purely biological viewpoint or because the zonation is of public health, agricultural, or economic significance. Stratifications have been noted among the communities of trickling filters receiving sewage and waste waters. Rock faces, too, have numerous microenvironments distinguishable with ease by the naked eye. Food products frequently contain small foci of intensive bacterial and fungal multiplication in an expanse of otherwise unspoiled material. The surfaces of roots, leaves, and stems, whether showing lesions resulting from the activity of pathogens or appearing totally healthy, harbor a variety of separate and distinct communities. Still, probably the greatest amount of attention has been paid to the microhabitats in the human and animal body.

MICROHABITATS IN THE ANIMAL

The animal body may be viewed as a vast array of biochemically different microenvironments. Infections by viruses, bacteria, protozoa, and fungi invariably are restricted to discrete tissues and organs, and each human and animal parasite is specialized for a given internal environment wherein it multiplies. Some parasites prefer cutaneous tissue, others the blood, and still others the liver, kidneys, or the epithelial surface of the upper respiratory tract. Histological studies, furthermore, provide evidence that not all sites in a single organ are extensively colonized, and microenvironments within the same organ support populations causing dissimilar types of lesions. Lesions at one point sustain active replication and an abundance of a certain parasite, while an immediately adjoining area sustains few of that kind of parasite or is inhabited by another species entirely.

The portion of the body where a parasite initially is deposited is often not crucial in determining its final localization. For example, *Diplococcus pneumoniae, Neisseria meningitidis,* and *Streptococcus* strains frequently have the same portal of entry; yet *D. pneumoniae* multiplies extensively in the lungs, *N. meningitidis* in the membranous coverings of the nervous system, and streptococci in the pharynx. The reasons for the localizations and the properties of the tissues or organs allowing for or favoring colonization of indigenous or pathogenic microorganisms are poorly understood, though it is clear that physical conditions, chemical constituents, and the supply of antibodies and phagocytes are not homogeneous in the innumerable internal microenvironments suitable for microbial proliferation.

Spatially separate parts of the digestive tract of man and animals have their own physicochemical peculiarities, and each supports a community adapted to that space. Although the domains of the individual species may overlap, distinct groups can still be isolated from different regions of the gastrointestinal system. Such a partitioning of a gross habitat into microenvironments is well illustrated in the communities residing on the epithelium of the gastrointestinal tract of the mouse, for here the microcosms within the esophagus, stomach, cecum, and colon contain regionally large populations of lactobacilli, anaerobic streptococci, fusiform bacteria, or yeasts, occasionally the microhabitat being so rich in one population that few other microbial types are observed (Savage and Dubos, 1967).

Small patches of microbial growth are also easily detected on the skin, whether the colonists are pathogenic or saprobic (Fig. 7.6). Fungi, bacteria, and yeasts grow at different locations on the skin of the foot, arm, trunk, auditory canal, or scalp, and the nails and hair contain their own communities as well. Many residents of the skin, nails, or hair, like *Trichophyton, Microsporum, Epidermophyton, Staphylococcus,* and *Saccharomyces,* are limited to the superficial areas and not only fail to move laterally to new microenvironments but are incapable of invading subcutaneous tissues (Marples, 1965). In the mouth, too, contiguous sites support dissimilar communities, and dental plaques represent highly localized concentrations of bacteria. Host response to microbial invasion may establish novel microen-

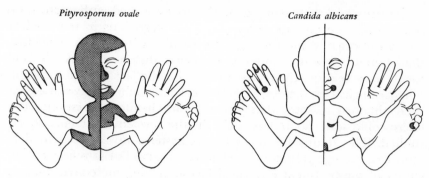

Pityrosporum ovale   *Candida albicans*

Figure 7.6   Favored habitats (shaded) of *Pityrosporum ovale* and *Candida albicans* on the human skin (Marples, 1965). Courtesy of Charles C Thomas, Publishers, Springfield, Ill.

vironments when the host erects fresh barriers, as in the deposition of fibrin walls about an abscess.

A dramatic instance of a set of microhabitats has been brought to light as a result of the use of chemotherapeutic agents, especially antibiotics, for the control of human disease. Treatment of patients with a drug effective against the pathogen causing the illness results in disappearance of symptoms of the disease, but after a time the communicable disease may recur. Contrary to expectations, the pathogen responsible for the relapse following an apparent clinical improvement is in some cases not resistant to the chemical employed but rather retains the drug susceptibility exhibited by the original parasite population. The capacity of drug-sensitive microorganisms to remain viable in vivo despite treatment with chemicals known to be effective in vitro poses a serious clinical problem in the control of diseases caused by species of *Staphylococcus, Diplococcus, Mycobacterium, Treponema,* and *Plasmodium.* The microorganism here appears to remain in a body locality where it exists in a state not susceptible to antimicrobial action, or it is lodged at a site where it is protected from contact with the administered drug (McDermott, 1958).

DETERMINANTS OF MICROENVIRONMENTAL PATTERNS

Profound physical and chemical heterogeneities characterize immediately adjacent microscopic regions in a high proportion of environments, and these heterogeneities are associated with microcosms suitable for one but not another species. Variations in the kind or amount of organic and inorganic nutrients, hydrogen ion concentration, osmotic tension, sunlight, moisture, dissolved gases, E_h, and toxins have been postulated to account for the lack of uniformity in microecological conditions, and some evidence, admittedly scanty, has been obtained to show that physical and chemical variations on a microscopic scale do indeed exist in the few environments so far studied. The results for the select ecosystems investigated to date show that average values for the habitat at large fail to reflect the intensity of the pertinent ecological factors in the microenvironment, and that the sphere of influence of many factors is narrow, extending for distances of the order of microns or millimeters but rarely greater.

Microscopically small sites, like their macroscopic counterparts, regularly display sharp gradients of several of the key determinants of microbial distribution. This is well illustrated by the microlocalization patterns of nutrients. Droppings of vertebrates and invertebrates, fragments of crop residues, conifer needles, and so on provide a locally high concentration of selected nutrients, each favoring no more than a small portion of the microorganisms indigenous to the macrohabitat. The surfaces of a plant root and of the human skin have large amounts of organic compounds serving as a food source at one point, and little or none nearby. Disease alters the kinds and quantities of normal metabolites, and exotic metabolites as well as antimicrobial substances appear at the site of microbial colonization, so that here, too, a microenvironmental influence is established.

Nonuniform microclimates typify the areas where heterotrophs and autotrophs proliferate. The unique localization of syphilitic lesions and of the growth of species of *Cryptococcus*, for example, seems to be affected by the temperature of particular regions of the body (Bessemans, 1938; Hollander and Turner, 1954; Marples, 1965), and the restriction of *Uncinula necator* to the upper and center portions of the grape vine has been ascribed to the temperature gradient about the plant, a gradient broader than the temperature tolerance range of the fungus (Delp, 1954). Moisture is another component of the microclimate that is of undoubted significance in governing microterritorial mosaics.

A relationship also exists between tissue chemistry and the localization in the animal body of pathogenic agents. For example, in cattle in which abortion resulting from infection with *Brucella abortus* is imminent, about three-fourths of the viable *B. abortus* cells are in the fetal placenta. The presence of brucellae in this microhabitat appears to result from the presence therein of a compound, erythritol, found in the fetal placenta but not in a variety of other tissues of the host, a substance markedly stimulatory to the growth of these bacteria (Pearce et al., 1962; Smith et al., 1961; Williams et al., 1962). This is a striking illustration of how environmental biochemistry accounts for microbial distribution in discrete microhabitats.

REFUGE

Two species may be unable to coexist in a homogeneous environment because of interspecific interactions that lead to the replacement or overt destruction of one by the second, as when rapid exploitation of nutrients, toxin production, predation, or parasitism by the first population excludes its neighbor, but coexistence of the same two species is feasible in a heterogeneous habitat. Complete elimination of the less favored organism may be avoided in the nonuniform area because the two populations, though coexisting in time, are spatially separated, and this isolation spares the individuals that otherwise would not have survived. Many species possessing limited capacities to endure the physical, chemical, or biological stresses imposed on the surroundings at large owe their existence in a heterogeneous milieu to the fact that they are present in a microhabitat not exposed to the rigors imposed at adjacent sites. In effect, they have found a *refuge*.

Environmental heterogeneity often accounts for species diversity in locations abundantly endowed with microenvironments dissimilar in physical and chemical properties. Each of the residents has become established in selected microcosms where it can grow, and it is not eliminated despite the occurrence in nearby microsites of organisms that, in highly uniform ecosystems, would prevent its continued replication (Fig. 7.7). These small-scale barriers and discontinuities, in effect, constitute refuges that explain in large part the apparently anomalous coexistence of interacting competitors, of a predator with its prey, of a parasite and its host, and of a toxin producer with a toxin-sensitive individual.

Refuges of innumerable sorts abound in nature. Communities in soil and aquatic sediments are characteristically separated by inanimate particles, while infected plant or animal materials placed below or remaining on the soil surface serve to protect pathogens that would fare poorly or would be destroyed if introduced into the soil proper. The clumping together of cells quite likely shields individuals within

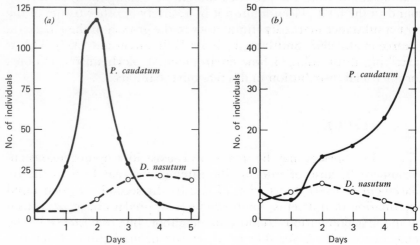

Figure 7.7   The survival of *Paramecium caudatum,* a prey of *Didinium nasutum,* in a homogeneous environment with no refuge (*a*) and in a heterogeneous environment (*b*) (Gause, 1934). By permission of The Williams & Wilkins Co., Baltimore, Md.

the aggregate from adverse conditions, and bacteria flocculating in waste water may thereby acquire a haven from marauding predatory protozoa. In the animal body bacteria probably find a suitable sanctuary in structures allowing for microbial penetration but not for the passage of phagocytes. Furthermore, bacteria located within cells of their animal hosts, including bacteria situated within but not destroyed by phagocytes, are protected against antimicrobial host reactions in their intracellular shelters.

The concept of refuge need not be restricted to an escape in space; it can readily be extended to an escape in time. Thus a lethal interaction among species occupying the same space is avoided whenever the interacting vegetative forms are separated in time. Bacteria similarly escape and persist by forming endospores, protozoa by encysting, and fungi by the production of resistant spores. The escape is successful provided that active development resumes before the resting structure uses up its stored reserves, becomes inactivated by abiotic factors, or is destroyed by neighboring parasites or predators.

References

REVIEWS

Bainbridge, R. 1957. The size, shape and density of marine phytoplankton concentrations. *Biol. Rev.*, **32**, 91–115.

Bisby, G. R. 1943. Geographical distribution of fungi. *Bot. Rev.*, **9**, 466–482.

Diehl, W. W. 1937. A basis for mycogeography. *J. Wash. Acad. Sci.*, **27**, 244–254.

Glover, R. S. 1961. Biogeographical boundaries: the shape of distributions. *In* M. Sears, Ed., *Oceanography*. American Association for the Advancement of Science, Washington, D.C. pp. 201–228.

Hedgpeth, J. W. 1957. Marine biogeography. *In* J. W. Hedgpeth, Ed., *Treatise on Marine Ecology and Paleoecology*, Vol. 1. National Research Council, Washington, D.C. pp. 359–382.

McDermott, W. 1958. Microbial persistence. *Yale J. Biol. Med.*, **30**, 257–291.

McLaren, A. D. 1963. Enzyme activity in soils sterilized by ionizing radiations and some comments on micro-environments in nature. *In* N. E. Gibbons, Ed., *Recent Advances in Microbiology*. University of Toronto Press, Toronto. pp. 221–229.

Pirozynski, K. A. 1968. Geographical distribution of fungi. *In* G. C. Ainsworth and A. S. Sussman, Eds., *The Fungi*, Vol. 3. Academic Press, New York. pp. 487–504.

Stamp, L. D. 1964. *The Geography of Life and Death*. Cornell University Press, Ithaca, N.Y.

OTHER LITERATURE CITED

Ajello, L. 1967. *Bacteriol. Rev.*, **31**, 6–24.

Aristovskaya, T. V. 1963. *Mikrobiologiya*, **32**, 663–667.

Becking, J. H. 1961a. *Plant Soil*, **14**, 49–81.

Becking, J. H. 1961b. *Plant Soil,* 14, 297–322.

Bessemans, A. 1938. *Ann. Physiol. Physicochim. Biol.,* 14, 944–970.

Blum, J. L. 1956. *Bot. Rev.,* 22, 291–341.

Delp, C. J. 1954. *Phytopathology,* 44, 615–626.

Emerson, R. 1941. *Lloydia,* 4, 77–144.

Furcolow, M. L. 1959. Seminar Report, U.S. Public Health Service, Vol. 4, No. 1.

Gause, G. F. 1934. *The Struggle for Existence.* Williams & Wilkins, Baltimore.

Hollander, D. H., and Turner, T. B. 1954. *Am. J. Syph.* 38, 489–495.

Hulburt, E. M., and Guillard, R. R. L. 1968. *Ecology,* 49, 337–339.

Marples, M. J. 1965. *The Ecology of the Human Skin.* Thomas, Springfield, Ill.

Meadows, P. S., and Anderson, J. G. 1968. *J. Marine Biol. Assoc. U.K.,* 48, 161–175.

Menzies, J. D. 1963. *Bot. Rev.,* 29, 79–122.

Pearce, J. H., Williams, A. E., Harris-Smith, P. W., Fitzgeorge, R. B., and Smith, H. 1962. *Brit. J. Exptl. Pathol.,* 43, 31–37.

Raymont, J. E. G. 1966. *Advan. Ecol. Res.,* 3, 117–205.

Ryther, J. H., and Kramer, D. D. 1961. *Ecology,* 42, 444–446.

Savage, D. C., and Dubos, R. J. 1967. *J. Bacteriol.,* 94, 1811–1816.

Scagel, R. F. 1963. *In* M. J. Dunbar, Ed., *Marine Distributions.* University of Toronto Press, Toronto. pp. 37–50.

Smayda, T. J. 1958. *Oikos,* 9, 158–191.

Smith, H., Keppie, J., Pearce, J. H., Fuller, R., and Williams, A. E. 1961. *Brit. J. Exptl. Pathol.,* 42, 631–637.

Stotzky, G., and Martin, R. T. 1963. *Plant Soil,* 18, 317–337.

Stotzky, G., and Post, A. H. 1967. *Can. J. Microbiol.,* 13, 1–7.

Tucker, C. M. 1933. *Mo. Agr. Expt. Sta. Res. Bull.* 184.

Webb, M. G. 1961. *J. Animal Ecol.,* 30, 137–151.

Williams, A. E., Keppie, J., and Smith, H. 1962. *Brit. J. Exptl. Pathol.,* 43, 530–537.

Zeidberg, L. D. 1954. *Am. J. Trop. Med. Hyg.,* 3, 1057–1065.

8

Natural Selection

Every species is endowed with biochemical and structural properties permitting its development in one region or another, traits allowing for its multiplication in the face of the frequent and assorted environmental stresses that it will surely encounter. The restriction of the organism to fixed habitats results from its specialization and its particular suitability for growth in the localities wherein it is found. The territorial affinity, microenvironmental distribution, and extent of spread are typically linked to the functioning of *natural selection*.

Natural selection operating during the course of time is the basis for evolution, and a vast wealth of information has been uncovered in this regard. Natural selection in space, that is, at specific sites, is a critical element in ecology, and the spatial aspect of selection is of especial relevance to the present chapter. The very fact that distinct communities are readily definable, despite the ease of dispersal of microorganisms, attests to the working of natural selection. So, too, do the changes that occur as a consequence of the interactions among the heterogeneous collection of individuals alighting on a previously uncolonized site.

Populations of unicellular or multicellular organisms have the ability to produce an enormous number of new cells or large masses of filamentous material in short periods of time. The realization of this capacity is rarely achieved in nature, except possibly in the early stages of colonization, where conditions, though initially advantageous to the pioneer, become progressively more unsuitable as the habitat gains additional residents. The action of natural selection is immediately evident long before the food resources available to a heterogeneous community are depleted. The presence of organisms with different properties is an essential ingredient of selection, but this community heterogeneity need not involve a variety of species,

165

for a single species containing phenotypically different cells, that may have originated by mutation or hybridization, is frequently subjected to selection pressure. The two or more different forms contest for a limiting factor or interact in some other manner in such a way that one loses out in the struggle for existence, and the favored type multiplies and assumes dominance.

Organisms differing in physiology or morphology have dissimilar chances of remaining active or of merely surviving in a community, but success is to a significant extent governed by prevailing conditions. Conservation of a species, or of a variant form produced by genetic modification of pre-existing cells, requires a favorable outcome in its struggle for existence with neighboring cells or filaments, the more fit bringing about the displacement of the less fit. Various members of a heterogeneous biota are suited to ambient circumstances at the particular place and time, and these survive and multiply, but those inhabitants that are less well suited have a low probability of survival in that ecosystem. Each subsequent community in a succession thus evolves in such a manner that the remaining inhabitants are strikingly attuned to their surroundings. The direction of change is unwavering: it is in favor of populations with superior fitness for the prevailing situation. The community is progressively remodeled, a transformation dictated by and in line with the environment, a remodeling resulting in a suppression of the ill-suited species. The habitat calls the tune, and the direction of community modification, as scored by the biotic and abiotic characteristics of the site, leads to the dominance of heterotrophs or autotrophs best able to cope with their milieu, the placement of less well favored populations in secondary positions, and the total elimination of groups not tolerating the various biotic and abiotic stresses. The populations cohabiting an ecosystem and constituting the climax are thus those that have been naturally selected—those in maximum harmony with their environment—from among the propagules that have reached the particular locality.

When a habitat is subjected to change, a small indigenous population, indeed a single viable cell, may have traits that are suddenly of enormous selective value, and hence its population size may increase until the species, previously unnoticed, assumes a dominant position. In some ecosystems the evidence suggests that the newly emergent group did not long inhabit the locality, but was derived instead from a mutant that appeared among the many other cells and was favored in the changed habitat. The same kind of mutant may have arisen earlier, but it had no ability to survive until the conditions were altered to permit its multiplication. Whether the cells were permanently resident in the area, were derived from a newly arriving propagule, or arose by mutation or hybridization, they possess traits of great selective value in the modified milieu and thus frequently are capable of suppressing or eliminating members of the pre-existing community. Similarly, among the variety of potential colonists gaining access to a previously uninhabited area, such as freshly exposed plant or animal tissues, a species represented by few propagules may assume a focal position in the developing community because it has physiological or morphological properties with selective value, whereas a species represented by hordes of cells or spores may not gain a foothold because it has no traits advantageous at that time and place.

Natural selection acts to eliminate species residing in the same microenvironment and occupying the identical niche, niche of course referring to function or activity and not spatial location. However, should the niches of two species be dissimilar, no matter how small the differences, both may be preserved and coexist indefinitely. By these means, selection discourages duplication of species performing the identical biochemical function in the same microenvironment but favors populations participating in community functions in somewhat dissimilar ways. Moreover, despite intense selection taking place all around, a fresh arrival can establish itself should a niche remain unoccupied.

Differential growth or colonization rates may determine success or failure in the struggle for existence, the selection being for the population with the greater multiplication rate in *that particular environment*. On the other hand, instances are known in which the selection appears to be governed by differential mortality rates among the interactants in a microbial community. Certain cells and filaments are eliminated because they are unable to cope with the stresses of their environs, and they disappear gradually or, occasionally, quite rapidly. Frequently, however, the bases for selection remain in doubt, and differences in growth rate, mortality rate, or other factors may be involved.

FITNESS

The identities of the attributes allowing the residents of an ecosystem to survive or multiply are difficult to establish for living things in general, but the problem of their characterization is compounded in microbiology because the number of microorganisms is large, their sizes small, their growth rates high, and the interspecific interactions numerous. Only a meager amount of knowledge is available, and rarely is the information adequate to argue strongly for a specific trait or a constellation of properties serving as the basis for an organism's prominence. Nevertheless, assessing the factors underlying fitness ought to be a fascinating field for biological exploration. Indeed, some studies have met with notable success, while others have created tantalizing leads though the conclusions remain equivocal. A few of the characteristics are immediately obvious, and in highly specialized habitats, such as where temperature extremes, high salinity, excessive sugar concentrations, prolonged drought, or the marauding action of phagocytes are of consequence, the fitness traits are simple to establish. More often than not, however, the overt features of a microorganism, its obvious physiological and morphological properties, are probably of less importance in natural selection than covert ones.

Traits underlying fitness endow their possessors with the capacity to cope with the physical, chemical, and biological stresses of the surroundings. These useful attributes are at the heart of success in the process of natural selection; species bearing them are favored and their prominence is accentuated during the course of selection. A fitness trait may be beneficial because it protects against the injurious effects of temperature, extremes of osmotic pressure, predators, parasites, synthetic or microbiologically formed toxins, or substances associated with host response to infection. The critical character may be a set of enzymes catalyzing a reaction sequence to make an exotic substrate available, or it might be a complex of physiological functions responsible for rapid growth or the intracellular storage of nutrient reserves.

The size, survivability, and activity of a population and its ability to leave behind daughter cells or to extend its filaments are excellent measures of fitness. Species with beneficial properties are conserved at the expense of those lacking them. Yet a feature advantageous in one place or at one time may be of no consequence elsewhere or at another time, inasmuch as the expression of fitness is related to the particular environment where the organism is found. A precise quantification of fitness, albeit desirable, is rarely feasible and is likely to be impossible in most natural environments.

Success in natural selection, furthermore, is frequently not attributable to an individual trait but to a multiplicity of cellular properties that have ecological relevance. Even where a distinct biological feature having potential selective value clearly typifies all inhabitants of an ecosystem, many alien cells fail to become established or do not appear in the region though they too possess the same physiological or morphological feature. Residence at a certain place may require the presence in the individual of a given quality, but the successful outcome of selection may depend on not just that one but instead a multitude of traits, a deficiency in any one of which might be sufficient cause for exclusion of the species. Despite the complexity of interspecific interactions and the variety of cellular attributes determining the outcome of selection, there is nevertheless great virtue in considering the individual factors separately, but remembering always that complexity and not simplicity is the usual rule. In addition, because of the scores of traits ultimately deciding the end result of selection in vivo, frequently it is not the fittest that constitute the indigenous species of the community; that is, it is not the organisms that have the greatest intensity of the fitness trait, be it growth rate, tolerance of toxins, resistance against attack by parasites, or the like. The fit, not necessarily the fittest, endure.

The acquisition of new traits is commonly linked with the loss of old ones, and the consequent modifications in the physiology or morphology or an organism that enable it to live in a given area often parallel the loss of characteristics potentially significant in another region. Thus, as additional properties are acquired by continued growth and by the natural selection of the cells more fit for a new set of circumstances, qualities that made the population successful in its former abode disappear, so that a strain invading and establishing itself in a habitat may, with time, no longer be able to multiply if returned to its original surroundings. This is particularly evident among parasites of animals and plants, the specialization leading to parasitism and the acquiring of new enzymatic capacities regularly resulting in the inability of the microorganism to cope with the co-inhabitants of the area it evolved from; the parasite then is totally reliant on the host for its continued existence. The same type of shift occurs among free-living heterotrophs, the gain of fresh traits needed for life in a new habitat being linked to the disappearance of attributes required for life in the old.

The *enrichment* or *elective culture* technique is designed to exploit fitness traits of microorganisms and illustrates well the operation of natural selection in vitro. As usually performed, a small amount of natural material that is heterogeneous in microbial composition— soil, sewage, river mud—is inoculated into a medium constituted so as to favor one physiological or nutritional type, an enrichment based on a trait allowing one sort of organism to assume dominance in the medium. The favored microbial types multiply preferentially, and the community changes from that of the original natural material to one in which the dominants are those selected out by the conditions established in the enrichment. The environment in vitro is deliberately manipulated in such a way that only the progeny of the fit cells multiply, while the remaining cells in the inoculum fail to develop or fare poorly. The species ultimately dominating in this artificial community has a selective advantage, one dictated by the particular environmental conditions created by the investigator (Schlegel and Jannasch, 1967; van Niel, 1955).

Any of a multiplicity of physiological properties may be the basis for an enrichment. Ammonium-oxidizing chemoautotrophs like *Nitrosomonas* have the advantage in a medium containing an ammonium salt, inorganic nutrients, and CO_2 as the sole carbon source, and provided that they are incubated in the dark to prevent photoautotrophs from appearing, these bacteria will grow extensively when such an inorganic medium is inoculated with an appropriate soil or water sample. Similarly, the introduction of water from many hot springs into an inorganic medium containing neither a nitrogenous compound nor a source of organic carbon will, if incubation is conducted at high temperatures and in the light, lead to the profuse development of blue-green algae, since these thermophiles are able to obtain their carbon from CO_2 and nitrogen from N_2. The selective trait in enrichments frequently is an enzyme system that allows the organism to make use of a substrate not utilized by other coinhabitants of the same environment; for example, fungi degrading cellulose or keratin are favored when cellulosic materials are the sole carbon source in a liquid medium or when hair, a keratin-rich material, is introduced into soil. Physiologically uncommon bacteria gain the ascendancy when biodegradable pesticides are the carbon sources, because these synthetic compounds are novel and are decomposed by but a few species (Fig. 8.1); such enrichments are not mere laboratory curiosities, moreover, in that the addition of the same nutrients to the natural habitat, a common occurrence, selects out similar groups from among the inhabitants.

The artificial environment can be altered in innumerable ways by changing its physical, chemical, or nutritional properties, as by the omission of nitrogen or organic carbon, addition of a novel carbonaceous substance, varying the pH, raising the temperature, or the introduction of an antibiotic or inorganic toxin. Inasmuch as it is likely that every organic compound formed biologically sustains the

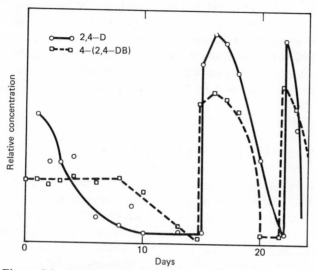

Figure 8.1 Enrichment for soil bacteria able to grow at the expense of 2,4-dichlorophenoxyacetate, 2,4-D, and 4-(2,4-dichlorophenoxy)butyrate, 4-(2,-4-DB). Soil was inoculated into solutions containing these pesticides as carbon sources. The drawing depicts the slow rate of disappearance of the chemicals initially added and the more rapid disappearance of the second and third increments of the compound added at 2 and 3 weeks (Whiteside and Alexander, 1960).

growth of some organism, each biologically synthesized molecule presumably can provide the basis for a natural selection, but the inoculum must of course contain a cell having the enzymes necessary to catalyze the decomposition. In all instances the species favored is the one in the inoculum that is fit for the conditions set up, although the circumstances may be far from ideal should the organism in question be examined in axenic culture.

Microorganisms offer many advantages in studies of natural selection. They multiply at a more rapid rate than higher plants and animals, and hence the influence of countless selective factors can be examined in a short period of time. A small site may contain large populations of diverse physiological groups, and because of the intensity and rapidity of selection, the ecologist has available to him notably useful working models. The small experimental areas required and the ability of the biologist to establish carefully controlled or uniform environmental circumstances also lend attraction to investigations of selection in microbial communities. Yet the microbiologist is prone to be led astray by the simplicity and ease of manipulating his artificial models, particularly because conclusions derived from laboratory investigations are frequently not directly applicable to natural communities. Granting that useful preliminary information is unquestionably obtained by examining the course of selection in vitro, the heterogeneity and fluctuations of the ecosystem in vivo dictate prudence in facile extrapolations from artificial systems.

INTERSPECIFIC SELECTION

Many of the ecologically significant features of natural environments may be forces in selection, and they, acting in conjunction with cellular characteristics, govern which of the resident or invading organisms will assume a focal position, which will maintain a minor role, and which will be overwhelmed in the struggle for continued existence. Only a few of these abiotic or biotic factors have been defined, however. Among the abiotic properties of the habitat, the quality and intensity of light, toxic chemicals, nutrient inadequacies, porosity, osmotic pressure, the availability of free water, pH, hydrostatic pressure, CO_2 level, and O_2 tension may exert a differential effect on the indigenous or invading populations. Biotic influences known to regulate the course of selection include the presence of toxin-producing microorganisms, predators, and parasites, the action of phagocytes on microbial cells gaining entry into the animal body, and plant- and animal-produced compounds differentially affecting tissue invaders.

Growth rate is often stated to be the ultimate determinant of selection, populations with higher rates of development allegedly overgrowing and displacing slow growers. A gradual assumption of dominance of this sort is readily demonstrable in cultures inoculated with two organisms that are nutritionally and physiologically similar. Rapid proliferation is indeed frequently beneficial to a species in vitro, and differential replication rates may account for the dominance of a reasonable proportion of fungi, bacteria, and algae. Yet too many physiological or morphological properties are responsible for ecological success to warrant the conclusion that growth rate is paramount. Countless bacteria, fungi, protozoa, and algae develop exceedingly slowly, but still these organisms, though sluggish growers, are both ubiquitous and abundant. Protective structures or pigments, the resistance to attack by predators, cell size, or additional traits may override the benefit of a potential for rapid development. Nevertheless, given that several species gain access to a colonizable environment, that the nutrients required by them are present therein, and that they are equally susceptible to the prevailing biotic and abiotic stresses, growth rate may be taken as a first approximation to be of considerable importance in selection and a critical factor associated with success in nature.

Evidence derived from investigations both in vitro and in vivo shows that in addition to growth rate a successful outcome in interspecific selection may result from the possession of one of a number of physiological and cytological properties. The presence of an enzyme permitting the cell or filament to make use of a substrate not available to coinhabitants of the ecosystem often is decisive. For example, only a small segment of the community of soils, rivers, and sewage treatment systems contains the requisite enzymes to degrade and grow at the expense of synthetic organic chemicals employed in agriculture as pesticides, used in the home as detergents, or introduced into streams with industrial effluents. Similarly, few autotrophs or heterotrophs are able to assimilate N_2, so that blue-green algae and strains of *Clostridium*, among others, may have a selective advantage over their neighbors in habitats poor in organic nitrogen, ammonium, and nitrate. The possession of unique physiological characteristics likewise underlies the occasional dominance of autotrophs in habitats containing little organic matter, a uniqueness associated with the capacity to multiply with CO_2 as sole carbon source and light, H_2, or reduced inorganic nitrogen, sulfur, or iron as the only source of energy.

The cells of certain species are endowed with structures protecting them against abiotic stresses to which others succumb, and some contain surface constituents shielding them from biotic stresses imposed by parasites, microbial predators, or host defense mechanisms. The prevalence of various groups results from their having morphological peculiarities or cellular inclusions fitting them for life in special localities, traits lacking in individuals entering into but not able to survive the very same circumstances. Particular morphological entities facilitating success in natural selection have been thoroughly studied, and a number of these will be considered briefly.

The outcome of the struggle for existence is quite evident in environments exposed to intense sunlight, and although elimination of the unfit has not been witnessed, distinctive characteristics of the victors provide clues to what undoubtedly transpired. Photosynthetic organisms require sunlight, but they vary in their capacity to make use of different wavelengths of the light they may receive. Because they, as well as heterotrophs, are light-sensitive, light may be considered to act as a selective force in two ways, by preferentially stimulating organisms on the basis of the wavelengths most effectively used in photosynthesis and by preferentially eliminating light-sensitive cells. The dual effect of sunlight is especially evident among the algae. Representative algae are prominent at the water's surface, on the exposed faces of rocks, and on unshaded portions of soil or higher plants, whereas many groups cannot tolerate full sunlight and are typical inhabitants of deep waters. That a significant component of this response to light results from a selection for light-tolerant types is suggested by the fact that algae whose natural habitat receives high light intensities appear to be far more tolerant to direct exposure to sunlight than species residing in areas where sunlight is not as intense (Biebl, 1956).

On the other hand, genera of aquatic algae benefit because the quality of light, rather than solely its intensity, is markedly altered with increasing depth of water. Water absorbs red light far more than light with wavelengths in the central portion of the visible spectrum, and this is the basis for a differential influence on algae. Groups bearing pigments capable of using the green and blue light that is available at some depth for photosynthesis thus have a distinct advantage in deep waters, while algae containing pigments suited for capturing red light are located at depths where the quality of light is suitable for their metabolism. In a typical study the rate of photosynthesis by *Enteromorpha linza,* a green alga, was observed to be rapid in red but slow in blue and green light, while that of two red algae, species of *Porphyra* and *Delesseria,* did not markedly decline in blue and green light; the first photoautotroph is located highest on the seashore while the remaining two live lower on the shore or below water (Klugh, 1931).

Pigmented organisms commonly seem to be shielded against inactivation by solar radiation (Swart-Fuchtbauer, 1957), and such protection should have a tremendous selective value in ecosystems exposed for prolonged periods to the action of the sun. Indeed, the heterotrophic communities of habitats exposed to bright light are usually dominated by pigment-producing strains, as is evident in the prevalence of carotenoid-containing bacteria in the air flora, on the surfaces of leaves (Stout, 1960), in salt ponds undergoing solar evaporation, and in related salty bodies of water receiving appreciable sunlight (Dundas and Larsen, 1962). Apparently, carotenoid-rich bacteria have a fitness trait not possessed by nonpigmented cells, which fail to establish themselves or are outgrown in time in these

Table 8.1

Selective Value of Carotenoids in Mixtures of *Halobacterium salinarium* and a Nonpigmented Mutant [a]

	Ratio of Red to Colorless Cells	
Time	Light-Grown Cultures	Dark-Grown Cultures
At start of experiment	0.11	0.11
End of first growth period	0.23	0.12
End of second growth period	0.86	0.12
End of third growth period	1.6	0.10
End of fourth growth period	6.8	0.12

[a] From Dundas and Larsen (1962).

habitats. Experiments with nonpigmented mutants derived from carotenoid-producing photosynthetic and heterotrophic bacteria or with cultures grown in the presence of compounds preventing the biosynthesis of colored carotenoids have established why such pigments are beneficial: the colored bacteria multiply in solutions incubated in the light whereas the carotenoidless mutants or cultures prevented from synthesizing carotenoids either divide far more slowly or are soon killed in the light (Dundas and Larsen, 1962; Kunisawa and Stanier, 1958; Sistrom et al., 1956).

A selection for pigmented cells is well illustrated in an investigation of a mixed culture containing the extreme halophile, *Halobacterium salinarium,* and a colorless mutant derived from the red parent cells. The mixture was incubated in the light or in the dark, and four consecutive transfers were made to fresh media incubated under identical conditions. At the time of each transfer and at the end of the experiment, the ratio of cell types was determined, and, as depicted in Table 8.1, the red cells were noted to have no advantage in darkness but were markedly favored and assumed dominance in media maintained in the light. Pigmented hyphae of some fungi are not killed, although colorless hyphae of the same species are inactivated in direct sunlight (Goldstrohm and Lilly, 1965), and a *Chlorella* mutant synthesizing no colored carotenoids is rapidly killed when placed in the light (Claes, 1954). It appears, therefore, that a key character determining fitness in habitats bathed with bright light is the ability to synthesize carotenoids, pigments safeguarding the cells against the deleterious action of sunlight and allowing the organisms to assume dominance over populations not having this fitness trait.

Other kinds of ecologically consequential light filters have been described among the fungi. Spores of diverse fungal genera are disseminated through the atmosphere and undoubtedly alight in the desert; yet few of the propagules survive the intense sunlight and drought of the dry area. The dominants in many desert communities, however, have a common attribute, dark pigments. These species apparently form melanin, presumably mainly in the spore and hyphal walls, and the melanin pigment serves a purpose similar to the carotenoids, namely, to allow for survival during prolonged exposures to sunlight (Borut, 1960; Durrell and Shields, 1960; Nicot, 1960). Fungi on the leaves of particular tropical trees are characterized by thick, black walls, which may filter out light to permit growth or survival of these but not potential leaf colonists not so endowed. Investigations of a large number of fungi have shown unmistakably that dark or thick-walled spores are far more tolerant of radiation than nonpigmented or thin-walled structures (Sussman, 1968), suggesting that melanization and wall thickness are important in the struggle for existence in light-saturated habitats.

Melanins confer a selective benefit totally unrelated to the germicidal effect of sunlight, a benefit attributable to their remarkable properties of protecting cell surface components from digestion by extracellular enzymes released by bacteria and actinomycetes. The mycelium of a high proportion of the fungi appearing in soil when organic nutrients are supplied is attacked by heterotrophs excreting enzymes that digest those polysaccharides serving as structural backbones of the filament, a digestion that leads to destruction of the hyphae and at the same time provides the enzyme producer with sugars it needs for multiplication. Fungi affected in this manner persist in soil because they form spores or other resting bodies. Nevertheless, though these resting structures, as well as the hyphae of occasional fungi, resist digestion by the soil parasites, they often contain in their surface layers the very same polysaccharides that are hydrolyzed by enzymes of the parasites. This anomaly was resolved when it was discovered that the resistant external layer of the hyphae of *Rhizoctonia solani*, the sclerotia of *Sclerotium rolfsii*, and the conidia of *Aspergillus phoenicis* contain melanin. This resistant polyaromatic material complexes with or overlays the labile polysaccharides of the fungal surface in a manner that prevents digestion. The hypothesis that it is melanin that accounts for survival was supported by the finding that the melanin-rich mycelial walls of *Aspergillus nidulans* were refractory to enzymatic digestion, whereas the walls of a melaninless mutant of the fungus were readily attacked (Bloomfield and Alexander, 1967; Kuo and Alexander, 1967; Potgieter and Alexander, 1966). Selection here is for the organism that possesses at some stage in its life cycle a constituent guarding against the ravages of microbial parasites.

Morphological features determining fitness and community composition are evident also among the populations inhabiting flowing waters. Cells and filaments suspended in the fluid—be it a stream, river, or estuary—are carried away with the moving waters, and selection favors organisms with some structural modification facilitating their retention in a relatively fixed position. Thus algae adhering to the surfaces of rocks in streams and rivers often have a means of attachment so that they are not borne away by the moving current, these attachment devices being in part the basis for their unique distribution. Marine algae living on rocks violently and incessantly battered by wave action are well suited to these dangerous habitats, from which other autotrophs are excluded, by virtue of the peculiar morphological characteristics they possess to withstand wave action and prevent their being torn from the rocks. Various bacteria indigenous to flowing rivers or water pipes are fit for this precarious existence because they can attach themselves and adhere to solid surfaces, thereby resisting the dislocating action of the fast-flowing waters that wash away most propagules. Even among the parasitic protozoa inhabiting the alimentary tract, one finds organelles serving to attach the microorganisms to the walls of the intestine, organelles not common to protozoa residing in blood or within the cells of the animal body; these attachment organelles probably confer a selective advantage on their possessors inasmuch as they retain the protozoa in a favorable microhabitat and prevent their elimination with the stream of food traversing the gastrointestinal system.

An interspecific selection based on an entirely different set of characteristics is conspicuous in communities of surface waters. The inhabitants of such locations must be buoyant, motile, or sink relatively slowly, lest they settle downward to sites where, for many of the groups involved, conditions are inhospitable. For algae, by way of example, sinking below the zone where active photosynthesis is possible would be disastrous. Buoyancy mechanisms are of immense importance, because the mean specific gravity of the protoplasm of most cells exceeds that of the usual aquatic environment, and individuals whose protoplasm is dense would be unable to maintain themselves in surface waters. The successful species have compensations for the high specific gravity of many cellular constituents, devices that insure the individuals continued flotation or slow sinking. These include motility, the accumulation of intracellular fat droplets, the possession of gelatinous capsules, the production of gas bubbles in the cytoplasm, and the formation of mucilage. Natural selection eliminates populations or propagules unable to cope with the problem of low specific gravity of the surrounding fluid and favors those possessing appropriate flotation devices (Fogg, 1965).

A striking outcome of selection pressure resulting from the physical characteristics of the surroundings becomes obvious when the cell size of residents of certain ecosystems is taken into consideration. Soil diatoms, as a case in point, are usually appreciably smaller than their aquatic counterparts, and terrestrial strains of a single species are frequently smaller than sister strains living in water. Small flagellates and rhizopods dominate the soil protozoan community, whereas the large-celled ciliates are relatively rare, and here, too, the soil dwellers typically are tinier than closely related strains inhabiting lakes, ponds, and streams. Similarly, the dimensions of the sand-dwelling ciliates are related to the size of the spaces between the sand grains; at the extreme, where the interstices between sand grains are minute, the existence of a ciliate fauna is precluded (Faure-Fremiet, 1951). The dimensions of the cell or filament unquestionably are of ecological consequence, and a reasonable number of environments clearly are more suited for small than large organisms. In ecosystems where the interstices and pores are tiny, as in sand or soil, the physical barriers encountered by larger individuals seem to explain their comparative rarity. Smallness also has biological value inasmuch as little cells often have a greater capacity to absorb nutrients that are present in low concentrations, as is the case in many ecosystems. Individuals with microscopic dimensions have a high surface area per unit volume, and this may be advantageous because of the high metabolic activity generally associated with cells having high surface-to-volume ratios. Largeness, on the other hand, may be of value in that such organisms are resistant to small predators; indeed, protozoa are noted for feeding discriminatingly on prey of particular size ranges, and thus dimensions may be crucial not only to the predator in its feeding habits but to those groups escaping destruction by marauding protozoa. The data are too limited as yet, however, to permit anything except speculation as to the possible causes of most kinds of selection linked with cell dimensions.

Interspecific selection regularly results from differentials in tolerance to toxins in the environment. Antimicrobial agents are found in animal and plant tissues, and the success of many parasites and the failure of closely related bacteria or fungi in colonizing these tissues are attributable to the parasites' resistance to the host's inhibitors. Antimicrobial substances in serum, saliva, food, and soil undoubtedly have a differential effect and aid in the course of selection, and investigations too numerous to cite have revealed the greater tolerance of members of the indigenous communities of such ecosystems than of alien species. The presence of various metallic ions and organic chemicals in heavily polluted rivers adversely affects a major segment of the indigenous algae, particularly the dominants, but they are soon replaced by species more tolerant of the pollutants in the water. The deliberate addition of chemical preservatives to food and manufactured products likewise has a depressing effect on much of the community, but when spoilage takes place, the common incitants are discovered to be resistant to the preservatives employed. In this regard it is of interest that staphylococcal poisoning is particularly prevalent in foods deleterious to many bacterial genera but not to the growth of *Staphylococcus* strains. In these instances tolerance to the toxins of the habitat is an obligatory fitness trait.

Considerable attention has been focused recently on the possible contamination of Mars by microorganisms derived from the earth and carried on vehicles used for space exploration, and on the colonization of earth by components of a Martian biota clinging to the spacecraft or astronauts returning from planetary exploration. If life exists on two or more planets of this solar system, the differences in the biospheres may be minor or they may be enormous, but it is likely that a struggle for existence between the invaders and the residents will ensue, provided that they have common attributes or make use of the same environmental resources. Selection determines that the fit will survive and, in time, assume dominance, but no reason can be cited a priori to preclude the possibility that an extraterrestrial organism, one having no prior contact with the earth's biota, or at least not with the currently existing species, will not be eliminated by selection. In view of the marked dissimilarity of the terrestrial and Martian surfaces, it is likely that tolerance to the multitude of stress factors will be of paramount importance in determining the successful transfer of living things from the earth to Mars or in the reverse direction.

At temperatures more suitable for one genus, species, or strain than for another, the population with the greater multiplication rate will frequently displace the slow growers. Because of this displacement the dominants arising during the progress of natural selection

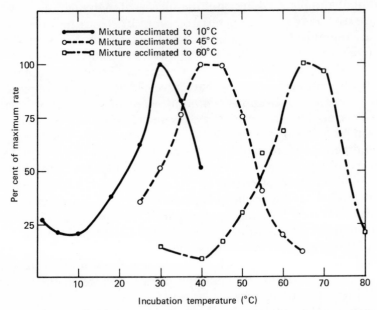

Figure 8.2   Temperature ranges and optima for glucose assimilation by a heterogeneous mixture of microorganisms acclimated to 10°, 45°, and 60° C (Allen and Brock, 1968).

in one region often show a different temperature range or optimum for development than the dominants in a similar ecosystem exposed to somewhat different temperatures (Fig. 8.2). Sometimes populations of the climax community established through selection are obligate in their temperature relationships, strict thermophiles or psychrophiles being common in hot or cold environs. The ultimate outcome when temperature is the selective force is evident in the copious blooms of cold-loving algae on mountain snows and their absence farther down the mountain slopes, the abundance of thermophilic blue-green algae in hot springs and their absence from cold spring water, and the frequency of cold-tolerant bacteria and fungi in foods spoiling in refrigerators and their paucity in edible products kept at room temperatures. Similarly, bacteria and yeasts multiplying readily at or having an obligate dependence on low temperatures are numerous in Antarctic soils (Straka and Stokes, 1960; diMenna, 1960), and algae tolerant of freezing-thawing cycles are more frequent in Antarctic than in temperate areas (Holm-Hansen, 1967). A shift to a higher optimum temperature for growth is evident among tropical as compared with temperate strains of the fungus *Glomerella* (Brown and Wood, 1953) and among nitrifying chemoautotrophs of southern in comparison with northern soils of the Northern Hemisphere (Mahendrappa et al., 1966). Even among the protozoa it has been observed that species parasitizing warm-climate amphibians are more tolerant of heat than are protozoa inhabiting amphibians residing in cold climates (Poljansky and Sukhanova, 1967). The temperature-governed exclusion undoubtedly results often from the complete inability of invaders to proliferate under the prevailing warm or cold conditions, but not uncommonly the organism, though able to live at the existing temperature, cannot cope with neighbors that are benefited by the prevailing temperature.

Interspecific selection favors microorganisms possessing one or possibly several qualities that allow them to gain prominence in saline environments. Unique communities fit for an active existence in highly saline circumstances are located in the Dead Sea, Great Salt Lake, salt marshes, brine solutions, saline soils, salt-treated food products, solar evaporation ponds of salt works, and in NaCl-treated olive, cucumber, or cabbage fermentations. The communities are made up principally of species remarkably similar, both morphologically and physiologically, to counterparts in nonsalty habitats, such as species of *Micrococcus, Sarcina, Pseudomonas,* and *Flavobacterium* among the bacteria, the protozoa *Paramecium* and *Colpoda,* the yeast *Debaromyces,* and numerous algae. Most organisms reproducing in saline ecosystems are more resistant to high salt concentrations than isolates from analogous nonsaline environments, but all have the capacity to multiply in the presence of the ecologically important salt. Some populations are obligately halophilic, while others are halotolerant and can develop vegetatively in dilute solutions.

Many studies, especially of bacteria and algae, have been devoted to ascertaining the physiological or structural characteristics responsible for fitness in salt-rich habitats, interest centering on salt-requiring or salt-tolerant enzymes, cell permeability, and the presence of organelles with osmoregulatory functions. It is noteworthy that a large percentage of marine bacteria and fungi as well as microorganisms native to other salt-rich situations exhibit a requirement for sodium ions, a need rare among residents of dissimilar habitats. The necessity for this cation cannot be explained simply by stating that it is present in the surroundings, since a host of ions are constituents of ecosystems inhabited by microorganisms but few have any nutritional role. A population requiring sodium is not at a disadvantage in the sea, for the cation is always present, and the fact that the oceans are dominated by sodium requirers, despite the continual injection of terrestrial species into the waters, suggests that dependence on sodium is ecologically advantageous, selection benefitting populations with a sodium need.

Selection for halophilic or salt-tolerant populations is immediately detectable in previously sterile or sparsely colonized areas and in waters that become increasingly saline through evaporation. In solar evaporation ponds, salted foods undergoing spoilage, and NaCl-treated fruits or vegetables subjected to lactic fermentation, the fit species multiply and assume an ever rising importance, while those not able to cope with the salty conditions fail to become established or decline in relative significance, so that the climax contains largely or solely halophilic or salt-tolerant populations. Thus strains of tolerant *Staphylococcus* frequently supplant the less suited bacteria in meats with high salt concentrations (Kelly and Dack, 1936), salt-resistant yeasts are prominent in lactic fermentations, and halophilic algae predominate in salt-producing ponds (Carpelan, 1964). Algae growing on the hulls of ships passing back and forth between rivers and the open sea and those indigenous to estuaries survive a particularly striking selection process operating on residents of both the sea and fresh water, the algae representing types not only able to proliferate at both salinity levels but also resistant to abrupt changes in salt concentration.

Of the countless propagules reaching a particular site, the few that survive and multiply occasionally reflect, in a quite obvious way, the abiotic determinants contributing to their surviving the weeding-out process. For example, parasites growing in acid plant tissues, such as immature apples or lemons, are quite resistant to acid (Brown and Wood, 1953), and soils of low pH are richer than neutral soils in acid-tolerant actinomycetes (Fig. 8.3). Because the hydrostatic pressure over large areas of the ocean floors exceeds 300 and occasionally 1000 atmospheres, an essential fitness trait among the bacterial denizens of the deep is the ability to endure and multiply at high pressures; conversely, most species characteristic of shallow water and other surface regions of the earth fail to multiply at the extremes of pressure (Morita, 1967). Furthermore, the distribution of fungi resistant to CO_2, a gas inhibitory to a number of organisms, may mirror the CO_2 partial pressure in the environment; this is evident in the finding that the localization of strains of *Rhizoctonia solani* on plants and in soil is correlated with the relative tolerance of the individual fungus strain to CO_2, suggesting that the CO_2 content of the particular environment contributes to the selection among *R. solani* populations

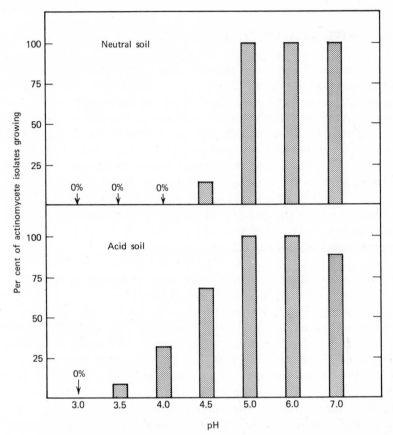

Figure 8.3 Percentage of actinomycetes from acid and neutral forest soil growing in vitro at different pH levels (Corke and Chase, 1964).

(Durbin, 1959). The selective force exerted by these abiotic properties of natural environments is easily demonstrable, but far more difficult is the establishment of the remaining fitness traits, characteristics that undoubtedly exist because numerous thermophiles, psychrophiles, barophiles, or acid-tolerant cells probably gain access to warm or cold environments, the sea bottom, or acid regions, yet few survive and but a scant number assume a prominent position in the community. Some of the additional factors contributing to success in natural selection will serve as subjects for consideration in subsequent chapters.

HOST AS A SELECTIVE FACTOR

Parasites have novel physiological qualities which, when the appropriate host is present and the environment permits the expression of these characteristics, give them an enormous selective advantage. Whereas saprobes commonly are in perennial competition for a limited nutrient supply, the parasite has a rich food source largely to itself, namely, the host on which it lives. As a rule, facultative parasites do not fare well when forced to interact with other microorganisms, but they are spared from harm because of their ability to obtain nutrients from the tissues of viable hosts. The tremendous value to the parasite of the unique ecosystem in which it resides shows itself when the vast number of bacterial, fungal, or protozoan propagules are released from the diseased host tissue: the parasite population declines, usually precipitously, and after several days or weeks not a single propagule may be detectable by even the most sensitive of microbiological techniques.

The biochemical bases for the selective advantage possessed by parasites are poorly understood, despite the impressive biochemical achievements in human, animal, and plant pathology in recent years. The picture is made more complex because often just one of two closely related heterotrophs lives parasitically, Similarly, only one of two very similar animal species or plant varieties is susceptible to parasitic colonization. Subtle differences in the composition and metabolism of host tissues and in the physiology of the parasite are undoubtedly involved, but far more work is necessary before a clear view will emerge of the specific fitness traits of the microorganisms concerned.

The virtue of a parasitic mode of life is totally abolished if the microorganism kills the host through unbridled pathogenicity. The host is lost, but so too is the benefit accruing to the microorganism, and often the latter disappears, for its novel properties have become ecologically irrelevant. To maintain itself, the parasite population must perpetuate a degree of balance with its host, and occasionally this host-parasite balance evolves in elegant ways. The very existence of certain microbial species is dependent on a timing in their growth and morphogenesis in complete harmony with the behavior of the animal or plant harboring them, and serious discord between the two organisms can result in the demise of either one or both. The close relationship between the life cycle of a parasite and its host's activities is nicely illustrated by the ciliates harbored by various crustaceans. Cysts of the ciliates are found on the gills of crabs, for example, and molting of the crustacean provides the sole stimulus for conversion of the cyst to an active form of the protozoan. The active cells feed rapidly on components of the skin shed by the crab, and these cells soon give rise to free-living cysts, which in turn produce daughter cells that can reinvade and encyst on the crustacean; the cysts remain there in a resting stage until molting is reinitiated. The entire sequence is triggered by and requires the molting process (Trager, 1957).

INTRASPECIFIC SELECTION

Although natural communities frequently are dominated by a single species and not a few are monospecific, nevertheless intense selection may occur within the dominant or the sole species of the biota. The selection expresses itself in the emergence of strains, variants, or mutants fit for the prevailing circumstances and in the decline of less fit populations. Communities in diseased tissues are generally monospecific or are dominated by a single species, the organism responsible for the ill effects, and the changes taking place in these microfloras, either spontaneously or under the influence of chemotherapeutic agents, clearly demonstrate the course of intraspecific selection.

Dramatic examples of this kind of selection come from human medicine, and the community modifications are readily discernible in patients treated with a drug nominally effective against the pathogen causing the communicable disease. Studies in vitro have revealed that populations of bacteria and fungi, when grown in culture media supplemented with gradually increasing doses of toxic compounds, show a rise in their tolerance to the antimicrobial chemical. Frequently the increase in resistance is gradual, as with bacteria cultured in the presence of penicillin, but occasionally the resistance of the population alters abruptly, as when some bacteria are exposed to streptomycin. Cultures of drug-resistant *Diplococcus, Mycobacterium, Neisseria, Plasmodium, Pseudomonas, Staphylococcus, Trichophyton,* and *Trypanosoma,* to mention but a few pathogens, have thus been found. The ease of obtaining a drug-tolerant from a sensitive culture depends on both the species and the chemical. Parasites like *Entamoeba histolytica* and *Treponema pallidum* rarely give rise to tolerant populations, while a few pathogens do so with reasonable frequency. In the presence of some inhibitors, tolerant strains often arise, but resistant cultures rarely appear following the exposure of microorganisms to other chemicals.

With most and possibly nearly all drugs that lose effectiveness as tolerant individuals appear, variants either initially present in the otherwise susceptible population or those arising after the initiation of drug treatment are favored in the environment containing the toxic agent, and these variants reproduce and eventually displace the sensitive individuals in the population. The antibiotic or synthetic compound is therefore the agent of selection, and the fitness trait is the physiological characteristic of the resistant cells that makes them tolerant to the drug and provides their tremendous advantage in environments subject to drug stress.

The decline in effectiveness of a chemotherapeutic agent previously useful for controlling a pathogen may be explained in one of two ways: either (a) a decrease in relative abundance of sensitive strains in nature due to the widespread use of the drug or (b) the acquisition, during therapy, of resistance by the parasites residing in the host. Both phenomena occur in nature, and both reflect the operation of natural selection. Regarding the first explanation, a steady increase in the drug resistance of many clinically important microorganisms has been noted. Coinciding with this increased tolerance of human pathogens is a parallel rise in treatment failures, so that severe limitations have been imposed on the usefulness of antibiotics and related drugs that were once quite important in medicine. Indeed, penicillin-resistant *Staphylococcus,* sulfonamide-resistant *Streptococcus* and *Neisseria gonorrhoeae,* and tryparsamide-tolerant trypanosomes became clinically prominent some years back, but similar problems continue to emerge. A rise in the relative abundance of drug-resistant strains in patients following the widespread use of penicillin and tetracycline antibiotics is now amply documented. The results of a typical investigation are presented in Fig. 8.4. Evidently selection enforced by the physician has altered the prominent parasitic strains with regularity and on a nearly global scale.

Regarding the second explanation, sensitive strains of several bacterial pathogens have been observed to be replaced by resistant populations when the microorganisms were passed through experimental animals receiving subcurative levels of drugs. The parasites include *Diplococcus pneumoniae, Neisseria meningitidis,* and *Mycobacterium tuberculosis.* In a typical study all *Staphylococcus aureus* isolates recovered from untreated mice inoculated with the bacterium were found to be sensitive to less than 31 μg of streptomycin per milliliter,

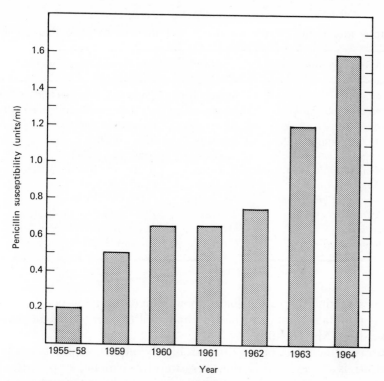

Figure 8.4 Increase in penicillin resistance among *Neisseria gonorrhoeae* strains not eliminated by penicillin levels usually used in therapy (Thayer et al., 1965).

whereas 27% of the isolates from streptomycin-treated mice showed significantly greater tolerance and one isolate was inhibited only by 16,000 μg/ml (Bliss and Alter, 1962). The development of resistance during the course of treatment of human infections caused by *Staphylococcus, Streptococcus, Salmonella, Pseudomonas, Hemophilus, Brucella,* and *Histoplasma* has also been described.

Considerable attention has been given to establishing the particular fitness traits underlying the selection of drug-resistant variants. One or more physiological modifications in a cell may make it fit for growth in an environment containing a drug toxic to related strains or to the wild type from which a mutant originated. The tolerant cell may become impermeable to the drug so that it no longer reaches the intracellular site of action. The resistant cell may contain an enzyme catalyzing destruction of the drug, as in the classical case of

Table 8.2

Bacteria and Fungi Whose Virulence Is Enhanced by Host Passage

Microorganism	Host	Reference
Brucella suis	Guinea pig	Jones and Berman (1951)
Diplococcus pneumoniae	Mouse	MacLeod and Bernheimer (1965)
Fusarium solani f. phaseoli	Bean	Burkholder (1925)
Phytophthora infestans	Tomato	Mills (1940)
Pseudomonas pseudomallei	Mouse	Nigg et al. (1956)
Pseudomonas solanacearum	Tobacco	Averre and Kelman (1964)
Salmonella gallinarum	Chicken	Gowen (1951)
Salmonella typhimurium	Guinea pig	Page et al. (1951)
Staphylococcus aureus	Mouse	Beining and Kennedy (1963)
Urocystis spp.	Oats	Christensen and DeVay (1955)
Xanthomonas stewartii	Corn	Lincoln (1940)

bacteria producing penicillinase, an enzyme detoxifying penicillin, or the fit population may synthesize a metabolite countering the action of the drug (Moyed, 1964).

Intraspecific selection, again with highly important practical ramifications, is also evident in the alterations in virulence of parasites affecting plants and animals. The virulence of bacteria and fungi is often profoundly affected by passage through suitable hosts, the virulent cells having an advantage not possessed by the avirulent cells and hence becoming ever more prominent with time (Table 8.2). The environment at distinct localities within the host is selective and permits extensive proliferation of virulent cells derived either from a propagule present in a population that was heterogeneous at the time it encountered the host or from a virulent cell that arose after infection, as by mutation or from a heterokaryotic individual. Passage through the animal or plant may favor the virulent individuals by virtue of their greater tolerance to the host's defense mechanisms or because environmental factors in the tissues or body fluids are more conducive to their proliferation than to growth of the avirulent cells. Demonstration of the quantitative changes is simpler with bacteria than with fungi, owing to the relative ease of counting bacteria and the difficulty in assessing the quantity of fungal protoplasm in such an ecosystem. The results of a number of studies of bacterial parasites, one of which is summarized in Fig. 8.5, reveal the increase in relative

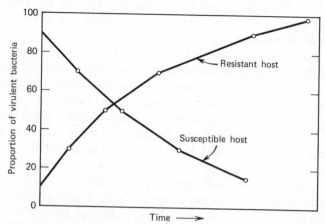

Figure 8.5 Effect of passage through corn plants on the relative abundance of virulent *Xanthomonas stewartii* cells. Resistant and susceptible plants were initially inoculated with a mixture of virulent and avirulent cells (Lincoln, 1940).

abundance of virulent cells developing in an appropriate host and the culling out of avirulent cells. The figure also shows an interesting change observed in certain plants, namely a disease-resistant variety favors virulent strains while a susceptible plant variety favors avirulent individuals.

Despite the presence of both avirulent and virulent propagules in the environs, an animal or plant of one species supports avirulent types while another host, a close or distant relation of the first, is colonized by the virulent microorganism and comes down with disease. Should the avirulent population be subject to mutation at a reasonable frequency, one host may function as a reservoir for a microbial species that invades another host and, following mutation and selection, gives rise to a virulent population. Moreover, disease-resistant crop varieties developed by plant breeders often serve to select especially virulent strains and races of pathogens, these being the only kinds successfully parasitizing the new, resistant varieties. In addition, when microorganisms are passed through hosts of moderate susceptibility, selection may favor the emergence of populations able to parasitize hitherto immune hosts (Christensen and Daly, 1951; Christensen and DeVay, 1955). Not only is the selection dictated by the particular animal or plant, but individual host tissues may be beneficial to either the virulent or avirulent type of the same pathogenic microorganism, which then reigns supreme in the community of that tissue (Chandler et al., 1939; Platt, 1937).

Little is known regarding why virulent cells assume dominance in such largely or totally monospecific communities, despite the many pathogens in which this phenomenon has been recorded. Studies in vitro show that some virulent organisms grow more readily than the avirulent strains because distinct metabolic products preferentially enhance their development, and analogous effects may occur in the living animal (Firshein and Braun, 1960). On the other hand, the avirulent forms of certain microorganisms are preferentially inhibited by discrete metabolic products, and possibly the intraspecific selection in vivo may sometimes result from a greater sensitivity of the avirulent strains to toxic metabolites (Braun, 1946, 1958).

Among the bacterial parasites of man and animals, the greater fitness of the virulent cells may be related to the fact that they are less susceptible to phagocytosis than avirulent counterparts. Thus the fate in the animal body of *Diplococcus pneumoniae, Bacillus anthracis,* and *Staphylococcus aureus* is correlated with the production of a capsule or other kind of surface-localized defense mechanism against phagocytes; the naked, avirulent cells are readily destroyed whereas encapsulated, virulent bacteria are shielded from attack by phagocytes. Presumably the displacement of the avirulent by the virulent cells of these species results from selection in vivo for individuals producing the phagocyte-resistance characteristics. Likewise a nonencapsulated, avirulent mutant of the yeast *Cryptococcus neoformans* is far more susceptible to phagocytosis than the virulent, encapsulated strain (Bulmer and Sans, 1967). Components of the capsule of some organisms are harmful to the host's phagocytes, but toxicity apparently does not account for the lack of destruction of other parasites by phagocytes. Factors in addition to surface components may also be important in protecting the pathogen from destruction. Regardless of the precise mechanism, such resistance is obviously a critical fitness trait associated with intraspecific selection among animal pathogens and their triumph in pathogenesis.

Intraspecific selection in a monospecific, albeit artificial, community is also frequently observed when pathogenic bacteria, fungi, or protozoa are repeatedly transferred in laboratory media. In the process of *attenuation* the attributes of the culture change progressively with time as the proportion of avirulent cells, or organisms of low virulence, increases with serial transfers, until ultimately a population is established that has little or absolutely no virulence. Such declines in virulence following initial isolation from diseased tissue have been found in strains of *Clostridium, Diplococcus, Pasteurella, Pseudomonas, Staphylococcus,* and *Xanthomonas* among the bacteria, in *Trichomonas* and *Trypanosoma* among the protozoa, and in the fungus *Fusarium,* to mention a few. The avirulent or less virulent cells may appear as a result of a mutation during growth of the parent culture, or they may have been present in the original inoculum. These cells thrive better in vitro and, on account of their selective advantage, ultimately displace the cells of the virulent population. Attenuation may also occur in vivo. Many years ago, for example, Pasteur and Thuillier (1883) noted that rabies virus passed through guinea pigs had decreased virulence for rabbits, and analogous attenuations have been reported for plant viruses as well. The selection for avirulent types in vivo is illustrated in Fig. 8.5. This kind of attenuation in vivo is known to take place occasionally when animal parasites are passed through normally resistant hosts (Braun, 1956) or plant parasites are passed through normally susceptible hosts (Christensen and DeVay, 1955).

ADAPTATION

Scores of heterotrophs and autotrophs acquire the faculty to grow in regions inhospitable to the original population, and modest modifications in the concentration of drugs, poisons, or salts, shifts in light intensity or temperature, or the appearance of a new substrate or host may effect a change in the population such that new physiological or morphological properties come to the fore. Microorganisms respond in different ways to an alteration in the environment, and those modifications permitting survival in the changed habitat are termed *adaptive.* In the process of *adaptation* or *acclimation* the individual or population of individuals adjusts to new circumstances so that it is more in harmony with the altered habitat. The adaptation, or adaptive trait, represents an advantageous quality, one that has survival value and permits the bearer to live in the new surroundings.

The adaptive trait may be obscure and expressed subtly, but its existence is manifested by the capacity of the species, strain, or race to tolerate or make use of a set of conditions different from that previously encountered. Adaptation at times results from the synthesis of a new enzyme or set of enzymes, or it may on occasion be linked to the formation of additional morphological features. In the last analysis, properties of the altered environment determine the precise biological characteristics, be they covert or overt, that a cell or filament must possess in order to remain alive or multiply. The forces of natural selection operating in the changed habitat select the individuals with advantageous properties, and these cells multiply and supplant the populations not possessing the beneficial traits.

For the survival of a high proportion of species, a degree of adaptability is essential to guarantee survival in a varying habitat. *Adaptability* refers to an organism's capacity to adjust to a state of greater fitness, each organism having a certain *adaptive capacity* or *plasticity* that enables it to respond to alterations in the surroundings. Adaptability is widespread among saprobes; yet the range encompassed by a particular adaptation rarely is overly broad in nature, particularly in communities composed of many populations, and the breadth of this range governs the sites potentially inhabited by the species. Nevertheless occasional species not only are readily adaptable but have a far wider range of potential responses than that of neighboring organisms. Still, countless species fail entirely to adapt under conditions allowing neighbors to acclimate with ease, as noted, for example, in attempts to increase the resistance of heterotrophs to toxic chemicals.

Two procedures are commonly employed to demonstrate acclimation in vitro. In one, successive serial transfers are made in media with ever increasing intensity of a factor—a drug, poison, hydrogen ion concentration, temperature, and so on. In the second technique, a large population is subjected to an extreme condition for as long as is needed to allow for the emergence and proliferation of a variant able to cope with the stress in question. Certain species tolerate no more than a modest change imposed in the first procedure, but the new population, once acclimated, can adapt to a second and a third similar stress, the end result being a population capable of developing under circumstances far different from those permitting activity of the original cells. The second procedure is useful for species tolerating an abrupt shift in the intensity of the particular environmental factor. The length of exposure of the cells to the environmental stress before acclimation is evident varies with the microorganism, the factor or compound in question, and the conditions prevailing during the exposure, the time elapsing ranging from a few hours to several weeks or, occasionally, months.

Acclimation to one factor sometimes results in the simultaneous acquisiton of a second biochemical attribute. Thus adaptation to increasing resistance to one drug often leads to a concomitant rise in tolerance to another. Conversely, acclimation to a new environmental factor may lead to the disappearance of an apparently unrelated capacity or function; for example, loss of pathogenicity at times parallels the adaptation of a parasite to altered circumstances. The loss may be in a physiological property analogous to the one gained, as shown by the numerous reports that acclimation of a population to higher than usual levels of some factor takes place at the same time as the capacity to grow at low levels of the factor disappears; for instance, acquisition of the ability to multiply at increasingly high salt concentrations commonly results in an inability to grow at previously suitable, low salt levels.

Adaptive capacity is of considerable ecological consequence. Though many populations are eliminated or relegated to a minor position when conditions are altered, adaptable organisms acclimate to the prevailing temperature, salinity, light intensity, toxic agents, or newly available organic nutrients, thereby resisting elimination. Table 8.3 summarizes a few of the factors to which microorganisms adapt, factors of appreciable importance to both saprobes and parasites. The occurrence of such phenomena in nature cannot be disputed, although far too little attention has been given to acclimations in natural habitats. The adaptation to tolerate toxic chemicals—including antibiotics, synthetic drugs, and fungicides—is widespread among pathogenic bacteria, fungi, and protozoa in vitro, and sufficient observations have been made to suggest that analogous responses occur in vivo. For most of the factors tabulated in Table 8.3, however, the information is derived solely from laboratory investigations.

Morphological changes induced by environmental modification or arising as an outcome of movement or growth of an organism from one site to another, by contrast, are seen both in vitro and in vivo. Responses of this sort have been observed in a variety of heterotrophs, but particular interest has been taken in the fungi, especially those showing a dimorphism involving a mycelial and a yeastlike phase as the fungi respond to their surroundings. Dimorphism is evident among species of *Candida, Blastomyces, Histoplasma,* and *Sporotrichum,* organisms frequently pathogenic for animals and man, and the fact that the fungal reaction to their surroundings is accompanied by a conversion of one form to the second assumes added relevance inasmuch as only one of the two forms, usually the yeast stage except for species of *Candida,* is found in infected tissues. Thus acclimation to a tissue ecosystem is accompanied by an enormous alteration in the physiology of the pathogen, one unquestionably linked with a number of different enzymatic steps (Romano, 1966).

Table 8.3

Factors to Which Microorganisms Show the Ability to Adapt

Factor to which Organism Adapts	Microorganism	Reference
Temperature, increasing	Bacillus subtilis	Dowben and Weidenmüller (1968)
Temperature, decreasing	Paramecium caudatum	Poljansky and Sukhanova (1967)
Heat resistance	Bacillus subtilis	Davis et al. (1948)
Salinity, increasing	Colpidium campylum	Loefer (1939)
Salinity, decreasing	Marine bacteria	Pratt and Waddell (1959)
pH, increasing	Bacillus cereus	Kushner and Lisson (1959)
pH, decreasing	Thiobacillus ferro-oxidans	Golomzik and Ivanov (1965)
Ultraviolet intensity, increasing	Neurospora crassa	Goodman (1958)
New bacterial host	Bacteriophage	Hewitt (1954)
Lignin utilization	Polyporus abietinus	Gottlieb et al. (1950)
Lysozyme tolerance	Micrococcus lysodeikticus	Brumfitt et al. (1958)
Tolerance to chemicals		
Actidione	Sclerotinia fructicola	Grover and Moore (1961)
Arsenite	Ustilago zeae	Stakman et al. (1946)
Copper	Mycobacterium avium	Kushner (1964)
Organic arsenicals	Trypanosoma spp.	Bishop (1959)
Pentachloronitro-benzene	Rhizoctonia solani	Elsaid and Sinclair (1964)
Phenylmercuric acetate	Escherichia coli	Kushner (1964)
Streptomycin	Anacystis nidulans	Kumar (1964)
Undecylenic acid	Trichophyton mentagrophytes	Vilanova and Casanovas (1950)

MECHANISMS OF ADAPTATION

A population may adapt to a fresh set of conditions in several possible ways. Nongenetic modifications are temporary and can be readily reversed by a return to the originally prevailing circumstances; these are designated *phenotypic adaptations. Genotypic adaptations,* by contrast, require a genetic modification; the change is in-

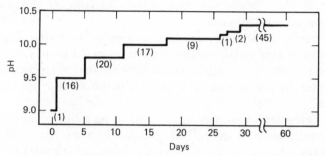

Figure 8.6 The development of resistance to alkalinity in *Bacillus cereus* by serial transfers in media of increasing pH. The figures in parentheses indicate the number of subcultures made (Kushner and Lisson, 1959).

herited by daughter cells, and it may persist for some time after the environment is restored to its original state. Phenotypic adaptations, moreover, differ from those having a genetic origin not only in being induced by a feature of the environment but also in constituting a direct response to that provocation. Those acclimations having a genetic basis must be expressed phenotypically in order to be ecologically of consequence. These genotypic adaptations result from mutation, the transfer of genetic material between cells, or heterokaryosis. The variant cells or the variant nuclei in the filaments of a heterokaryotic fungus, provided that they exhibit superior fitness to their progenitors, are then acted upon by natural selection so that a population emerges having the ability to proliferate in the altered habitat. In Fig. 8.6 is depicted the adaptation of a strain of *Bacillus cereus* to alkaline conditions effected by repeated subculture of the bacteria in media of gradually increasing pH. The pH values are the highest permitting growth of the *B. cereus* strain.

Microorganisms probably respond to short-term fluctuations in the ecosystem by physiological adaptations, whereas genetic modifications are likely to be more common if the change in the surroundings is of long duration or is notably severe. Rarely, however, have microbiologists distinguished between the selection of variants appearing in vivo by genetic mechanisms and those arising by nongenetic means, with the notable exception of the emergence of antibiotic-resistant bacteria. In many instances of interspecific selection in nature, as one might expect, a drastic or enduring alteration of the environment is not countered by the emergence of a variant derived from a previous dominant in the community, but rather—because of the ubiquity, ready dispersal, and physiological versatility of microorganisms—by the appearance of a newly arrived species or one that previously played a minor role in the community. In infected tissues or in other largely or totally monospecific communities, major stresses may indeed be met by the arising of a variant originating from the prior dominant.

Phenotypic adaptation may be effected by an adjustment in the enzymatic composition of the cell in response to a specific substance or factor in the environment. The factor bringing about the change may be a substrate that the organism can metabolize. When a substrate brings about a response in an entire population such that enzymes for utilization of the substrate, and for certain products formed in the metabolism of that compound, appear in sizable amount, the substrate is termed an *inducer* and the enzymes appearing are said to be inducible enzymes. In natural communities a small population reacting to the presence of an inducer, like an organic nutrient, will proliferate and assume a greater role in the ecosystem. The progeny of the cells quickly lose the adaptation of their forebears once the substrate serving to induce the formation of the enzymes is wholly consumed.

Alterations in phenotypic behavior also occur with no detectable change in the inorganic or organic nutrient supply. For example, differences are noted among cultures grown in media of the same composition but at different temperatures, osmotic pressures, pH levels, or light intensities, and analogous effects in nature can have survival value. A rise or fall in pH may bring about the formation of large amounts of one enzyme and diminish the production of another, changes that influence the ecosystem and, at the same time, affect the ability of the cell to maintain itself (Gale and Epps, 1942). With enzymes induced by a fresh substrate, improved fitness is readily apparent inasmuch as the nutrient source, otherwise useless to the species, now can be utilized for development. Among the algae clinging to the hulls of ferry boats making repeated trips from fresh to seawater in ports along the seacoast, adaptation allows the organism to adjust itself to the extreme and rapid variations in salinity. Other algae acclimate to different light intensities and alter the amount of photosynthetic pigments in the cell as they are exposed to more or less intense light (Steeman Nielsen and Jorgensen, 1968), while selected heterotrophs, too, show responses to sunlight, some of which may affect their very survival. Nonhereditary structural modifications of a variety of kinds are widespread and commonly make the organism more fit for the new conditions. In the last analysis, to be sure, the extent of phenotypic adaptation is governed by the genotype, each genotype allowing for a limited range of biochemical reactions and behavioral responses, and that organism endowed with the potential to adapt quickly or to diverse environmental factors is said to possess *phenotypic plasticity*.

Genotypic adaptation reflects the existence of a nonhomogeneous population, an assemblage of organisms containing a few variant cells that normally have low fitness but which occasionally come to the forefront through natural selection. In the presence of a powerful toxic agent, some other sort of extreme environmental stress, or a nonutilizable nutrient, the bulk of the population cannot multiply, but the rare variant finds the new situation to its liking, multiplies, and attains a dominant position. Selection thereby favors the fit genotype. The characteristic of *genotypic plasticity* is especially advantageous, allowing the possessor to occupy a reasonably large number of habitats or to cope with modified surroundings.

Mutation is a common means of genotypic adaptation, a mechanism in which an entirely new heritable property or physiological activity is suddenly introduced into a population. Though mutants probably arise continually, most fail to survive in an environment that is reasonably constant because the existing populations are already well adapted to the ecosystem. It is when a change occurs that the rare mutant is favored, since it has greater fitness than the parent cells. The available information indicates that the extent of these adaptations is generally small, and several mutations usually are required to bring about a major alteration in the population, as, for example, in the gradual rise in toxin resistance of heterotrophs as they are repeatedly exposed to increasing concentrations of the inhibitor. There is evidence, sometimes excellent but often equivocal, that the appearance of the following results from the selection of a mutant arising in the original population: antibiotic-resistant bacteria, drug-tolerant protozoa, strains of fungi and bacteria with altered virulence, and fungi and yeasts tolerating heavy metal inhibitors.

Genotypic changes of possible ecological value may arise in ways other than by mutation. The temporary fusion of two individuals and the transfer of hereditary determinants between them occur in a large number of microbial types. Transformation takes place in certain bacteria; in transformation, a portion of the DNA released upon the death and lysis of a cell enters another bacterium of the same or a closely related species, and the DNA combines with the hereditary material of the recipient and alters it genetically. In transduction, a bacteriophage infecting one bacterium acquires a small amount of DNA from this cell and injects the nucleic acid into a second bacterium, which the virus infects but does not lyse.

The various processes involving the transfer of hereditary determinants have been explored exhaustively in vitro, but their significance in adaptations in vivo is largely unknown. Hybridization between races of *Ustilago hordei* has been observed during their passage through a susceptible plant host (Tapke, 1944). Transformation has been found to occur in mice infected with two genetically distinct strains of *Diplococcus pneumoniae,* and it has been proposed that new strains of this bacterium may be generated during pneumococcal pneumonia, strains which, if possessing different antibiotic susceptibilities or antigenic properties, could affect the course of disease (Conant and Sawyer, 1967). Transduction has been suggested as the cause of the multiplicity of distinct *Salmonella* serotypes and the development of new ones (Bailey, 1956). Nevertheless, the evidence that a transfer of genetic material between populations is important in nature remains skimpy and largely inconclusive.

Heterokaryosis involves two or more genetically dissimilar nuclei within a single hypha or spore. The fungal heterokaryon is potentially quite variable as different nuclei associate or the nuclear ratios become altered and bring about different physiological activities. Attributed to heterokaryosis in fungi are the appearance of new races of pathogens, decreases or increases in virulence, gain or loss in the ability to utilize individual substrates, and adjustments of the individual to a slowly fluctuating environment. Most authorities agree that heterokaryosis does indeed occur in nature, but its frequency or significance remains in dispute.

DEADAPTATION

Upon the return of the environment to its original state, the population commonly deadapts, but the length of time needed to regain the original characteristics is extremely variable. Whether the period necessary for deadaptation is prolonged or brief is determined by the microorganism, the factor responsible for the initial acclimation, and the mechanism of the adaptation or deadaptation. Deadaptation of individual species to some factors is notably short, while a long time is necessary before the same species regains its original traits in the case of other factors. Different species, moreover, deadapt at markedly dissimilar rates to the same variable in the environment. Losses of drug resistance and fungicide tolerance have been well explored in this regard.

Restoration of the original behavioral pattern by a population following a phenotypic adaptation is rapid, as a rule, once the inducing substrate is metabolized completely or some other feature responsible for the adjustment is no longer present. On the other hand, a return to the original condition may require several successive transfers in vitro, as in the loss of arsenite tolerance by *Ustilago zeae* (Stakman et al., 1946) or chloramphenicol resistance by *Escherichia coli* (Herrmann and Steers, 1953). A reasonably rapid restoration of the initial microbial type seems to occur in nature, as witnessed by the reappearance with increasing frequency of drug-susceptible bacteria in hospitals or in human populations at large upon termination of the therapeutic use of antibiotics. Yet *Trypanosoma rhodesiense* can maintain its atoxyl resistance in excess of 24 years (Bishop, 1959),

Table 8.4

Relative Proportion of *Escherichia coli* Cells That Are Resistant to Chloramphenicol When the Bacteria Are Subcultured in Chloramphenicol-Free Media [a]

No. of Subcultures	Per Cent of Cells Resistant to Chloramphenicol	
	1.0 mg/ml [b]	2.0 mg/ml [b]
0	89	89
1	86	52
2	80	37
3	83	4.4
4	37	1
5	4.1	0
6	1.3	—
7	0	—

[a] From Herrmann and Steers (1953).
[b] Concentration of antibiotic the cells would tolerate. The culture was originally resistant to the drug.

and though few microbiological studies span a period as long as a quarter of a century, it is a common observation that no deadaptation is discernible during the course of investigation of numerous heterotrophs.

A population arising by genotypic adaptation may be at a disadvantage when the factor responsible for its selection is removed, and it may be replaced by organisms physiologically quite similar to the original cells from which the new group arose. This is particularly likely when the deadaptation is attributable to the emergence of a mutant that lacks the particular adaptive trait, a mutant whose growth is favored once the stress factor is gone; thus, when a chloramphenicol-resistant strain of *E. coli* is placed in antibiotic-free media, some of the sensitive mutants that arise have greater fitness to the new conditions—in this instance, fitness being the greater growth rate of the susceptible population—and hence displace the resistant cells on serial subculture (Table 8.4). A population acclimated to a stress condition is very likely at a disadvantage with respect to its predecessors in nonstress circumstances, because the processes resulting in its adaptation brought about additional physiological changes as well, for example, diminished rate of multiplication, greater growth factor requirements, or lower ability to cope with host defense mechanisms, and it is selected against—as it was previously selected for—when its peculiar adaptive trait no longer has survival value in interspecific or intraspecific interactions.

References

REVIEWS

Christensen, J. J., and Daly, J. M. 1951. Adaptation in fungi. *Ann. Rev. Microbiol.*, 5, 57–70.

Christensen, J. J., and DeVay, J. E. 1955. Adaptation of plant pathogen to host. *Ann. Rev. Plant Physiol.*, 6, 367–392.

Davies, R., and Gale, E. F., Eds. 1953. *Adaptation in Micro-organisms.* Cambridge University Press, Cambridge.

Grant, V. E. 1963. *The Origin of Adaptations.* Columbia University Press, New York.

Pardee, A. B. 1961. Response of enzyme synthesis and activity to environment. *In* G. G. Meynell and H. Gooder, Eds., *Microbial Reaction to Environment.* Cambridge University Press, London. pp. 19–40.

Person, C. 1968. Genetical adjustment of fungi to their environment. *In* G. C. Ainsworth and A. S. Sussman, Eds., *The Fungi*, Vol. 3. Academic Press, New York. pp. 395–415.

Schlegel, H. G., and Jannasch, H. W. 1967. Enrichment cultures. *Ann. Rev. Microbiol.*, 21, 49–70.

Schnitzer, R. J., and Grunberg, E. 1967. *Drug Resistance of Microorganisms.* Academic Press, New York.

van Niel, C. B. 1955. Natural selection in the microbial world. *J. Gen. Microbiol.*, 13, 201–217.

Williams, G. C. 1966. *Adaptation and Natural Selection.* Princeton University Press, Princeton, N.J.

OTHER LITERATURE CITED

Allen, S. D., and Brock, T. D. 1968. *Ecology*, 49, 343–346.
Averre, C. W., and Kelman, A. 1964. *Phytopathology*, 54, 779–783.

201

Bailey, W. R. 1956. *Can. J. Microbiol.,* **2,** 555–558.

Beining, P. R., and Kennedy, E. R. 1963. *J. Bacteriol.,* **85,** 732–741.

Biebl, R. 1956. *Protoplasma,* **46,** 63–89.

Bishop, A. 1959. *Biol. Rev.,* **34,** 445–500.

Bliss, E. A., and Alter, B. M. 1962. *J. Bacteriol.,* **84,** 125–129.

Bloomfield, B. J., and Alexander, M. 1967. *J. Bacteriol.,* **93,** 1276–1280.

Borut, S. 1960. *Bull. Res. Council Israel.,* **8D,** 65–80.

Braun, W. 1946. *J. Bacteriol.,* **52,** 243–249.

Braun, W. 1956. *Ann. N.Y. Acad. Sci.,* **66,** 348–355.

Braun, W. 1958. *J. Cellular Comp. Physiol.,* **52,** Suppl. 1, 337–369.

Brown, W., and Wood, R. K. S. 1953. *In* R. Davies and E. F. Gale, Eds., *Adaptation in Micro-organisms.* Cambridge University Press, Cambridge. pp. 326–337.

Brumfitt, W., Wardlaw, A. C., and Park, J. T. 1958. *Nature,* **181,** 1783–1784.

Bulmer, G. S., and Sans, M. D. 1967. *J. Bacteriol.,* **94,** 1480–1483.

Burkholder, W. H. 1925. *Am. J. Bot.,* **12,** 245–253.

Carpelan, L. H. 1964. *Ecology,* **45,** 70–77.

Chandler, C. A., Fothergill, L. D., and Dingle, J. H. 1939. *J. Bacteriol.,* **37,** 415–425.

Claes, H. 1954. *Z. Naturforsch.,* **9b,** 461–470.

Conant, J. E., and Sawyer, W. D. 1967. *J. Bacteriol.,* **93,** 1869–1875.

Corke, C. T., and Chase, F. E. 1964. *Proc. Soil Sci. Soc. Am.,* **28,** 68–70.

Davis, F. L., Wyss, O., and Williams, O. B. 1948. *J. Bacteriol.,* **56,** 561–567.

diMenna, M. E. 1960. *J. Gen. Microbiol.,* **23,** 295–300.

Dowben, R. M., and Weidenmüller, R. 1968. *Biochim. Biophys. Acta.,* **158,** 255–261.

Dundas, I. D., and Larsen, H. 1962. *Arch. Mikrobiol.,* **44,** 233–239.

Durbin, R. D. 1959. *Am. J. Bot.,* **46,** 22–25.

Durrell, L. W., and Shields, L. M. 1960. *Mycologia,* **52,** 636–641.

Elsaid, H. M., and Sinclair, J. B. 1964. *Phytopathology,* **54,** 518–522.

Faure-Fremiet, E. 1951. *Biol. Bull.,* **100,** 59–70.

Firshein, W., and Braun, W. 1960. *J. Bacteriol.,* **79,** 246–260.

Fogg, G. E. 1965. *Algal Cultures and Phytoplankton Ecology.* University of Wisconsin Press, Madison.

Gale, E. F., and Epps, H. M. R. 1942. *Biochem. J.,* **36,** 600–618.

Goldstrohm, D. D., and Lilly, V. G. 1965. *Mycologia,* **57,** 612–623.

Golomzik, A. I., and Ivanov, V. I. 1965. *Mikrobiologiya,* **34,** 465–468.

Goodman, F. 1958. *Z. Vererbungslehre,* **89,** 675–691.

Gottlieb, S., Day, W. C., and Pelczar, M. J. 1950. *Phytopathology,* **40,** 926–935.

Gowen, J. W. 1951. *In* Dunn, L. C., Ed., *Genetics in the 20th Century*. Macmillan, New York. pp. 401–429.

Grover, R. K., and Moore, J. D. 1961. *Phytopathology*, 51, 399–401.

Herrmann, E. C., and Steers, E. 1953. *J. Bacteriol.*, 66, 397–403.

Hewitt, L. F. 1954. *J. Gen. Microbiol.*, 11, 261–271.

Holm-Hansen, O. 1967. *In Environmental Requirements of Blue-Green Algae*. Pacific Northwest Water Laboratory, Corvallis, Ore. pp. 87–96.

Jones, L. M., and Berman, D. T. 1951. *J. Infect. Dis.*, 89, 214–223.

Kelly, F. C., and Dack, G. M. 1936. *Am. J. Public Health*, 26, 1077–1082.

Klugh, A. B. 1931. *Contrib. Can. Biol. Fisheries*, 6(4), 41–63.

Kumar, H. D. 1964. *J. Exptl. Bot.*, 15, 232–250.

Kunisawa, R., and Stanier, R. Y. 1958. *Arch. Mikrobiol.*, 31, 146–156.

Kuo, M.-J., and Alexander, M. 1967. *J. Bacteriol.*, 94, 624–629.

Kushner, D. J. 1964. *In* R. J. Schnitzer and F. Hawking, Eds., *Experimental Chemotherapy*, Vol. 2. Academic Press, New York. pp. 113–168.

Kushner, D. J., and Lisson, T. A. 1959. *J. Gen. Microbiol.*, 21, 96–108.

Lincoln, R. E. 1940. *J. Agr. Res.*, 60, 217–239.

Loefer, J. B. 1939. *Physiol. Zool.*, 12, 161–172.

MacLeod, C. M., and Bernheimer, A. W. 1965. *In* R. J. Dubos and J. G. Hirsch, Eds., *Bacterial and Mycotic Infections of Man*. Lippincott, Philadelphia. pp. 146–169.

Mahendrappa, M. K., Smith, R. L., and Christiansen, A. T. 1966. *Proc. Soil Sci. Soc. Am.*, 30, 60–62.

Mills, W. R. 1940. *Phytopathology*, 30, 830–839.

Morita, R. Y. 1967. *Oceanog. Marine Biol.*, 5, 187–203.

Moyed, H. S. 1964. *Ann. Rev. Microbiol.* 18, 347–366.

Nicot, J. 1960. *In* D. Parkinson and J. S. Waid, Eds., *The Ecology of Soil Fungi*. Liverpool University Press, Liverpool. pp. 94–97.

Nigg, C., Ruch, J., Scott, E., and Noble, K. 1956. *J. Bacteriol.*, 71, 530–541.

Page, L. A., Goodlow, R. J., and Braun, W. 1951. *J. Bacteriol.*, 62, 639–647.

Pasteur, L., and Thuillier, L. 1883. *Compt. Rend.*, 97, 1163–1169.

Platt, A. E. 1937. *J. Hyg.*, 37, 98–107.

Poljansky, G. I., and Sukhanova, K. M. 1967. *In* A. S. Troshin, Ed., *The Cell and Environmental Temperature*. Pergamon Press, Oxford. pp. 200–208.

Potgieter, H. J., and Alexander, M. 1966. *J. Bacteriol.*, 91, 1526–1532.

Pratt, D. B., and Waddell, G. 1959. *Nature*, 183, 1208–1209.

Romano, A. H. 1966. *In* G. C. Ainsworth and A. S. Sussman, Eds., *The Fungi*, Vol. 2. Academic Press, New York. pp. 181–209.

Sistrom, W. R., Griffeths, M., and Stanier, R. Y. 1956. *J. Cellular Comp. Physiol.*, 48, 473–515.

Stakman, E. C., Stevenson, F. V., and Wilson, C. T. 1946. *Phytopathology*, **36**, 411.

Steeman Nielsen, E., and Jorgensen, E. G. 1968. *Physiol. Plantarum*, **21**, 401–413.

Stout, J. D. 1960. *N. Z. J. Agr. Res.*, **3**, 214–223.

Straka, R. P., and Stokes, J. L. 1960. *J. Bacteriol.*, **80**, 622–625.

Sussman, A. S. 1968. *In* G. C. Ainsworth and A. S. Sussman, Eds., *The Fungi*, Vol. 3. Academic Press, New York. pp. 447–486.

Swart-Fuchtbauer, H. 1957. *Arch. Mikrobiol.*, **26**, 132–150.

Tapke, V. F. 1944. *Phytopathology*, **34**, 993.

Thayer, J. D., Samuels, S. B., Martin, J. E., and Lucas, J. B. 1965. *In* J. C. Sylvester, Ed., *Antimicrobial Agents and Chemotherapy—1964*. American Society for Microbiology, Ann Arbor, Mich. pp. 433–436.

Trager, W. 1957. *Biol. Bull.*, **112**, 132–136.

Vilanova, X., and Casanovas, M. 1950. *J. Invest. Dermatol.*, **15**, 161–162.

Whiteside, J. S., and Alexander, M. 1960. *Weeds*, **8**, 204–213.

Part 2

INTERSPECIFIC RELATIONSHIPS

9

Homeostasis

Populations rarely exist independently in nature, and the component members of multispecific communities are as a rule significantly affected by their neighbors. As a previously uncolonized region becomes inhabited, a variety of relationships appear, some beneficial, some detrimental. Because of the proximity of the individuals in natural communities, interspecific interactions are particularly pronounced and have a marked influence on the participating cells. It is these interactions and the interrelationships between microorganisms and their abiotic surroundings that regulate the composition of the established microflora and microfauna of the ecosystem.

Many communities are characterized by a degree of biotic balance, an equilibrium or steady-state condition in which the cell densities or masses of filaments of the indigenous groups are held reasonably fixed. This balance is largely maintained by biotic components of environmental resistance; that is, restrictions imposed by biotic factors in the environment that keep in check the number of individuals of each indigenous population. The biotic factors, as well as abiotic determinants, likewise serve to prevent the establishment of aliens that have gained access to the area by either accidental or deliberate introduction.

Communities are regularly exposed to a multitude of modest biological and nonbiological stresses that might be expected to upset the balance, but, surprisingly, the communities are usually largely unchanged, adjusting themselves to all but the more extreme environmental changes. The capacity to maintain community stability and integrity in a variable environment is termed *homeostasis*. In effect, when the environment changes, self-regulatory mechanisms or homeostatic reactions come into play to restore the relationships previously existing among the residents and to maintain the relative constancy in the makeup of the community. Biological alterations occur as a result of environmental stress, but the processes contributing to homeostasis tend to restore the initial steady-state situation.

Scores of regulatory interrelationships and coordinated processes contribute to preservation of the community's composition and to maintenance of the existing populations at reasonably fixed levels. Though not directly proven for microbial communities, it is likely that the tendency to maintain a balance among the resident organisms is enforced with increasing species diversity and with greater biological complexity of the community. An essential element of homeostasis, one enforced by species diversity, is *negative feedback*, which in the present context refers to a situation in which a change in one or more populations induced by a modification in some environmental variable brings about a response in other populations in such a manner that the original biological fluctuation is opposed or damped. Because positive feedback, a condition in which the initial population response brings about biological modifications that increase the extent of change, is rare or unknown in microbial ecology, the term *feedback* is generally used to indicate negative feedback.

SURVIVAL OF ALIENS

A characteristic of many environments is their accessibility to propagules of species alien to that ecosystem, an invasion that can have enormous public health, agricultural, or economic importance. The behavior of the aliens and the response of the community illustrate elegantly the impact of self-regulatory mechanisms.

Occasionally a foreign organism may enter an already inhabited environment and become established. The establishment of an organism suggests that a niche that it could occupy was not already filled, that natural selection has not been sufficiently rigorous to favor the most efficient exploitation of the niche in question (Slobodkin, 1961), or that the change in the habitat opens up new opportunities for the alien. More frequently, however, propagules of an alien alighting on an already inhabited area die out at a reasonably rapid rate. The species already making up the community have been sorted out through natural selection and are peculiarly suited for that locality. Selection tends to preserve the status quo because the invader generally is less fit than the species that have survived the prior struggle for existence. Since the community reflects the habitat, only a modification of the ecosystem will allow for appreciable establishment of a foreign species in a climax community.

The basis for the ephemeral existence of aliens has attracted great interest. Although occasionally the mechanism of elimination of alien species is quite evident, far more often the available data are sparse, and the limited information is the subject of disagreement. Biological activities, nonbiological factors, or both may be involved, and some investigators have emphasized the biological destruction of invaders, while others have stressed the abiotic contribution. It should be no great suprise, moreover, that frequently not one but rather the interplay of several determinants is responsible for the suppression of propagules introduced into a climax community, with the role of each factor varying in prominence with time.

For obvious reasons, the survival of parasites apart from their hosts has attracted considerable attention. Few pathogenic bacteria, fungi, protozoa, or viruses endure for long periods away from their hosts, whether the environment is soil, water, sewage, air, or the surfaces of animate or inanimate objects. Soil, the repository of diseased tissues throughout history, receives human and animal pathogens in the form of excreta and dead bodies, yet rare is the pathogen entering the soil which persists there. Despite its vast assortment of residents, soil harbors surprisingly few of the animal and human pathogens it receives, the aliens soon dying off. Similarly, a high proportion of plant parasites retain their viability in soil for remarkably short periods. Typical curves showing the disappearance of aliens deliberately introduced into soil are presented in Fig. 9.1. Among the parasites and saprobes demonstrated to be destroyed in soil—often in a few days, sometimes in a few weeks, but at times only after several months—are the following: species of *Agrobacterium, Brucella, Corynebacterium, Erwinia, Escherichia, Mycobacterium, Pseudomonas, Salmonella, Streptococcus,* and *Xanthomonas,* among the bacteria; species of *Blastomyces, Fusarium, Helminthosporium, Ophiobolus,* and *Pythium,* of the fungi; and the rickettsia *Coxiella burnetii.* The time of persistence varies with the soil texture, moisture level, pH, temperature, and depth, and it also depends on the organism in question, some being notably nonhardy, others quite durable. The extended survival of numerous parasitic fungi and of *Bacillus anthracis,* however, is not unexpected, since they form resting structures in soil, while many parasites owe their longevity to the protection afforded by buried dead host tissues. Nevertheless, the latter do disappear from soil as the diseased tissues undergo decomposition. On

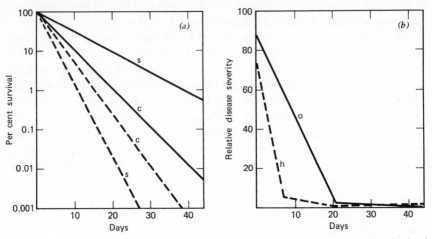

Figure 9.1 Survival in soil of bacteria and fungi. (a) Survival of coliform bacteria (c) and *Streptococcus* (s) of fecal origin. Dashed lines represent summer, and solid lines winter. (b) Survival of the plant pathogens *Ophiobolus graminis* (o) and *Helminthosporium sativum* (h). Survival is expressed in terms of relative disease severity in host plants seeded into the soil at various times after introduction of the fungi (Semeniuk and Henry, 1960; Van Donsel et al., 1967).

occasion, an animal or plant pathogen that is commonly considered to be rapidly destroyed below ground remains viable for astonishingly long periods, a tenacity of considerable concern, but this probably results from the localization of a few cells in microenvironments where they are protected from adverse conditions.

Enormous numbers of nonaquatic microorganisms enter the sea or inland waters with eroding land or with human and animal wastes. Countless cells move from lakes and reservoirs into streams, or from rivers into the sea. Increasing urbanization has fostered an ever-increasing local discharge of sewage, commonly inadequately treated, into streams, rivers, estuaries, and coastal waters, and often irrigation water, too, contains parasites capable of causing human or plant disease. These intrusions of alien microorganisms into aquatic environments are almost invariably followed by a rapid decline in the number of invaders, occasionally following a brief period of replication. Despite the rapid death rate of human pathogens in fresh and sea water, this water may still bear viable cells sufficiently long to pose a public health menace. With time, however, the so-called self-purification makes the polluted waters safe to drink.

Many studies have been conducted to assess the influence of environmental conditions and to establish the reasons for the decline in rivers, oceans, reservoirs, and well water of bacteria derived from human and animal feces, for example, *Escherichia coli, Streptococcus faecalis, Salmonella typhosa, Shigella sonnei,* and *Vibrio comma.* To a lesser extent, the fate of freshwater species in the sea and of lake plankton in streams has been examined. Affecting the population decline of one species or another are the degree of oxygenation, organic matter level, temperature, pH, season of year, and kind and concentration of salt. The cell densities of specific groups decrease at wholly dissimilar rates, and controversy has arisen among public health scientists because of these differences, a controversy originating in the fact not only that bacteria employed as indicators of the course of water purification, such as *E. coli* or *S. faecalis,* are generally destroyed at different rates, but that often the loss of viability by the indicator bacterium does not parallel that of pathogens of concern, like the enteric viruses.

Recent years have witnessed an upsurge in interest in attempts to ascertain the causes for the disappearance of aliens from water. The abiotic factors proposed to account for this decline include the settling out of cells, adsorption to particulate matter and subsequent death, inactivation by solar radiation, the unavailability of carbon sources usable by the foreigner, or conditions of temperature, salinity, aeration, pH, or E_h inimical to the invader. Among the possible biotic factors cited are the presence of biologically formed toxins, the action of predatory protozoa and lower metazoa, attack by bacteriophage or other parasites, and the inability of the foreigner to compete effectively with the indigenous community for a limited store of nutrients. In view of the diversity of organisms entering natural bodies of water, no single explanation is likely; yet even for the most extensively studied alien, *E. coli,* no one factor has been shown to be the sole or dominant reason for the decline (Pramer et al., 1963; Mitchell, 1968).

Ecosystems of a variety of types, in addition to the aquatic milieu, receive continual inputs of wide assortments of microorganisms, and though these locations can support life, as indicated by the heterogeneous and commonly large communities they contain, the foreigners fail to flourish. The community in each instance maintains its integrity, and the aliens are rejected. Sewage, for example, acquires hordes of organisms from feces, soil, water, and urban runoff; yet it is the exception, if any, that survives for long, and this highly specialized sewage biota has few of the human enteric pathogens or of the many saprobes that are always being introduced (Rudolfs et al., 1950). The eating habits of man and beast cause them to swallow vast numbers of heterotrophs; yet the intestinal tract's communities, except during overt disease, rarely accept the newcomers as functional inhabitants, most of the transients soon dying out. Experimental trials have shown in like fashion that bacteria and yeasts not indigenous to the bovine rumen are rapidly eliminated after their introduction (Adams et al., 1966), results analogous to those obtained when the bovine animal consumes feed and forage harboring a heterogeneous assemblage of bacteria, fungi, and yeasts. By its perennial contacts with contaminated surfaces of animate and inanimate objects and its exposure to the atmospheric biota, the skin receives a veritable flood of organisms, but the indigenous flora persists in the face of all these potential colonists, and the alien cells vanish. Even when large populations of strains of *Staphylococcus, Streptococcus, Serratia, Proteus, Pseudomonas, Klebsiella,* or *Escherichia* are applied to the skin, they fail to maintain themselves in appreciable densities, and few become established for any length of time.

Microbiologists, medical researchers, plant pathologists, nutritionists, agriculturalists, and laymen have deliberately inoculated one or another species, presumed or occasionally proven to be beneficial in vitro, into a natural and already well-populated habitat in order to alter the existing community so as to bring about some desirable process or to eliminate an unwanted population. Among the failures in attempts to alter a heterogeneous community by direct inoculation are the following: (a) application to soil or seed of N_2-fixing *Azotobacter* to overcome the paucity of these bacteria and satisfy the need of crops for appreciable nitrogen; (b) addition to soil or seed of strains of *Bacillus megaterium* to convert the organic phosphorus of soil to inorganic phosphate, which is directly assimilable by plants; (c) introduction of antibiotic-synthesizing or lytic microorganisms into soil to control soil-borne plant pathogens; (d) feeding of milk containing *Lactobacillus* in order to alter man's intestinal flora; (e) introduction into the mouth of species that, at least in vitro, inhibit lactic acid-producing bacteria; and (f) inoculation of cucumber and cabbage fermentations in the making of pickles and sauerkraut. Scattered successes have been reported, but they are far less common than the failures, and many of the alleged successes are questionable, transitory, or of marginal value. The added culture soon fades away as the cells lose in the struggle for existence with the entrenched inhabitants. Indeed, failure would have been predicted for many of the attempts, because microorganisms are generally cosmopolitan, and natural inoculation undoubtedly occurs repeatedly, so that the scarcity or absence of the species whose presence is sought reflects biotic or abiotic conditions hostile to its existence at that place and time. Microbial ubiquity dictates frequent, nondeliberate inoculations of many organisms, and it is the physical and chemical characteristics of the habitat, together with the specific community arising because of these abiotic properties, rather than the failure of a propagule to alight that regulates the species composition of a large proportion of habitats.

Nevertheless, inoculation with certain organisms accomplishes the desired end. Thus occasional plant pathogens inoculated into soil indeed cause disease, *Rhizobium* added to seeds of leguminous plants do invade the developing roots and then incite nodule formation, and inoculated strains of N_2-fixing blue-green algae such as *Tolypothrix, Nostoc,* and *Anabaena* unquestionably become established in paddy fields and apparently fix N_2 or increase rice yield (Fig. 9.2). The organism introduced in these instances, however, occupies an unfilled niche or is colonizing a habitat that is newly available for microbial growth, and organisms that can assume the identical niche or antagonists may be scarce.

COMPONENTS OF HOMEOSTASIS

Abundant information has now been collected to demonstrate that homeostatic mechanisms are critical in ascertaining the biological composition of different environments and in controlling the capacity of communities to regulate their own constitution. Evidence for this self-regulation has been obtained from studies in which life is totally or partially eliminated through a treatment that destroys all or most of the inhabitants of the site while having minimal effects on the physical or chemical properties of the habitat. Although innumerable aliens still are unable to proliferate despite the scarcity or absence of other microorganisms, presumably because suitable nutrients are lacking or abiotic factors are not conducive to their proliferation in the treated environment, a large number multiply vigorously. Many saprobic and parasitic bacteria, fungi, and actinomycetes added to sterile soil, for example, develop profusely; the same populations introduced into natural soil may persist briefly, but rarely do they replicate, and with time all individuals except possibly resting stages are generally eliminated. Similarly, aseptically cultured higher plants, when exposed to selected fungi, come down abruptly with disease, whereas the same microorganisms applied to plants

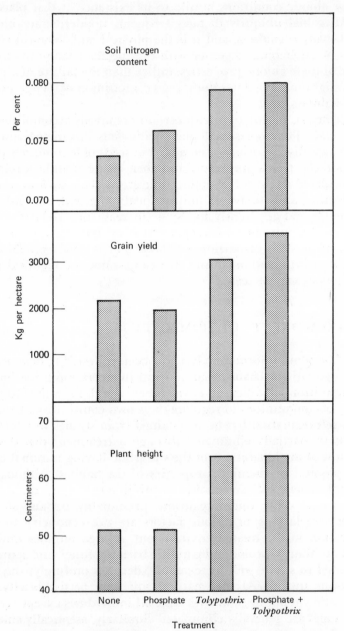

Figure 9.2 Effect of inoculation of soil with *Tolypothrix tenuis* on rice growth and soil nitrogen levels (Jha et al., 1963).

grown in identical but nonsterile circumstances have a much less deleterious influence or do no harm at all. Because of the marked protective effect of the natural community dwelling in or on plants or animals, the requirements of Koch's postulates often cannot be fulfilled if the parasite is applied to plants in a location teeming with nonparasites, while, on the contrary, no difficulties are encountered in establishing that a pathogen can do injury and the postulates are fulfilled when the parasite is inoculated into sterile tissue or the host is maintained free of microorganisms before meeting the parasite. Self-regulation is evident from studies of germfree animals, as aseptically grown individuals succumb to bacteria like *Escherichia coli* and *Bacillus subtilis*, which are not discernibly detrimental when inoculated into animals grown under normal circumstances (Luckey, 1963). Communities in food products likewise seem to govern their own composition; if the flora is large and active, bacteria like the staphylococci responsible for food poisoning may not become established, at least not to an extent sufficient to produce hazardous levels of their specific toxins (Casman, 1965).

Various lines of evidence, therefore, indicate that established and heterogeneous communities react in such a way as to reject newly introduced propagules or to hold their abundance in check. It is more likely that all the potential niches are occupied in the complex than in the simple community, and if natural selection has been intense, these niches are filled by populations efficient in performing the necessary function so that the potential invaders find it difficult to gain a foothold (Slobodkin, 1961). The biological bases for the rejection of aliens nevertheless require definition and deserve attention, but the processes involved undoubtedly vary from ecosystem to ecosystem and differ according to the interacting populations. Rivalry for a limiting nutrient or excretion of a toxic compound may rid an environment of an alien or hold down an indigenous population. Fungi are often seen to be destroyed by substances produced by cells active in the lysis of hyphae or spores. Predatory protozoa and parasitic bacteria, fungi, and viruses are widespread, each destructive in its own way to a limited spectrum of organisms. Beneficial interactions of a variety or sorts are ubiquitous also, mutual aid accounting for the frequency of many fastidious microorganisms.

Several schemes have been proposed for delineating the types of biological interactions. None is universally accepted, and all suffer from the disadvantage of creating more or less firm lines between the sorts of interrelationships of consequence in vivo. Nature recognizes few rigid boundaries. Nevertheless, a degree of categorization is necessary, if only to facilitate description and to discern similarities and dissimilarities in the behavioral patterns of microbial populations. Singling out interactions for separate consideration has the virtue of allowing for the definition of components of the whole, but the complex of relationships must always be borne in mind, particularly inasmuch as a single population may be affected simultaneously and, indeed, may rely for its existence on several kinds of associations.

Populations occasionally reside together in harmony, neither harming nor benefiting their neighbors; *neutralism* denotes the absence of interaction. *Commensalism* is quite common; this term describes an interaction providing benefit to one but having neither a detrimental nor a beneficial effect on the second of two interacting populations; in the absence of a suitable partner, however, the former may suffer because of its reliance on its associate. *Protocooperation* refers to an association mutually advantageous to the two cooperators; this interaction may be essential for the existence of the organisms in the particular ecosystem or for the operation of some metabolic process, but frequently the species contributing to this loose partnership change. The term *symbiosis*, or *mutualism*, describes an association involving two different species having an obligate reliance on each other for life in the ecosystem or for the performance of some reaction. In symbiosis, as in protocooperation, the relationship is beneficial to the two populations concerned.

Harmful as well as beneficial associations contribute to homeostasis. The term *antagonism*, or *interference*, is generally used for interspecific interactions in which one of the interacting species suffers because of the activities of its neighbors. Competition, amensalism, parasitism, and predation are categories of antagonism. In *competition* the two species are in rivalry for a limiting factor in the environment—a nutrient, for example—and though both are to some extent adversely affected, one commonly is less vigorous and loses out in the struggle. In *amensalism* one population is suppressed because of toxins synthesized by a second; the toxin producer is not directly benefited, but it may gain indirectly by virtue of the elimination of a potential competitor for a limited nutrient supply. *Parasitism* and *predation* involve a direct attack and feeding by one organism on another; the parasite or predator profits thereby and often depends for its existence on the association, while the host or prey population is harmed and possibly eliminated. If 0 is taken to indicate no effect, + a benefit, and − some degree of harm, a simple scheme can be established to characterize these interactions (Table 9.1).

Table 9.1

Types of Interactions between Two Species [a]

| Interaction | Effect on Growth and Survival of | |
	Species A	Species B
Neutralism	0	0
Commensalism	+	0
Protocooperation	+	+
Symbiosis	+	+
Competition	−	−
Amensalism	−	0
Parasitism	+	−
Predation	+	−

[a] From Odum (1959).

Intraspecific relationships—that is, between individuals of the same species—are of ecological consequence, particularly in regard to competition for an inadequate supply of nutrients or possibly limited space. Yet little is known of intraspecific interactions of microorganisms in nature. To date most interest has centered on characterizing interspecific associations, those involving individuals of different species.

ECOLOGICAL UPSETS

Although self-regulatory mechanisms operate to maintain the composition of the community under reasonably stable conditions, under certain circumstances an *ecological upset* or *ecological explosion* occurs, and the abundance of a single species rises dramatically (Elton, 1958). These population changes may take place abruptly, or the density of a species may slowly rise and just as slowly decline. Such biological modifications seem to suggest an escape from the homeostatic processes normally controlling the abundance of the particular population, but in fact the upset usually comes about because the environment or a force holding the organism in check has itself been perturbed in some manner.

With a physical or chemical alteration in the habitat or a variation in the behavior of a host serving as the domicile of a parasite, strange

or additional selective stresses are introduced and the original homeostasis is disturbed, so that new species come to the fore, displacing
those with less fitness. The imposed change may alter part of the community, with the appearance of new dominants, or all species in the
area may be affected to a significant extent. Among the stresses shown
to bring about ecological upsets of consequence are nutrients added
by man or through natural processes, antibiotics employed in chemotherapy, chemical or physical treatments leading to elimination of
large numbers of organisms, toxic pollutants introduced into water,
and variations in the physiology of animal or plant hosts.

A clear and unequivocal disturbance of the community is evident
upon the introduction of fresh or different microbial nutrients. Both
the types and numbers of microorganisms are typically affected. A
minor organism is frequently favored and multiplies in response to
the substrate, an occurrence routinely noted upon the addition of carbonaceous materials to water, soil, and sewage, or a modification in
the diet of the ruminants or young children. The introduction of a
plant variety or species into an area where it previously was unknown,
as is often done in agriculture or landscaping, may be regarded as
providing a new substrate for parasites. The genesis of certain kinds
of algal blooms, such as the so-called red tides or the occasional massive growths occurring in inland waters, offers a spectacular demonstration of the explosive development of a hitherto obscure population
in a sparse, reasonably stable community. The sudden appearance of
these algal masses often results from the upwelling of nutrient-rich
waters from deep ocean currents, an overturn in a lake in the spring
or autumn season, or the contamination of water with nutrients derived from terrestrial sources, the nutrients suddenly inserted into
the lighted zone in the surface water promoting the abrupt onset of
algal proliferation.

The use of antimicrobial agents in medicine, veterinary practice,
and agriculture or the release of toxic materials into flowing waters
constitutes an ecological perturbation that destroys the original homeostatic mechanisms and commonly establishes conditions suitable
for the dramatic rise in numbers of a species hitherto uncommon in
the locality. At times dire effects ensue. Such microfloral changes and
their consequences are well illustrated by the therapeutic use of antibiotics. The mouth, upper respiratory tract, intestine, and skin are
inhabited by an assemblage of species living in harmony, but this
balance is destroyed when an antibiotic is administered. For example,
the *Lactobacillus* population of the intestine is eradicated in mice
treated with penicillin while the numbers of enterococci and gram-

negative enterobacilli rise markedly (Dubos et al., 1963), and the ubiquitous gram-positive bacteria situated on the skin are eliminated by treatment with neomycin plus aluminum chlorhydroxide, with a concomitant parallel proliferation of gram-negative bacteria (Shehadeh and Kligman, 1963). Similarly, local treatment of burn infections with antibacterial drugs sometimes alters the microflora in a manner that facilitates the development of fungi that would not normally be found were the bacteria not suppressed (Livingood et al., 1952), and prolonged oral administration of selected antibacterial drugs reduces substantially the density of distinctive bacterial groups in the intestinal tract, so that *Candida albicans* flourishes with little hindrance and causes moniliasis (Eisman et al., 1961). Antibiotics inhibiting the normal intestinal bacteria sometimes allow for the proliferation of strains of *Staphylococcus, Proteus,* and *Pseudomonas,* microorganisms that would not have been prominent in the absence of the chemical; such organisms in turn cause infections that probably would not have been evident in untreated patients.

The widespread and often indiscriminate administration of antibiotics has been accompanied by an increase in the prevalence and severity of clinical infections by fungi, yeasts, and bacterial genera not susceptible to the antibacterials employed. Many of the bacteria, such as *Escherichia,* the *Klebsiella-Aerobacter* group, *Alcaligenes, Proteus,* and *Pseudomonas,* are indigenous to man. Bacteremias caused by these organisms were rare and the microorganisms were not considered to be highly pathogenic before the antibiotic era. The situation is now totally different, for infections by these bacteria are quite common, an unfortunate outcome of the necessity for antibiotic therapy in human medicine (Fig. 9.3). The weight of available evidence suggests that administration of the antimicrobial agent creates a stress that destroys the antibiotic-sensitive populations previously residing in the mouth, respiratory tract, intestine, or on the skin, and a new biota emerges made up of drug-resistant microorganisms, some of which may be pathogenic. Members of this new microflora could not cope effectively with the homeostatic processes operating before the ecosystem was put under antibiotic stress, but they can multiply extensively when the original inhabitants of the site are suppressed or eliminated.

Chemical or heat treatment of soil for the purpose of eradicating plant pathogens rarely sterilizes the soil, and hence the stage is set for the explosive development of species previously held down by self-regulatory processes typical in the terrestrial community. Minor fungi and bacteria suddenly assume prominence. The application of chem-

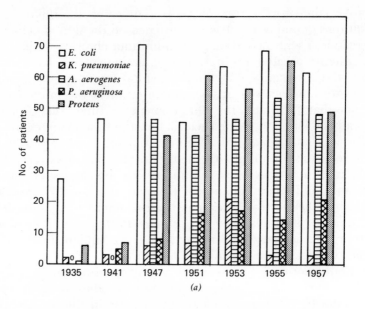

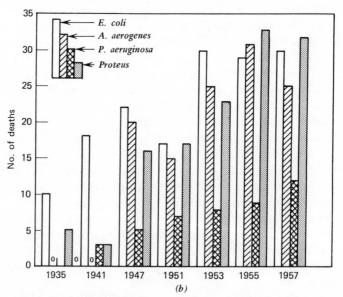

Figure 9.3 Organisms involved and deaths resulting from bacteremia in patients of Boston City Hospital before and following the introduction of antibiotic therapy. (*a*) Occurrence of gram-negative bacilli in blood cultures. (*b*) Deaths in patients with gram-negative bacilli in blood cultures (Finlay et al., 1959).

icals to lakes for the control of excessive algal growths and the inadvertent release of acids, phenolic materials, metallic ions, or other toxic wastes into rivers have analogous influences. The upset may be so extreme that in the early stages of recolonization a single or just a few fungal species in soil or algal species in lakes and rivers may be dominant.

An ecological explosion involving a parasite often results from a disturbance in the physiology of the host. Potential disease agents, for instance, reside indefinitely in thickly populated parts of the human body without doing harm. Once the host ecosystem is perturbed, the previously innocuous population multiplies, invades suitable tissues, and makes its insidious effects felt. Infections associated with perturbations in host physiology are known to occur after surgery, following X-radiation, or during corticosteroid therapy. There is little doubt that numerous pathogen flare-ups occur in direct response to a lowering of the body's defense mechanisms, but it is likely, too, that the cause is frequently a disturbance in the balance among the indigenous microbial populations.

Ecological explosions characterize situations in which a particular species only rarely encounters a given type of habitat it can colonize. The infrequent encounters may be a consequence of the organism's rarity or its poor dispersal. Countless micro- or macroenvironments are occupied by notably few species, at least for short periods or, alternatively, for the length of time that the habitat exists. As stated above, a fresh arrival has a good likelihood of becoming established when niches have yet to be exploited or are exploited inefficiently because of the lack of biological heterogeneity. Pathogens thus regularly find niches that have not yet been taken advantage of, in the sense that they utilize tissue constituents not readily available to existing residents. The occasional explosive outbreak of a disease suggests a nonuniversal distribution of the microbial parasite and rare encounters between the pathogen and its potential host or susceptible host tissues. Upsets of this sort occur repeatedly, as in the catastrophic outbreaks of typhoid fever among humans or of plant diseases caused by selected fungi.

References

REVIEWS

Elton, C. S. 1958. *The Ecology of Invasions by Animals and Plants*. Methuen, London.

Finlay, M., Jones, W. F., and Barnes, M. W. 1959. Occurrence of serious bacterial infections since introduction of antibacterial agents. *J. Am. Med. Assoc.*, **170**, 2188–2197.

Langley, L. L. 1965. *Homeostasis*. Reinhold, New York.

Mitchell, R. 1968. Factors affecting the decline of non-marine micro-organisms in seawater. *Water Res.*, **2**, 535–543.

OTHER LITERATURE CITED

Adams, J. C., Hartman, P. A., and Jacobson, N. L. 1966. *Can. J. Microbiol.*, **12**, 363–369.

Casman, E. P. 1965. *Ann. N.Y. Acad. Sci.*, **128**, 124–131.

Dubos, R., Schaedler, R. W., and Costello, R. 1963. *Federation Proc.*, **22**, 1322–1329.

Eisman, P. C., Weerts, J., Jaconia, D., and Barkulis, S. S. 1961. *In* P. Gray, B. Tabenkin, and S. G. Bradley, Eds., *Antimicrobial Agents Annual: 1960*. Plenum Press, New York. pp. 224–230.

Jha, K. K., Ali, M. A., Singh, R., and Bhattacharya, P. B. 1963. *J. Indian Soc. Soil Sci.* **13**, 161–166.

Livingood, C. S., Nilasena, S. N., King, W. C., Stevenson, R. A., and Mullins, J. F. 1952. *J. Am. Med. Assoc.*, **148**, 334–339.

Luckey, T. D. 1963. *Germfree Life and Gnotobiology*. Academic Press, New York.

Odum, E. P. 1959. *Fundamentals of Ecology*. Saunders, Philadelphia.

Pramer, D., Carlucci, A. F., and Scarpino, P. V. 1963. *In* C. H. Oppenheimer, Ed., *Symposium on Marine Microbiology*. Thomas, Springfield, Ill. pp. 567–571.

Rudolfs, W., Falk, L. L., and Ragotzkie, R. A. 1950. *Sewage Ind. Wastes*, 22, 1261–1281.

Semeniuk, G., and Henry, A. W. 1960. *Can. J. Plant Sci.*, 40, 288–294.

Shehadeh, N. H., and Kligman, A. M. 1963. *J. Invest. Dermatol.*, 40, 61–71.

Slobodkin, L. B. 1961. *Growth and Regulation of Animal Populations*. Holt, Rinehart and Winston, New York.

Van Donsel, D. J., Geldreich, E. E., and Clarke, N. A. 1967. *Appl. Microbiol.*, 15, 1362–1370.

10

Commensalism and Protocooperation

Despite the extremely close physical proximity of two species co-existing in space and time, they may fail to affect one another either beneficially or adversely. Neutralism is evident in vitro when the rates of growth and final densities of two populations are the same in mixed as in axenic culture. Data suggesting the occurrence of neutralism in laboratory media have been obtained in dual cultures involving protozoa (Burbanck and Williams, 1963), algae (Parker and Turner, 1961), and bacteria (Lewis, 1967). Demonstration of neutralism in vivo is far more difficult, and its frequency or significance in natural microbial communities has yet to be assessed. It is likely, however, that neutralism is of consequence when population densities are low, the food supply is abundant, and the requirements for growth of the two populations are sufficiently different that they do not interact for limiting nutrients or common needs. Species having dissimilar demands and behavioral patterns are less likely to interact than those with similar requirements and activities. During the early colonization of a habitat, when few cells are present relative to the capacity of the region to support life and at a time when the extent of possible toxin accumulation is negligible, interactions may be inconsequential. Still, when the environment is modified or when the supply of nutrients becomes low, adjacent species may begin to affect one another.

Beneficial relationships have been described for innumerable groups of microorganisms. The extent of benefit, the degree of reliance of one population on the second, and the mechanism involved in the association vary from organism to organism and are frequently altered by environmental conditions. For the sake of convenience,

224

commensalistic, protocooperative, and symbiotic relationships will be considered separately, although in fact lines of demarcation are sometimes far from distinct.

COMMENSALISM

Mutual relationships range from loose associations to closely knit ways of life. Because of their usual crowding in areas rich in nutrients, microbial populations are constantly interacting, and some of these interactions are unquestionably advantageous to the participants. In commensalism one of the partners profits by living together with a second species, but the latter receives neither good nor harm from the organism it favors. The microorganism deriving the benefit, the *commensal,* is provided with conditions helpful or essential for its well-being. The relationship between the two companions is frequently, but not necessarily, casual, in that a number of species can provide the commensal with the factors it requires, and there is often little specialization of either partner for its associate. The dependencies, whether loose or obligatory, lead to a degree of coordination among the residents of an ecosystem, and commensalism effectively provides the community with a definite integration in the activities and the organisms inhabiting a particular site.

Various kinds of commensalism involving microorganisms have been described. A few involve a close physical linkage between the organisms, while many require no direct contact between the individuals. The following commensalistic relationships have received the greatest attention:

(a) One population converts a substrate unavailable to a second population into a product that is assimilated and serves as a macronutrient for the latter.

(b) One species excretes a growth factor essential for the proliferation of its commensal.

(c) One destroys toxins or removes inhibitory factors from the environment, thereby allowing for multiplication of its associate. Destruction of organic or inorganic toxins, lowering or raising of the pH or E_h, removal of O_2, the creation of light-shading effects, reducing the osmotic pressure by microbial metabolism of sugar or salts, and changing the inorganic nutrient level to favorable concentrations are means by which the benefit is conferred.

(d) A macro- or a microorganism provides a surface that is particularly suitable for colonization by the commensal, a surface giving the partner a marked ecological advantage over free individuals of the same or different species. An association of this sort is often designated *phoresis,* the supporting associate being known as the host.

(e) One individual provides nutrients, protection, or shelter to another that is living within it, the commensal doing neither harm nor good to the organism in which it resides. Such relationships occasionally verge on parasitism.

The unilateral stimulation of an alga, bacterium, fungus, protozoan, or actinomycete in a two-membered association has been widely noted, in vitro at any rate. Medical microbiologists have extensively inquired into the changes brought about in mixtures of organisms because of the significance of heterogeneous communities in infection and the enhancement of virulence by nonpathogens. For example, the virulence of *Proteus vulgaris* to mice is markedly promoted by *Staphylococcus aureus* (Arndt and Ritts, 1961), and the abundance of *Borrelia vincentii* in the human mouth is increased by components of the mixed microflora characteristic of certain inflammatory processes (Nevin et al., 1960). The stimulation of pathogenesis is quite noteworthy with *Entamoeba histolytica;* when introduced into germfree guinea pigs, the protozoan does not cause amebic infections or lesions, whereas disease develops when *E. histolytica* is inoculated either into germfree animals together with several bacterial species or into nongermfree guinea pigs (Phillips, 1968). Such findings are not unexpected, because alleviation of this amebic disease sometimes is noted following the use of bactericidal drugs. Thus organisms seemingly harmless in axenic culture may reveal their pathogenicity when coexisting with a neighbor. Similar phenomena occur in plants as well as in animals, the resistance of plants to fungal infection being overcome, for instance, when the tissues encounter selected products of microbial metabolism (Linderman and Toussoun, 1968). A number of other activities that are more pronounced in two-membered than in axenic cultures are listed in Table 10.1.

Microorganisms representing a large collection of taxonomic categories are responsible for the appearance of organic and inorganic substances in natural habitats. These products either result from the activities of extracellular enzymes or are formed intracellularly and are excreted by the cell. Whether generated intra- or extracellularly, the metabolites are potent insofar as their ecological impact is concerned inasmuch as they become part of the milieu of adjacent populations. The growth and very survival of the latter may be governed

Table 10.1

Stimulation by One Population of an Activity of a Second

Process	Transformation		Reference
	Responsible Microorganism	Stimulator of Process	
Acid production	Streptococcus thermophilus	Candida albicans	Peppler and Frazier (1942)
Fruit infection	Diplodia natalensis	Penicillium digitatum	Reichert and Hellinger (1932)
Cellulose decomposition	Trichoderma sp.	Rhizopus sp.	Waksman and Hutchings (1937)
Phenolsulfatase production	Fusobacterium polymorphum	Bacteria	Schultz-Haudt and Scherp (1955)
Nitrogen fixation	Clostridium pasteurianum	Bacillus circulans	Emtsev (1960)
Toxin formation	Corynebacterium diphtheriae	Bacteriophage	Freeman (1951)

entirely by the availability of these compounds, many of which, from the viewpoint of the producer, are merely metabolic wastes (Lucas, 1961). Commensalistic relationships thereby arising between producer and recipient of metabolic products are ubiquitous.

Notwithstanding its reliance on neighbors to provide essential nutrients, the recipient is at no disadvantage so long as it coexists with its commensalistic partner. Indeed, the increased dependence may allow for greater efficiency in other aspects of cellular metabolism or greater fitness in ways not related to the nutritional dependency. Although the relation between the two commensalistic associates is generally fortuitous, an advantage gained by the commensal over nearby species in the habitat may be favored by natural selection, so that in time the commensal assumes greater prominence. Simultaneously, however, the factor or chemical elaborated by the producer assumes ever-increasing ecological importance as selection favors organisms relying on that substance (Dales, 1957).

An appreciable number of compounds serve as the basis for commensal relationships in nature. A few of the compounds are produced in large amounts and serve as energy, carbon, or nitrogen sources. Some are released by the cell in small quantities, but these traces suffice to satisfy the demand of the cohabitants, as with the vitamins, amino acids, purines, and pyrimidines. If the chemical in question is of the sort that is in constant demand as a nutrient, its concentration in the environment probably will be low, despite the fact that copious amounts may be formed. The concentration detected at any particular time represents the steady-state level between the rate of production and the rate of utilization of the nutrient by commensals as well as by other organisms, and, as stated in Chapter 5, there is a continuous turnover of nutrients of this kind in natural ecosystems.

Owing to their pioneering role in aquatic ecosystems, algae have been the focus of tremendous interest. Axenic cultures of greens, blue-greens, and diatoms, both marine and freshwater genera, liberate an array of water-soluble substances during proliferation in laboratory media. These products are not derived from autolytic breakdown or decomposition of dead cells by heterotrophs, but come from metabolically active cells. Among the compounds excreted in vitro are saturated and unsaturated fatty acids, amino acids, free sugars, sugar acids, high-molecular-weight polysaccharides, nucleic acids, and both simple and complex polypeptides. Glycolic acid is a major excretion of several species during active photosynthesis, and red and brown seaweeds generate large amounts of mucilaginous polysaccharides. The magnitude of the excretion depends on the organism, its age, and prevailing conditions, but it may be appreciable; less than 2% to more than 50% of the organic carbon synthesized during photosynthesis is known to be liberated extracellularly.

Extrapolation from laboratory data to the field is difficult in regard to either the kinds or amounts of organic compounds excreted. Yet both the colonization of algal surfaces by heterotrophs and analyses showing typical algal products in sea and fresh water supporting blooms of these organisms suggest that excretion by algae not only occurs but is of considerable consequence in nature. Moreover, evidence for excretion by natural, healthy phytoplankton populations has been obtained; in one region, for example, from 5 to 16% of the photoassimilated carbon was excreted into the water (Hellebust, 1967). Furthermore, microscopic examination reveals that bacteria develop alongside growing algae, sometimes adhering to the algal surfaces, sometimes proliferating in solution in the immediate vicinity of the photoautotrophs. Multiplication of the bacteria is sustained by the carbonaceous and/or nitrogenous products released by their nutritionally less demanding neighbors. Studies in controlled conditions have revealed that large numbers of bacteria can be supported by simple algal excretion products, whether the compounds are supplied in mixtures or individually, the bacteria consuming and getting

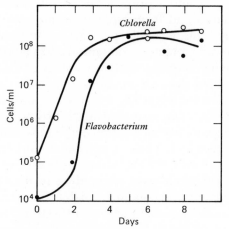

Figure 10.1 The commensal growth of a *Flavobacterium* sp. in an inorganic medium inoculated with *Chlorella pyrenoidosa*. Cultures were incubated in the light (Vela and Guerra, 1966).

energy from the algal metabolites (Fig. 10.1). The complex polysaccharides of seaweeds, too, are utilized by marine bacteria as carbon sources. The heterotrophs in turn may be able to generate and excrete trace substances necessary for the replication of aquatic auxotrophs. On account of the simultaneous production and utilization of organic molecules, the concentration of readily oxidizable carbonaceous substances in oxygenated waters is almost invariably low, and probably only the more resistant compounds will generally be found in significant levels.

Many bodies of water not only contain little organic carbon to support heterotrophs but are nitrogen-deficient as well. Certain bluegreen algae, however, are uniquely able to function in commensalistic relationships, since they excrete a variety of organic molecules containing nitrogen. These organisms assimilate N_2 during active development and liberate surprisingly large amounts of the N_2 fixed as simple, soluble compounds. In culture, from 5 to 60% of the N_2 assimilated may appear outside the cell. Part of the nitrogen is in the peptide form, and part appears as free amino acids (Stewart, 1963). Such algae include species of *Nostoc, Calothrix, Anabaena, Tolypothrix,* and *Anabaenopsis*.

A microorganism unable to use a given substrate may live commensally because it obtains assimilable products that the heterotrophic associate generates during the degradation of the substrate in question. Fungi, bacteria, and actinomycetes act on a fantastic assortment of organic molecules and form from them an impressive collection of sugars, organic acids, alcohols, peptides, amino acids, phosphorus- and sulfur-containing organic compounds, ammonium, phosphate, and so on. Nevertheless only a small percentage of the resident species generally metabolize the carbonaceous substrates entering an ecosystem, the rest developing at the expense of products generated by the microflora metabolizing the incoming substrates. Associations in which one cell grows by utilizing compounds synthesized by its neighbors are widespread. Thus in soil, water, and rumen and in the decomposition of plant remains, successive waves of different kinds of organisms appear, each wave responding to metabolites excreted by components of the previous microbial assemblage.

Such kinds of cooperation function in the transformations of carbon, nitrogen, phosphorus, and sulfur. By reason of the role of carbonaceous substances in providing energy to heterotrophs, carbon transformations have been explored most thoroughly. An excellent example is provided by the decomposition of plant polysaccharides. Heterotrophs unable to use cellulose and hemicelluloses proliferate in environments receiving these carbohydrates because they are preceded by and rely entirely on bacteria and fungi elaborating extracellular enzymes that hydrolyze the polymers, and the products of the hydrolysis support the polysaccharide utilizer and the nonutilizer simultaneously. Analogous biological relationships occur in the decomposition of proteins in nature, the amino acids formed in the degradation serving as energy sources for nonproteolytic cells. Similarly, fat-decomposing bacteria provide hydrolysis products that sustain heterotrophs incapable of oxidizing the lipids in the environment, and commensalism is also evident between the non-lactose-utilizing yeasts and the lactose-destroying bacteria inhabiting fermented milk. Much of the organic nitrogen and phosphorus in water is unavailable to algae, but they live as commensals and obtain the benefit from bacterial enzymes cleaving ammonium and phosphate from the organic complexes. Aquatic red and green photosynthetic sulfur bacteria are aided in a similar fashion when the sulfide they require is made by bacteria reducing sulfate to H_2S.

It is relatively easy to find multistage kinds of commensalism in which a commensal receiving benefit from its partner in turn creates a product that is an essential nutrient for succeeding species. For example, in the anaerobic decomposition of plant remains in swamps, the cellulose is attacked by bacteria that form large amounts of organic acids, which in turn are acted upon by and support the growth of methane producers. The methane escaping from the swamp passes through an oxygenated zone near the surface, in which unique aerobic bacteria proliferate at the expense of the methane. Multistage commensalism also characterizes steps in the nitrogen cycle; thus heterotrophs convert amino acids to the ammonium that *Nitrosomonas* needs as an energy source, and *Nitrosomonas* then generates the nitrite that is the sole energy source for *Nitrobacter,* which in its turn makes the nitrate that is an electron acceptor for denitrifying bacteria.

The unmasking of shielded or coated substrates by one species may be considered a form of commensalism, in that a second species gains as the previously inaccessible nutrient is exposed. Unmaskings undoubtedly occur frequently during the selective decomposition by microorganisms of individual structural components of plant residues. Similarly, the rotting of fruit by fungi effects a collapse of the structural integrity of the tissues, and yeasts and other organisms incapable of colonizing the intact fruit thereby are exposed to an environment more conducive to their replication.

Commensalism involving the excretion by one population of a growth factor needed by a second has been repeatedly demonstrated in vitro. The observations are of several sorts. (a) On nutritionally inadequate agar media, colonies of an organism of interest appear only in the vicinity of colonies of a physiologically dissimilar species, the latter synthesizing and excreting the nutrient needed by the former. Such relationships are often noted when samples from nature are plated in order to isolate a particular organism, which is found as a satellite colony adjacent to unwanted contaminants, or when auxotrophic mutants are found developing on agar near parent strains that excrete the diffusible nutrient the auxotroph is incapable of synthesizing. (b) A fastidious species thrives in mixed culture in liquid media but cannot be freed from contaminants because the solution is nutritionally deficient insofar as the fastidious cells are concerned, and the contaminants have the useful though exasperating function of supplying the missing growth factor. (c) Some cultures are maintained in the laboratory only if grown together with another species, such two-membered cultures still being essential for the propagation in vitro of a variety of heterotrophs, notably some protozoa. (d) Occasional cultures multiply more rapidly or attain higher cell densities if provided with the filtrate of a liquid medium previously supporting another organism than when fresh medium is employed. (e) Specific large marine algae, like *Ulva lactuca,* do not grow normally in bacteria-free circumstances, as the bacteria appar-

Table 10.2

Microbial Excretions Supporting the Proliferation of Commensals in Vitro

Compound	Species Requiring or Stimulated by Compound	Producer of Compound	Reference
Acetyl phosphate	Borrelia vincentii	Diphtheroid	Nevin et al. (1960)
Nicotinic acid	Proteus vulgaris	Saccharomyces cerevisiae	Shindala et al. (1965)
Purine	Lactobacillus	Saccharomyces cerevisiae	Challinor and Rose (1954)
RNA	Tetrahymena	Colpidium campylum	Stillwell (1967)
Thiamine	Phytophthora cryptogea	Bacterium	Erwin and Katz-nelson (1961)
Vitamin K	Bacteroides melaninogenicus	Staphylococcus aureus	Gibbons and Macdonald (1960)

ently generate compounds essential for the natural pattern of algal development.

The identities of several of the growth factors involved have been established, and representative results showing the kinds of compounds serving as the basis for commensalism are shown in Table 10.2. Bacteria, actinomycetes, algae, fungi, and protozoa from a broad range of genera and isolates derived from a wide cross-section of habitats are now known to excrete purines, organic acids, amino acids, and B vitamins in vitro. Individual species and even strains of the same species vary tremendously in the quantity of growth factors they excrete. The magnitude of the release is almost invariably governed by the age of the culture and growth conditions. Moreover, strains may excrete one, several, or an appreciable number of these compounds.

The profusion of growth factors probably accounts for the ability of large numbers of heterotrophs and photoautotrophs of differing degrees of fastidiousness to develop in natural bodies of water, soil, and around roots. Although autolysis, microbial decomposition of dead cells, and the release of essential metabolites from living hosts or their decaying tissues may contribute to the regeneration of growth factors that have been assimilated by auxotrophs, probably so too does the excretion of such substances by viable microbial cells, an excretion that underlies the commensalism between producer and consumer of the metabolite. The potential for such a relationship involving cellular excretions is common to residents of diverse environments, and indeed it has been deemed ecologically significant in soil, fresh water, the sea, the human mouth, and about plant roots. Nevertheless, the fact that laboratory studies demonstrate the ubiquity of growth factor-excreting populations is no guarantee that they have a role in nature. Still, in regions where organic matter is generated largely by microbial biosynthetic reactions, as in freshwater or marine ecosystems, the multiplication of auxotrophs suggests the importance of a tie between producer and consumer of growth factors. Commensalistic associations involving growth factors are probably loose and fortuitous as a rule, and the commensal need not have a specific or unique partner but rather only a companion capable of liberating the required metabolite in sufficient yield to support the commensal while at the same time producing no compounds harmful to it.

Heterotrophs excreting growth factors are abundant and ubiquitous. For example, bacteria synthesizing and excreting biotin, thiamine, nicotinic acid, and vitamin B_{12} constitute a significant portion of the marine community (Burkholder, 1963), and a high percentage of the soil and root-surface bacteria excrete, in culture media at least, the same four vitamins as well as riboflavin, pantothenic acid, folic acid, pyridoxine, and amino acids (Payne et al., 1957; Rouatt, 1967). The isolates frequently set free more than a single factor. Similarly, bacteria excreting stimulatory compounds are abundant in saliva (Scrivener et al., 1950). Though bacteria have received the greatest scrutiny, other microbial groups undoubtedly contain sizable numbers of individuals active in growth factor excretion, but systematic investigations of these groups are largely lacking.

One population may affect another by forming extracellular chelating agents that keep trace elements available or make insoluble nutrients more readily assimilable. Microbiologically generated chelators are known to promote algal growth in laboratory media. In nature the availability of micronutrients is frequently low despite the abundance of the element in question, and chelating agents elaborated by one species could thus enhance the development of its neighbors. Particularly attractive as ecologically important chelators are the slowly degradable organic compounds, which, because of their persistence, may have a prolonged influence. However, the contention that commensalism results from a mechanism involving the excretion of chelating agents affecting microbial nutrition is not supported by a body of direct evidence.

Destruction of toxins and elimination of deleterious factors in the environment by resistant individuals can aid adjacent sensitive organisms. The ability to destroy organic inhibitors formed by residents of the microbial community or by higher plants and animals is widespread. The highly toxic H_2S emanating from decomposing carbonaceous materials or sulfur springs is detoxified as it is oxidized in water by photosynthetic sulfur bacteria. Yeasts tolerant of high osmotic pressure metabolize the sugar added as a preservative to food, and they thereby facilitate the establishment of groups sensitive to high osmotic pressure. Food preservatives like benzoate and SO_2 are likewise destroyed microbiologically, and the effectiveness of $CuSO_4$ employed to control aquatic algae is reduced by the microbial formation of complexing agents, activities that allow sensitive populations to gain the ascendancy. A variety of heterotrophs cleave H_2S from sulfur-containing organic compounds and *Desulfovibrio* forms H_2S from sulfate, and the sulfide evolved in turn ties up mercury-containing germicides and permits bacterial proliferation (Bachenheimer and Bennett, 1961; Stutzenberger and Bennett, 1965). By assorted mechanisms, drug-tolerant strains permit the growth or enhance the resistance of antibiotic-sensitive organisms, a commensalism that may be of considerable moment in infections involving two or more populations; enzymes destroying antibiotics used in chemotherapy or the production of metabolites antagonizing the action of drugs are some of the protective devices responsible for this sort of commensalism (Cornforth and James, 1956; Sanders et al., 1962). A change in light intensity or quality resulting from algal blooms near the surfaces of lakes benefits underlying photoautotrophs, and raising the E_h by algal metabolism favors strict aerobes. Conversely, the removal of O_2 and lowering of the E_h by metabolic activities of facultative anaerobes benefit O_2-sensitive and strictly anaerobic populations; this type of commensalism takes place in soil, water, sewage, silage, fermented foods, infected tissues, deep wounds, and dental plaques, and it is the basis for the proliferation of many O_2-intolerant bacteria and protozoa, including pathogens, in habitats that would not support them were it not for the presence of helpful neighbors.

One population may alter the tolerance range of a second to temperature. For example, proliferation of a mesophilic *Bacillus* at 65°C is made possible when the bacterium is cultured together with a thermophile (Oates et al., 1963), and *Arthrobacter citreus* grows at 37°C in the presence of any of a variety of fungi or bacteria but not in pure culture (Chan and Johnson, 1966). The heat resistance of some microorganisms is also enhanced by accompanying organisms. The reasons for these broadenings of tolerance range remain nebulous.

A striking and repeatedly observed feature of the joint culture of two species is the morphological alteration induced by metabolic products of a partner in the interspecific association. Among the morphogenetic processes affected are (a) endospore formation by *Bacillus* and *Clostridium*, (b) sporulation of yeasts, (c) development of aerial mycelium by *Streptomyces*, (d) excystment of protozoan cysts, (e) germination of spores of the fungal genera *Agaricus, Entomophthora,* and *Tilletia,* and (f) formation of perithecia, appressoria, chlamydospores, sclerotia, rhizomorphs, sporangia, oospores, zoospores, and conidia by an array of fungi. Sometimes the morphological change occurs in axenic culture, but it is either slow or rare; with other organisms responding morphologically to the presence of dissimilar species, the structure does not appear in axenic culture but it is formed when a suitable associate is nearby. The partners responsible for promoting the structural response include strains of *Chlamydomonas* and *Streptomyces*, as well as numerous bacteria and fungi. Yet not all isolates tested bring about a morphological modification in their cultural companions, and the degree or extent of the change varies among the active populations. Frequently stimulation is first detected in agar at places where two colonies come close together, but often the response is seen initially in liquid media. Hypotheses to account for these phenomena include a reduction in the concentration of selected nutrients, depletion of O_2, accumulation of inhibitors, or the production of substances specifically triggering the morphogenetic sequence. In a rare few instances the compound— or a substance having an equivalent effect—causing or abetting the structural alteration in the neighboring population has been identified; organic acids, alcohols, hydrocarbons, B vitamins, CO_2, and organic phosphates have so far been implicated.

Commensalism requires that one of the companions benefit, and it is not difficult to postulate how a species responding to its neighbor's activities by forming structures having a reproductive, dispersal, or survival function gains an advantage. Were the commensal induced to enter into a resting stage when nutrients are depleted or when inhibitors accumulate, it could survive whereas the unmodified cells or filaments might not. Should germination of the resting stage be enhanced when conditions improve, moreover, the chances for the species to colonize the habitat successfully are greatly increased. By favoring the formation of structures concerned in dispersal, the partner may facilitate the commensal's establishment at a new site. Although data showing that these phenomena are prominent in nature are meager, it seems quite likely that the germination of resistant structures, the initiation of sporulation, the number of fungal spores formed, and stages of vegetative growth are affected by microbial metabolites in vivo as well as in vitro.

ECTOCOMMENSALISM AND ENDOCOMMENSALISM

Not a few species live as *ectocommensals* on the exterior surfaces of higher plants, animals, or microorganisms, the unique location usually being highly advantageous. On occasion the microscopic commensals possess peculiar structures that facilitate their remaining in place, though many microorganisms have no morphological features especially suiting them to their particular habitat. Many of these phoretic relationships are nonobligatory and, at best, are quite loose; little or no host specificity is evident. Conversely, certain ectocommensals are almost invariably linked with given host species, suggesting the existence of a degree of selectivity between the participants. A few ectocommensals probably use their host merely as a support and a base for growth, obtaining their foods from the surrounding fluid. Occasionally a host may shelter its commensal from deleterious factors in the environment or serve as a vehicle of transport. In other relationships, however, metabolic products of the host are unquestionably vital to sustenance of the adhering populations, influencing and determining the species composition of the dependent community, as with the rhizosphere and *phyllosphere* microfloras dwelling on the exteriors of roots and leaves, respectively.

Bacteria and fungi living on the surfaces of plants are known as *epiphytes*, whereas the protozoa residing on outer portions of animals are designated *epizoites*. The more general term *epibionts* is employed to include both epiphytes and epizoites. Many protozoa are epibionts, as in the case of the ciliates found on marine and freshwater animals. Often the protozoa are found on an aquatic animal in locations where they seem to benefit from the feeding or respiratory currents set up by the host, the microorganisms metabolizing the particulate matter brought to their immediate vicinity by the currents of water but obtaining no nutrients directly from the metazoan. Many

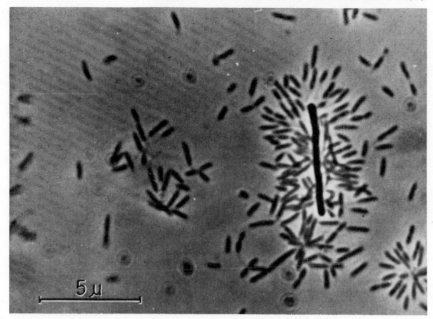

Figure 10.2 *Bacillus subtilis* cells surrounded by cells of *Caulobacter* sp. (Poindexter, 1964).

of these ectocommensals can have an independent existence in nature, but some are obligate commensals that do not survive apart from the host animal. This obligate dependency is nicely illustrated by *Opercularia,* which dies if detached from its freshwater animal host (Faure-Fremiet, 1906).

Microscopic examination of apparently viable fungal hyphae and algal filaments frequently reveals the presence of hordes of bacteria, and these may live as ectocommensals by using excretions of the filamentous hosts. Seaweeds are notable in having enormous numbers of bacteria on their surfaces, but even unicellular marine algae are accompanied by countless bacteria. An especially interesting ectocommensal is *Caulobacter;* not only is its host a microorganism, but it also has a specific means of attachment, one relying on an adhesive substance present at the end of the *Caulobacter* cell. This bacterium may thus become fixed to aquatic green and blue-green algae, diatoms, and unrelated bacteria (Fig. 10.2), from whose excretions it may acquire nutrients (Poindexter, 1964).

Microbial *endocommensals* reside in the gastrointestinal systems of marine animals and terrestrial vertebrates. The resident bacteria utilize a portion of the ingested foodstuff passing through the alimentary tract, while a high proportion of endocommensal protozoa probably consume the bacteria rather than the metazoan's nutrients. Some of these microorganisms are never found apart from their host and survive for very short intervals when separated from it. The gain to the microorganism by its presence in the unique habitat is obvious. Though at times the metazoan derives benefit as well, various populations in the gastrointestinal system have no known helpful or stimulatory influence on their host, and hence they may be considered commensals rather than symbionts. In truth, a degree of competition may occur between commensal and host for food, but the competition is rarely sufficiently severe for these microorganisms to be deemed parasites.

The relationship between the bacterial cell and the temperate bacteriophage it harbors is essentially a special case of endocommensalism. The association is stable, and the virus dwells in apparent harmony with the bacterium for long periods of time. The bacteriophage is transmitted to daughter bacteria during cell division, and only rarely does the virus cause lysis, as when a sensitive strain is made available. The frequency of occurrence of this relationship remains unknown, but nearly all members of a few categories of bacteria have been found to bear temperate viruses that maintain themselves in seeming perpetuity without destroying their hosts.

PROTOCOOPERATION

Microorganisms regularly participate in mutual relationships entailing the simultaneous cooperation of two species. Although in protocooperation the association is loose, nevertheless the existence of each partner in that locale requires the presence of a companion species, or of a population having an activity equivalent to that of the companion. The very stability of climax communities argues for the existence of interdependent relationships. In a heterogeneous community containing a multitude of potential nutrients, it is probably true that completely independent species sacrifice something in order to be self-reliant, whereas the development of a two-membered association that behaves as an independent unit likely allows each species to function in ways that it could not were it compelled to fend entirely for itself. The apparent frequency of protocooperative interactions argues for the view that these relationships are ecologically advantageous and that natural selection in many habitats favors organisms with protocooperative ties.

In protocooperation, which is sometimes termed nonobligatory mutualism, the two partners aid one another, and each is a critical component of the immediate environment of the other. The relationship is nonobligatory or nonspecific in the sense that the identities of the companions are not fixed; a species capable of effecting the particular change that is the basis for participation of one of the interactants can replace that partner. This contrasts with the legume-*Rhizobium* symbiosis, in which the symbionts are reasonably specific. Though the individuals acting jointly in protocooperation coexist under the prevailing conditions, they can and frequently do proliferate independently in different environments.

Since the early days of microbiology, it has been known that two species may together effect a change that neither could bring about alone or that would occur at a slower rate in axenic culture. The reaction might involve the synthesis of a product, the formation of a distinctive structure, or the induction of a pathological response. This phenomenon, *synergism,* frequently is the outcome of some form of protocooperation. Only selected examples need be cited to illustrate the variety of ways in which synergism can be expressed.

(a) Complex products are formed by two populations acting together but not singly, as in the synthesis of prodigiosin by two nonpigmented strains of *Serratia marcescens* (Siddiqui and Peterson, 1965) or the formation of a heat-stable antifungal toxin by a mixture of two fungi cultured together (Redmond and Cutter, 1951).

(b) Gaseous products are evolved in the fermentation of mono- or disaccharides by joint participation of two species, neither of which can generate gas from the test sugars in axenic culture (Castellani, 1953). Among the gases produced in this fashion are CO_2, H_2, and CH_4. The transformations leading to gas formation probably are performed by a sugar-decomposing, non-gas-producing culture and a gas producer that is inactive on the sugars. The latter uses excretions of the former; hence the volatile compound is liberated by the mixture of organisms. These associations represent, in fact, not protocooperation but rather commensalism.

(c) A polysaccharide like cellulose is degraded more rapidly in mixed than in pure cultures of cellulose-utilizing fungi, or it is destroyed when two species individually unable to decompose the polysaccharide are grown together (Oates et al., 1963; Rege, 1927).

(d) Sterilized plant residues are decomposed more quickly or fruit rots more rapidly when incubated with mixed inocula than with monocultures (Savastano and Fawcett, 1929; Waksman, 1931).

Table 10.3

Mortality of Mice Infected with Mixtures of *Staphylococcus aureus* and *Proteus vulgaris* [a]

	No. of Mice Dead/No. of Mice Infected	
S. aureus Strain	With P. vulgaris	Without P. vulgaris
Virulent strain 11096	13/15	0/10
Virulent strain 11462	9/15	0/10
Avirulent strain 11305	0/15	0/10
Avirulent strain 10827	0/15	0/10

[a] From Arndt and Ritts (1961).

(e) Mixed cultures fix and utilize N_2 in nitrogen-free solutions containing a carbon source unavailable to the N_2 fixer. The unavailable organic compound, however, is converted by one component of the microbial assemblage, itself unable to assimilate N_2, to a product that is suitable for the N_2 user. The latter, in turn, makes nitrogen compounds that sustain its partners. If the initial carbonaceous substrate is a polysaccharide or CO_2, a polysaccharide-utilizing heterotroph or an autotroph is necessary (Rubenchik, 1963).

(f) Harmless organisms, through their cooperative efforts, produce disease, or two weakly virulent pathogens do more harm than either does alone (Table 10.3). Among the animal pathogens, such synergistic actions may involve either a virus and a bacterium or two dissimilar bacteria, and it is now widely recognized that certain animal or human diseases are more prevalent than usual during epidemics caused by other pathogens, as with bacterial pneumonia and influenza (Arndt and Ritts, 1961; Loosli, 1968). Synergism is likewise evident among plant pathogens, be they viruses or fungi (DeVay, 1956; Kassanis, 1963). The extent to which these effects result from an influence of one microorganism on another or, alternatively, arise because of some structural or physiological modification in the host, predisposing it to a new infection, is difficult to assess.

(g) Mixtures of different bacteria found in dairy starter cultures are frequently more active in acid production than individual populations (Bautista et al., 1966).

(h) Two-membered associations involving strains of *Corynebacter-*

ium and *Chromobacterium* oxidize the manganous ion, but neither bacterium alone performs the oxidation (Bromfield, 1956).

The underlying causes of most described instances of protocooperation either have not been sought or have eluded the best efforts of the investigator. However, the basis for the interdependent association has unquestionably been unraveled in several dual-culture laboratory models or, conversely, the data strongly suggest the reason for the mutual dependency in populations of natural communities. At the present time the kinds of protocooperation for which there are documented explanations can be divided into four categories: (a) one associate provides an energy source to its partner, and the latter supplies the first with some other essential nutrient; (b) an aerobic heterotroph produces CO_2 for an alga, which in its turn evolves the O_2 necessary for the heterotroph; (c) each partner excretes a growth factor without which its associate cannot live; and (d) one population destroys a toxin suppressing its neighbor, and the species that is freed of the inhibitor supplies a needed compound to the detoxifier. Brief illustrations of each category are presented below.

In a variety of protocooperative relations, one partner—either a photosynthetic organism or a heterotroph—supplies an energy source to the species furnishing it with an essential nutrient. This seems to occur in the two-membered association of the sort cited above between a N_2-fixing bacterium such as *Azotobacter,* which uses solely simple organic compounds, and a species that cannot use N_2 but can convert an organic molecule unavailable to *Azotobacter* into an assimilable form. Another common protocooperation of this kind is the relationship between a photoauxotroph and bacteria growing epiphytically on its surface or immediately nearby; the former assimilates CO_2 and releases organic molecules that the latter must have, and the heterotrophs in turn excrete growth factors required by the photosynthetic cells. This is probably what occurs between certain marine algae and bacteria residing on their surfaces, the bacteria liberating a vitamin that is essential for the algae. There need not be a physical tie between the interactants, but the organisms must be sufficiently close to allow the respective metabolites to reach the partner before they are destroyed biologically. Protocooperation of this sort may also account for the persistence of bacteria in cultures of auxotrophic marine algae.

A different kind of protocooperation involves mutual exchanges of gases. In sewage oxidation ponds receiving large amounts of carbonaceous wastes, heterotrophs decompose the organic matter and release CO_2 that the associated algae need, while the algae contribute to their partners the O_2, generated in photosynthesis, that the hetero-trophic aerobes rely on. The net effect is that the organic carbon coming into the pond is decomposed at a faster rate than if no algae were present. This explanation for the enhanced degradation of or-ganic matter is supported by experimental findings showing that *Chlorella*, a dominant genus inhabiting oxidation ponds receiving domestic sewage, is incapable of efficiently utilizing components of the sewage and that indigenous heterotrophs are responsible for the deg-radation of the organic materials. The alga develops better in the sewage when bacteria are present, a stimulation of *Chlorella* that can be effected just as well if the photoautotroph is provided with sup-plemental CO_2 (Provasoli, 1961).

Unquestionably the most thoroughly investigated category of proto-cooperation is that entailing a bilateral exchange of growth factors. As usually observed in vitro, two auxotrophic species are found to be incapable of growing in deficient media in axenic culture, but multi-plication proceeds in two-membered cultures in the same medium because each supplies the other's needs for a particular vitamin or amino acid. Such mutual feeding by dissimilar auxotrophs is termed *syntrophism*, a relationship in which two or possibly more popula-tions are able to develop in nutrient-deficient circumstances not suit-able for the proliferation, or allowing for poor development at best, of either. In some laboratory experiments the two populations, though in the same deficient medium, are physically separated by a dialysis membrane so that they are not in direct contact, yet they de-velop well as long as the partner is present in the culture vessel and is able to excrete the appropriate diffusible substances. The metab-olite being exchanged need not always be a growth factor itself, inasmuch as intermediates in growth factor biosynthesis occasionally suffice; for example, a bilateral stimulation of two thiamine-dependent marine bacteria occurs in thiamine-free media if each excretes intermediates in the biosynthesis of this vitamin (Burkholder, 1963).

The phenomenon of growth factor exchange with a resulting mu-tual benefit to auxotrophs is probably common in environments initi-ally containing little readily available organic nutrients. Such habitats ultimately become enriched with fastidious microorganisms, many of which are likely to obtain their growth factors from the metabolites liberated by nearby microbial populations. As shown in Table 10.4, syntrophism has been reported for both bacteria and fungi. Syntroph-ism is also typical of cultures containing yeasts and auxotrophic bacterial mutants. Amino acids, B vitamins, and precursors in their biosynthesis are implicated. Moreover, although a significant percent-

Table 10.4

Syntrophy between Microorganisms in Vitro

First Associate		Second Associate		
Identity [a]	Metabolite Excreted	Identity [a]	Metabolite Excreted	Reference
Lactobacillus plantarum	Folic acid	*Streptococcus faecalis*	Phenylalanine	Nurmikko (1956)
Marine bacterium	Riboflavin	Marine bacterium	Pantothenate	Burkholder (1963)
Polystictus adustus	Biotin	*Nematospora gossypii*	Thiamine	Koegl and Fries (1937)
Proteus vulgaris	Biotin	*Bacillus polymyxa*	Nicotinic acid	Yeoh et al. (1968)

[a] Auxotrophic for metabolite excreted by the partner.

age of the indigenous bacteria of soil, water, and the rhizosphere excrete growth factors, many need growth factors too; hence the extensive distribution of bacteria both exporting and importing growth factors suggests a widespread occurrence of syntrophy in nature.

Ubiquitous microorganisms generate appreciable quantities of organic autoinhibitors that prevent their further multiplication, and a second species that uses the toxin as an energy source not only itself acquires an advantage because of the organic nutrient it has available for oxidation but also provides a benefit to the inhibitor producer. The lactic acid bacteria in milk, as one instance, tend to poison themselves except where a lactic acid-utilizing yeast or related organism metabolizes the acid and prevents a drastic fall in pH. In addition, the accumulation of metabolic products during the decomposition of cellulose often could deter the further progress of degradation were it not for the intervention of non-cellulose decomposers; the detoxifying cells gain from the interaction in that they obtain a usable carbon source, while destruction of the polysaccharide and further proliferation of the cellulolytic organisms are promoted as the inhibitors are removed (Fahraeus, 1949). *Hemophilus pertussis* participates in a similar mutually beneficial interaction; this bacterium produces autoinhibitory long-chain fatty acids that are metabolized by a diphtheroid, which thereby relieves the inhibition of *H. pertussis* (Pollock, 1948).

References

REVIEWS

Dales, R. P. 1957. Interrelations of organisms. A. Commensalism. *In* J. W. Hedgpeth, Ed., *Treatise on Marine Ecology and Paleoecology,* Vol. 1. National Research Council, Washington, D.C. pp. 391–412.

DeVay, J. E. 1956. Mutual relationships in fungi. *Ann Rev. Microbiol.,* **10,** 115–140.

Hellebust, J. A. 1967. Excretion of organic compounds by cultured and natural populations of marine phytoplankton. *In* G. H. Lauff, Ed., *Estuaries.* American Association for the Advancement of Science, Washington, D.C. pp. 361–366.

Loosli, C. G. 1968. Synergism between respiratory viruses and bacteria. *Yale J. Biol. Med.,* **40,** 522–540.

Lucas, C. E. 1961. Interrelationships between aquatic organisms mediated by external metabolites. *In* M. Sears, Ed., *Oceanography.* American Association for the Advancement of Science, Washington, D.C. pp. 499–517.

Orenski, S. W. 1966. Intermicrobial symbiosis. *In* S. M. Henry, Ed., *Symbiosis,* Vol. 1. Academic Press, New York. pp. 1–33.

Robert, D. S. 1969. Synergic mechanisms in certain mixed infections. *J. Infect. Dis.,* **120,** 720–724.

OTHER LITERATURE CITED

Arndt, W. F., and Ritts, R. E. 1961. *Proc. Soc. Exptl. Biol. Med.,* **108,** 166–169.

Bachenheimer, A. G., and Bennett, E. O. 1961. *Antonie van Leeuwenhoek J. Microbiol. Serol.,* **27,** 180–188.

Bautista, E. S., Dahiya, R. S., and Speck, M. L. 1966. *J. Dairy Res.,* **33,** 299–307.

Bromfield, S. M. 1956. *Australian J. Biol. Sci.,* **9,** 238–252.

244

Burbanck, W. D., and Williams, D. B. 1963. *In* J. Ludvik, J. Lom, and J. Vavra, Eds., *Progress in Protozoology.* Czechoslovak Academy of Sciences, Prague. pp. 304–307.

Burkholder, P. R. 1963. *In* C. H. Oppenheimer. Ed., *Symposium on Marine Microbiology.* Thomas, Springfield, Ill. pp. 133–150.

Castellani, A. 1953. *Proc. 6th Intern. Congr. Microbiol.,* 1, 188–189.

Challinor, S. W., and Rose, A. H. 1954. *Nature,* 174, 877–878.

Chan, E. C. S., and Johnson, M. B. 1966. *Can. J. Microbiol.,* 12, 581–584.

Cornforth, J. W., and James, A. T. 1956. *Biochem. J.,* 63, 124–130.

Emtsev, V. T 1960. *Mikrobiologiya,* 29, 529–535.

Erwin, D. C., and Katznelson, H. 1961. *Can. J. Microbiol.,* 7, 945–950.

Fahraeus, G. 1949. *Lantbruks-Hogskol. Ann.,* 16, 159–166.

Faure-Fremiet, E. 1906. *Compt. Rend. Soc. Biol.,* 61, 514–515, 583–585.

Freeman, V. J. 1951. *J. Bacteriol.,* 61, 675–688.

Gibbons, R. J., and Macdonald, J. B. 1960. *J. Bacteriol.,* 80, 164–170.

Kassanis, B. 1963. *Advan. Virus Res.,* 10, 219–255.

Koegl, F., and Fries, N. 1937. *Z. Physiol. Chem.,* 249, 93–110.

Lewis, P. M. 1967. *J. Appl. Bacteriol.,* 30, 406–409.

Linderman, R. G., and Toussoun, T. A. 1968. *Phytopathology,* 58, 1431–1432.

Nevin, T. A., Hampp, E. G., and Duey, B. V. 1960. *J. Bacteriol.,* 80, 783–786.

Nurmikko, V. 1956. *Experientia,* 12, 245–249.

Oates, R. P., Beers, T. S., and Quinn, L. Y. 1963. *Bacteriol. Proc.,* 44.

Parker, B. C., and Turner, B. L. 1961. *Evolution,* 15, 228–238.

Payne, T. M. B., Rouatt, J. W., and Lochhead, A. G. 1957. *Can. J. Microbiol.,* 3, 73–80.

Peppler, H. J., and Frazier, W. C. 1942. *J. Bacteriol.,* 43, 181–191.

Phillips, B. P. 1968. *In* M. Miyakawa and T. D. Luckey, Eds., *Advances in Germfree Research and Gnotobiology.* Iliffe, London. pp. 279–286.

Poindexter, J. S. 1964. *Bacteriol. Rev.,* 28, 231–295.

Pollock, M. R. 1948. *J. Gen. Microbiol.,* 2, XXIII.

Provasoli, L. 1961. *In* Algae and Metropolitan Wastes. U.S. Public Health Service Publication SEC-TR-W61-3, Cincinnati. pp. 48–56.

Redmond, D. R., and Cutter, V. M. 1951. *Mycologia,* 43, 723–726.

Rege, R. D. 1927. *Ann. Appl. Biol.,* 14, 1–44.

Reichert, I., and Hellinger, E. 1932. *Hadar,* 5, 203–206.

Rouatt, J. W. 1967. *In* T. R. G. Gray and D. Parkinson, Eds., *The Ecology of Soil Bacteria.* Liverpool University Press, Liverpool. pp. 360–370.

Rubenchik, L. I. 1963. *Azotobacter and Its Use in Agriculture.* Israel Program for Scientific Translations, Jerusalem.

Sanders, A. C., Pelczar, M. J., and Hoefling, A. F. 1962. *Antibiot. Chemotherapy,* **12,** 10–16.

Savastano, G., and Fawcett, H. S. 1929. *J. Agr. Res.,* **39,** 163–198.

Schultz-Haudt, S. D., and Scherp, H. W. 1955. *J. Bacteriol.,* **69,** 665–671.

Scrivener, C. A., Myers, H. I., Moore, N. A., and Warner, B. W. 1950. *J. Dental Res.,* **29,** 784–790.

Shindala, A., Bungay, H. R., Krieg, N. R., and Culbert, K. 1965. *J. Bacteriol.,* **89,** 693–696.

Siddiqui, M. A. Q., and Peterson, G. E. 1965. *Antonie van Leeuwenhoek J. Microbiol. Serol.,* **31,** 193–202.

Stewart, W. D. P. 1963. *Nature,* **200,** 1020–1021.

Stillwell, R. H. 1967. *J. Protozool.,* **14,** 19–22.

Stutzenberger, F. J., and Bennett, E. O. 1965. *Appl. Microbiol.,* **13,** 570–574.

Vela, G. R., and Guerra, C. N. 1966. *J. Gen. Microbiol.,* **42,** 123–131.

Waksman, S. A. 1931. *Arch. Mikrobiol.,* **2,** 136–154.

Waksman, S. A., and Hutchings, I. J. 1937. *Soil Sci.,* **43,** 77–92.

Yeoh, H. T., Bungay, H. R., and Krieg, N. R. 1968. *Can. J. Microbiol.,* **14,** 491–492.

11

Symbiosis

Dissimilar species sometimes coexist intimately and exert mutually beneficial effects. To such associations the term symbiosis is given. Symbiosis denotes a reasonably long-lasting relationship in which two or occasionally more different species live in immediate proximity and derive reciprocal benefits from their interactions. Some biologists employ the word mutualism to refer to the close interspecific tie that provides advantage to both partners, using symbiosis instead to include all relationships involving dissimilar organisms that are in close proximity on a permanent or semipermanent basis, whether good or harm results; such a definition for symbiosis includes commensalism, protocooperation, and parasitism as well as the more restricted usage of the term symbiosis adopted here.

Some symbioses involve two entirely unrelated microorganisms. Others include a microorganism on the one hand and a higher plant or animal on the other; the microscopic component in these instances is designated the *microsymbiont,* and the larger organism, the *macrosymbiont.* Occasionally three organisms may be linked together, although one may be harmed, as in the case of the orchid that symbioses with a fungus that is in turn parasitic on a tree. Not uncommonly, one—or possibly both—of the associates is an obligate symbiont, always requiring its neighbor to maintain an active existence in nature. The obligate symbiont clearly gains enormously from the interaction, on which its very survival depends, but the extent of benefit may be grossly different for the two partners. One species may be able to maintain itself independently with little difficulty in certain habitats, whereas the second is totally reliant and is only found together with the first.

Each of the symbionts is thus a helpful or an essential part of the microenvironment of its partner. These microenvironments are at

247

times so tightly intermeshed that it is difficult, if not impossible, to determine where one organism ends and the second begins. Because of the close physical and physiological interrelations, moreover, a change in one symbiont more likely than not will have a significant and prompt influence on its companion. Various symbioses involving microorganisms are associated with distinct and easily recognizable structural entities, and little imagination is required to visualize them as separate and functional units. This morphological feature houses the symbiosis, and it is here that the functional interplay takes place, an interplay that usually has as its basis an exchange of metabolites.

A component of permanency is inherent in all symbioses in the sense that the association lasts for a major part of the life of the participating individuals. A component of specificity apparently is also an invariable characteristic, in that the choice of partner is not random. A particular symbiont is only capable of developing this kind of relationship with one given type of organism. This permanency and specificity serve to distinguish symbiosis from protocooperation. That the relationships are eminently successful is attested to by their abundance and ubiquity in freshwater and marine ecosystems, within the bodies of animals of diverse phyla, on leaves, on subterranean portions of plants, and in soils of tropical, temperate, and polar regions of the earth.

Symbiosis entails not only mutual aid but also, and probably quite commonly, a degree of mutual exploitation. Each symbiont may make use of facets of the environment, metabolism, or structure of the other, and though the net effect may be beneficial, a measure of parasitism seems to be at the heart of some of these relationships. The parasitism results from the struggle between the two partners for life. Indeed, at certain stages of the interaction or when environmental conditions change, the covert parasitism may be expressed overtly so that the condition of mutual gain terminates, and a parasite-host situation develops.

INTERMICROBIAL SYMBIOSIS

An outstanding illustration of a symbiosis composed of two wholly dissimilar microorganisms is the lichen (Fig. 11.1). In this association between an alga and a fungus, known as the *phycobiont* and *mycobiont,* respectively, many different species can participate. At the present time approximately 20 genera of green algae, almost a dozen

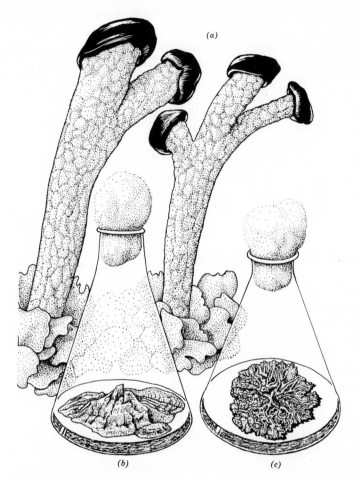

Figure 11.1   The lichen *Cladonia cristatella*. (a) Composite lichen. (b) Mycobiont in culture. (c) Phycobiont in culture (Ahmadjian, 1967a).

blue-greens, and a single representative of Xanthophyta, namely, *Heterococcus*, have been reported as lichen symbionts. The most common alga is *Trebouxia*, a green alga that sometimes is likewise found to be free-living, but *Trentepohlia* and *Nostoc* are reasonably abundant as well (Ahmadjian, 1967b; Hale, 1967). The mycobiont is an Ascomycete or, less frequently, a Basidiomycete. Occasionally two unrelated algae are involved in the lichen, although most lichens contain a single phycobiont and a single mycobiont species.

The external structure of the lichen bears little kinship to that of the isolated partners. The symbiotic unit has an independent existence, and because of the structural uniqueness, physiological distinctiveness, and ecological autonomy of the union, the practices of considering the lichen as a separate organism and classifying it as such are widely accepted. Moreover, in clearly unalike lichens, considered as different species, essentially the same type of *Trebouxia* serves as the phycobiont, emphasizing the distinctness of the symbiotic combination.

The mycobiont's hyphae and cells of the phycobiont exist in close proximity, the fungus comprising the bulk of the lichen protoplasmic material. The mycelium completely surrounds the algal cells, which either reside in a narrow region below the surface or are dispersed uniformly throughout the structure. The hyphae may in some instances penetrate the algal cells or, alternatively, be attached to them. Many authorities consider that the heterotroph actually parasitizes its photosynthetic mate, some of the algal cells being killed thereby but many still surviving. A contending view holds that there exists a bilateral parasitism, while a third school of opinion believes that the association is essentially harmonious.

Initiation of the symbiosis appears to occur under adverse circumstances, and, interestingly, conditions notably suitable for growth of the respective symbionts, such as environments rich in nutrients, tend to destroy the partnership. The predilection of lichens for adversity is evident in their distribution—their prominence in the antarctic flora, on forbidding surfaces near the North Pole, and on desert rocks exposed to intense sunlight. The essentiality of the symbiosis for the existence of the fungus and alga in various environments is manifest inasmuch as neither of the two can establish itself separately in most habitats where the association is notably prominent.

The lichens are notorious for their excruciatingly slow growth. For example, gross observations of antarctic lichens have failed to reveal any signs of spread during a 30-year period. This sluggish development may result from the rigorous conditions and inadequate nutrient levels prevailing at sites where lichens are conspicuous. Water is often limiting, and the temperature may be far from ideal. In addition, however, the symbionts individually may multiply slowly, and the fact that the net CO_2 fixation by the lichen is not rapid could contribute to its inability to proliferate quickly.

Microbial assemblages resembling primitive lichens are found in the sea. Some of these alga-fungus associations indeed appear to be lichens, but others possibly are more akin to relationships that are truly parasitic. Numerous genera of algae and fungi are implicated in these relationships (Johnson and Sparrow, 1961).

A second prominent type of intermicrobial symbiosis involves algae and protozoa. The photoautotrophs live and maintain themselves indefinitely as symbionts within the animal cells. Such symbioses have been described for several genera of ciliates and flagellates and for a large number of marine and freshwater rhizopods (McLaughlin and Zahl, 1966). A high percentage of these microscopic animals live in aquatic habitats in complexes with green- or brown-colored algae and, particularly when the photosynthetic partner makes up a large part of the protozoan cell, sometimes assume the algal color. Although few algal cells reside in some protozoa, the number of green algae in ciliates may range up to 3000 per cell, and they may occupy from 10 to 56% of the mass of the animal (Sund, 1968). In the dark the symbiosis may break down, and the algae are then digested by their former companions.

Paramecium bursaria together with its included *Chlorella* has been the most popular experimental model (Fig. 11.2). Scores of *Paramecium* and *Chlorella* strains can participate in this relationship, but the place of the photoautotroph can also be occupied, in vitro at least, by other algae. Thus *Scenedesmus* will infect alga-free *P. bursaria* in experimental trials. In nature, freshwater *P. bursaria* may contain hundreds of *Chlorella* cells within its cytoplasm, and the colorless ciliate assumes the color of the symbiont it harbors as the green cells invade and become more numerous in the animal's body. The process of ingestion of the chlorellae by the paramecium resembles that of solid food particles, but for unknown reasons the algae resist digestion and become successfully established (Karakashian and Karakashian, 1965).

Additional intermicrobial symbioses, or relationships that seem to be symbiotic in function, have been described. Occasional genera of algae are known to have dissimilar algae in their cytoplasm. Several species of the planktonic diatom *Rhizosolenia* contain inclusions of blue-greens, and other algae harbor still different symbionts. In a few of these associations the host has lost the capacity to synthesize its own chlorophyll, so that the colored cells play a critical role in the nutrition of the organism containing them. Bacteria have been reported to live inside protozoa and, in the case of the intracellular bacteria borne by flagellates indigenous to the termite, to contribute enzymes essential for the well-being of the protozoan. The flagellate

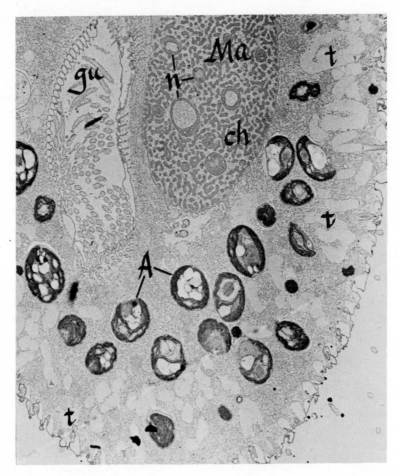

Figure 11.2   Electron micrograph showing *Chlorella* (*A*) in the cells of *Paramecium bursaria*. *Ma, gu, n, ch,* and *t* designate the macronucleus, gullet, nucleoli, chromatin, and trichocysts of the protozoan (Karakashian et al., 1968).

Crithidia oncopelti also contains a bacterium living symbiotically within its cell (Ball, 1969).

SYMBIOSIS INVOLVING HIGHER PLANTS

Symbiotic relationships having a microorganism and a higher plant as associates are widespread, common, and often of enormous ecological, economic, and agricultural importance. They are found

Table 11.1

The Extent of Nodulation among the Leguminosae [a]

Subfamily	Species in Subfamily	Species Examined for Nodulation	Species That Are	
			Nodulated	Non-nodulated
Mimosoideae	1,500	146	127	19
Caesalpinioideae	1,400	108	26	82
Papilionoideae	10,000	1,024	959	65
Total	12,900	1,278	1,112	166

[a] From Allen and Allen (1961).

in the tropics, in the temperate zones, and in the arctic, too, and each continent has its distinct assemblage of these intriguing partnerships. The plants come from a broad cross-section of orders and families, and the microorganisms also represent unrelated taxonomic categories, representatives being found among the bacteria, actinomycetes, algae, and fungi. Some of the symbioses are fortuitous, albeit of great value, while others clearly are essential for the life of one or both of the symbionts in nature and even, quite frequently, in vitro as well.

Undoubtedly the most thoroughly explored of these plant-microorganism complexes is that composed of a legume and a member of the genus *Rhizobium*. The only microsymbionts known to couple with legumes in the formation of the unique root nodular structures resulting from this symbiosis are bacteria of this single genus, and the only macrosymbiont that the rhizobia invade and function with is the leguminous host. However, not all *Rhizobium* strains can enter into such partnerships; similarly, the capacity to form nodules either is absent or is not expressed in many leguminous species (Table 11.1), although less than 10% of the approximately 13,000 legume species have been surveyed for the characteristic. Little attention has been given to the frequency of nodulation in two of the three subfamilies of Leguminosae, the family that encompasses leguminous plants. The third subfamily, Papilionoideae, has attracted greatest interest because of the economic or agricultural value of many of its representatives.

Legume root nodules are of various sizes and shapes, but usually they are small and either spherical, branched, or club-shaped. Despite the sizable collection of different rhizobia found regularly around roots, each nodule typically contains a single bacterial strain, suggesting that the infection is initiated by one or occasionally a few of the surrounding microorganisms. The morphological events in the infection process have been thoroughly documented, and the sequence of steps presented below is customary insofar as indicated by the available information. Nevertheless, given the almost 1300 species known to be nodulated and the uncounted rhizobial strains, no one series of events is likely to be universal.

Prior to nodulation the rhizobia begin to multiply around the roots, but this stimulation is not specific for the potentially invasive microorganism because a variety of rhizobia and unrelated bacteria also proliferate in the vicinity of the root in response to its excretions. The identities of a part of the excretion products that promote bacterial growth have been established; they include amino acids and B vitamins, as well as several organic compounds that may be energy sources. The rhizobia in turn liberate substances, possibly by merely transforming constituents of the root excretions, that have an influence on the root, causing it to become deformed and curl. Of the exuded substances, prime interest has been centered on tryptophan, which is released by the host and acted on by the microorganism to yield the auxin, indoleacetic acid; indoleacetic acid may contribute to the stimulation of root hair growth, but it is not the sole compound involved in the initial stages of infection, because noninfective rhizobia and members of different genera also form the auxin from tryptophan. In addition, the evidence for a functional or dominant role of indoleacetic acid in the curling, despite years of study, remains equivocal. Regardless of the precise mechanism, however, the first overt sign of an interaction is the enhanced growth of root hairs and their pronounced curling. The percentage of root hairs that curl depends greatly on the host and varies from very small to reasonably high.

In the early initial stage of the infection proper, the root hair invaginates, but the bacteria fail to penetrate into the cytoplasm of the hair. The physiological changes responsible for the invagination and subsequent steps in the infection have been carefully scrutinized. One hypothesis states that the rhizobia, by virtue of the polysaccharide they make in abundance, induce the plant to form polygalacturonase, an enzyme presumably acting on the pectin of the host's cell wall so that bacterial penetration is possible. Attractive as the hypothesis is, the data of a number of investigators fail to support the idea of selective induction of polygalacturonase formation.

Passage through the root hair is accomplished by means of an infection thread developing from the site of invagination (Fig. 11.3). This thin tube is composed of cellulosic materials of plant origin and contains the microsymbionts. Only a small proportion of the root hairs show infection threads, often from 1 to 5%. In a few legumes, like species of *Lotus* and *Anthyllis*, nodules appear but no infection threads are evident. Formation of the nodule itself appears to be initiated as an infection thread approaches a tetraploid cell of the host's cortical tissue, and this and adjacent diploid cells undergo division to yield a large mass that becomes the juvenile nodule. The host cells of the central region of the nodule characteristically contain twice the number of chromosomes as cells of non-nodular tissue. Ultimately the bacteria leave the infection threads and invade tetraploid cells in the center of the nodule, multiplying therein and assuming unique shapes, known as bacteroids, typical of the association. The bacteria in nodules of at least certain hosts are enclosed in membranous envelopes, the function of which remains obscure.

Bacteroids are generally irregular in shape and exhibit club, X, Y, or L shapes and even more bizarre forms. These morphological oddities can be induced from the usual rod shape of *Rhizobium* not only by components of the nodule microenvironment but also by glucosides and alkaloids added to laboratory media. Nevertheless the bacteria in occasional hosts fail to develop the bizarre shapes, retaining instead their original morphology.

The operation of the legume-*Rhizobium* symbiosis is expressed by the fixation of N_2, the N_2 assimilated satisfying the need of the organisms for this element. Neither the bacteria nor the legume alone effect this process, but the two together actively bring about the reaction. Studies employing isotopic nitrogen, ^{15}N, show unequivocally that the site of fixation is exclusively the nodular tissue, but until recently it was not certain whether the bacterial or the plant components of the nodule assimilated N_2. Some investigators have proceeded on the assumption that the bacteroid rather than the regularly shaped cell is active in the process, and so they established conditions in vitro that led to bacteroid formation, but these cells, too, were inactive. The assumption is tenuous, moreover, for several hosts containing only regularly shaped microsymbionts assimilate N_2 also, whereas bacteroid-filled nodules of many legumes are totally inactive.

A striking feature of nodules fixing N_2 is their redness. This color results from the presence of a protein, designated leghemoglobin, resembling the hemoglobin of blood. Not all nodules contain leghemoglobin, and these nonred nodules fail to bring about fixation.

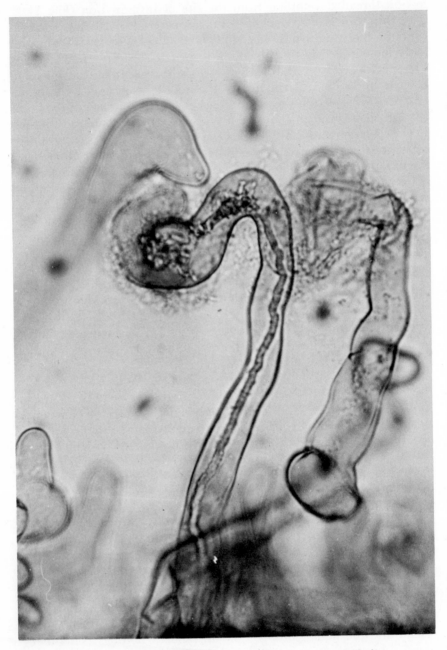

Figure 11.3   Infection thread developing through a legume root hair.

Table 11.2

Occurrence of Root Nodules among Nonleguminous Angiosperms [a]

Family	Genus	No. of Species in Genus	No. of Species Found to Bear Nodules	N_2 Fixation Verified
Betulaceae	Alnus	35	25	+
Casuarinaceae	Casuarina	45	14	+
Coriariaceae	Coriaria	15	12	+
Elaeagnaceae	Elaeagnus	45	9	+
	Hippophae	1	1	+
	Shepherdia	3	2	+
Ericaceae	Arctostaphylos	40	1 [b]	
Myricaceae	Myrica (Comptonia)	35	11	+
Rhamnaceae	Ceanothus	55	30	+
	Discaria	10	1	
Rosaceae	Cercocarpus	20	1	+
	Dryas	4	1 [b]	
	Purshia	2	2	+

[a] From Bond (1967).
[b] Identity of root structures as nodules not confirmed.

This trait has been the basis for the common field practice of using the color as a guide to the ability of the association to act on N_2. An appreciable proportion of rhizobia, though capable of prompting nodule formation, are typically unable to participate in the symbiotic fixation, and the nodules they produce are devoid of hemoglobin. Such bacteria are termed ineffective and contrast with effective cultures that, jointly with the macrosymbionts, convert N_2 into compounds usable by the two partners (Stewart, 1966).

A symbiosis localized in root nodules is also prominent among nonleguminous angiosperms. The genera bearing these structures and participating in the partnership are given in Table 11.2. The plants come from different families and thus contrast with the legumes, all of which are members of a single family; yet certain taxonomic affinities exist among the macrosymbionts. Thus Elaeagnaceae contains three and Rhamnaceae two genera with root nodules, and these two families are related taxonomically and are placed in the order Rhamnales. The remaining genera, however, are taxonomically distinct. Not all species of these genera have been examined as yet to determine whether they possess nodules, but some groups are clearly nodule-free. The distribution of nodulated species is global, however, and they are situated on river banks, mountain slopes, coastal soils, and eroded areas, and in forests, pastures, and acid bogs.

The nonleguminous nodule functions in the fixation of N_2, and hence it is physiologically analogous to the comparable legume structure. No part of the plant other than the nodule contains the N_2-fixing enzyme system. As indicated in Table 11.2, N_2 fixation has been verified by critical tests for species representing most of the genera, and future trials presumably will reveal all genera to be active in this way. Such plants, if nodulated, grow well in nitrogen-free solutions in the greenhouse, whereas non-nodulated plants may develop to the extent allowed by the seed nitrogen but they soon die out. The N_2 that is fixed is translocated from the nodule to the rest of the plant, as shown by the appearance of ^{15}N in tops after exposure of the roots to $^{15}N_2$. That this fixation is beneficial to the plant is evident from its improved growth in nitrogen-deficient soils once the symbiosis is established. Because it is unknown in nature apart from the nodule, the microorganism likewise benefits.

Not only cannot the microsymbionts be isolated from soils supporting the macrosymbionts, but there is widespread doubt as to whether they have ever been successfully cultured in vitro. Many claims have been made for the isolation of a microorganism from nonleguminous nodules of the sorts herein discussed, its growth in laboratory media, and the ability of the isolate to reinfect the host. Such isolates have included a number of actinomycetes, bacteria, and fungi, but *Rhizobium* is unquestionably not implicated. To date, none of these reports has been confirmed, and it is likely that many of the isolates were derived from the nodule surface. Sometimes a culture is capable of initiating an infection, but the process of nodule formation does not go to completion (Wollum et al., 1966). In view of the apparent failures of attempts to obtain the microsymbionts in culture and to establish normal infection with isolated organisms, information on their identity rests entirely on cytological grounds. The cytological investigations reveal that the microorganisms have a hyphal structure and that the filaments are commonly thin, branched, and septate and possess no distinct nuclear membrane. Characteristics of these sorts, which have been noted to pertain in varying degrees to the organisms residing in nodules of *Alnus, Ceanothus, Discaria, Myrica,* and *Purshia* species (Bond, 1967), indicate that the endophyte is probably an actinomycete.

In addition to the plants cited, several gymnosperms are nodulated, and they may be capable of assimilating N_2. The nodulating habit is particularly prominent among genera of Podocarpaceae, a family that includes trees important as timber. Although a number of genera of conifers are nodulated, *Podocarpus* has received the most attention. The microsymbiont in this instance also has not been cultured, but microscopic evidence reveals it to be a nonseptate Phycomycete-like fungus. The symbiosis of *Podocarpus* and the filamentous organism metabolizes N_2, but the rate of fixation may be insufficient to satisfy entirely the plant's demand for the element (Becking, 1965).

A scattering of tropical and subtropical gymnosperms classified among the Cycadales apparently symbiose with algae. The association is localized in nodules borne on the roots. The plants are species of *Bowenia, Ceratozamia, Cycas, Dioon, Encephalartos, Macrozamia, Stangeria,* and *Zamia,* and the incitants of nodule formation are strains of *Anabaena* and *Nostoc* of the blue-greens and a green alga, *Chlorococcum* (Allen and Allen, 1965). The blue-greens can be grown independently of the plant, and they themselves assimilate N_2 (Douin, 1953); hence those nodules that contain *Anabaena* and *Nostoc* presumably utilize N_2. Such a physiological activity has indeed been demonstrated for the root nodules of species of *Ceratozamia, Cycas, Encephalartos,* and *Macrozamia* (Bond, 1967). Blue-green algae sometimes live jointly with liverworts and ferns, and at least in the liverwort *Blasia pusilla,* which has cavities containing *Nostoc,* N_2 fixation is significant (Bond and Scott, 1955).

Tropical plants classified in the families Rubiaceae and Myrsinaceae frequently have nodules on their leaves. These nodulated plants are native to Africa, Australia, and South and Central America, and those most thoroughly investigated are species of *Psychotria, Pavetta,* and *Ardisia.* A single leaf may have up to 200 small nodules scattered about the leaf surface, along the leaf margin, or adjacent to the midrib. Not all species of the nodule-bearing genera exhibit these protuberances, the frequency of occurrence varying from genus to genus. The causative agents in this instance can be isolated and cultured with ease, and they have been identified as strains of *Klebsiella* and *Chromobacterium.* The bacterial microsymbionts, by contrast with those of legumes, fix N_2 independently of their host (Bettelheim et al., 1968; Centifanto and Silver, 1964).

A plant-microorganism symbiosis entirely different both morphologically and physiologically from the foregoing is that designated a *mycorrhiza.* This fungus-root association is readily recognizable to the naked eye, the mycorrhizae often appearing as short, thick, well-branched lateral roots, frequently with a distinct hyphal sheath. The

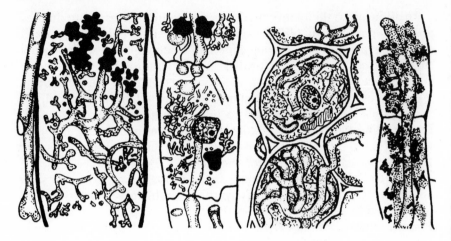

Figure 11.4   Hyphal development in endotrophic mycorrhizae (Boullard, 1968).

morphology is not stereotyped and depends on the particular host and the fungus concerned as well as prevailing environmental conditions. Two broad categories of mycorrhizae are distinguishable: the endotrophic and the ectotrophic types. In the former a portion of the hyphae penetrates into the host's cells, while in the latter the hyphae ramify between the plant cells and frequently extend in a network on the outside of the root (Fig. 11.4). A significant part of the mycelium grows out into the soil, and often the mycelium insulates the root surface so extensively that the nutrient-absorbing organ is the mycorrhiza rather than host tissue per se. The extension of the microorganism away from the seat of the partnership is in marked contrast to the nodule symbioses, wherein the microorganisms are entirely restricted to this single structural unit.

Mycorrhizae are characteristic of diverse plant groups. They appear on a wide range of perennials and annuals and are found on a high percentage of individuals of tropical, temperate, arctic, and alpine floras, including ferns and liverworts. Considerable attention has been given to forest trees and orchids, which have prominent and physiologically important symbiotic relationships in nature. The mycorrhizal fungi that have been investigated most thoroughly are classified among the Basidiomycetes, some 100 species of which have been reported to form mycorrhizae. *Amanita, Boletus,* and *Lactarius* are typical participants in the ectotrophic type and *Armillaria* in the endotrophic type of association. In addition, some fungi other than Basidiomycetes enter into a symbiosis of this sort. It is not uncommon, moreover, to find that an individual tree is infected by more than one kind of fungus. These heterotrophs, though often not culturable in laboratory media, are ubiquitous in soil, as evidenced by the widespread occurrence of mycorrhizae (Harley. 1959, 1968).

Environmental factors have a dramatic impact on the development of this symbiosis. High phosphorus and nitrogen levels, for example, generally suppress mycorrhiza formation in forest trees, and light intensity, too, affects the onset of the relationship. Age of the host is a factor in its response to the fungus, the mycorrhizae usually appearing only at particular periods of plant growth. The mutual benefit accruing to each of the partners is seen in improved growth and nutrient uptake by the plant and the extensive proliferation of the microorganism, which is difficult or impossible to isolate from the surrounding soil. Yet on occasion the association verges on parasitism, a fungus harming its symbiont, or sometimes a host, like the orchid, behaving as a parasite on its fungal partner.

SYMBIOSIS INVOLVING HIGHER ANIMALS

Nonpathogenic microorganisms are intimately linked with all phyla and possibly all classes of animals, ranging from the simple protozoa to higher invertebrates and vertebrates. Though frequently the significance of microscopic life to the well-being of the animal is unknown, unequivocal evidence exists to show that occasional populations not only gain from the animal harboring them but also contribute to and sometimes are essential for the health or very existence of the macrosymbiont. Dependencies of animals on the microbial communities contained within or upon their bodies are widespread, and study of these biological ties has occupied the energies of an international group of scientists.

The site where microbial symbionts become established varies considerably from animal to animal. The alimentary tract is commonly extensively colonized; many of its normal inhabitants apparently do no injury and often are so well integrated into the host's physiology that digestion cannot proceed without their presence and activity. The intestine, the rumen, and the cecum generally contain high microbial cell densities, but the Malpighian tubes or the lymph may be inhabited likewise by hordes of heterotrophs (Buchner, 1965).

An interesting and unique group of algae has the capacity to develop within marine and freshwater animals. These algae are subdivided on the basis of their color, and one speaks of zoochlorellae, zooxanthellae, and cyanellae to designate the algae that are pale to bright green, yellow to greenish brown, and blue-green, respectively. The microorganisms are classified taxonomically among the green algae, dinoflagellates, and the blue-greens. The zoochlorellae are common in freshwater animals, while the zooxanthellae are abundant in the marine fauna, particularly of warmer waters; for example, most species of coelenterates in the tropical seas contain zoochlorellae. In addition to the algal-protozoan relationship already mentioned, symbiotic algae—or at least species that appear to behave symbiotically—are found in Coelenterata, Porifera, Platyhelminthes, Mollusca, and other phyla (Droop, 1963; McLaughlin and Zahl, 1966). Immense numbers of the algae may live in conjunction with giant clams. Particular attention has been given to the biologically important association in the corals; in the massive coral *Favia*, for instance, a dinoflagellate forms a dense cover that effectively captures nearly all the light that reaches the coral (Halldal, 1968).

The metazoan-inhabiting photoautotrophs frequently participate in digestion or food transport. They are localized in the endodermal epithelium of coelenterates, in amebocytes of sponges, or in phagocytic cells in the blood of giant clams. The microorganisms may exist free in the body cavity of some animals, or, among the Turbellaria, they may occur extracellularly in spaces of the subepidermal parenchyma. Abundance of the algae is governed by the particular host; sometimes their population density is low, and on occasion it is quite high. Such associations range, moreover, from casual linkages among certain invertebrates, wherein the two symbionts may be found apart as well as together, to permanent relationships involving symbionts that in aquatic habitats rely on one another for the essentials of an active life.

The most thoroughly characterized type of microorganism-metazoan symbiosis, from both the biological and biochemical points of view, is that situated in the gastrointestinal system. It occurs in an appreciable number of vertebrates and invertebrates and requires, as microsymbionts, genera of protozoa and bacteria. The relationships involve interactions yielding a common benefit, and both the macro- and many of the microsymbionts are mutually dependent. Such dependency typifies most, but usually not all, of the heterotrophs of the alimentary tract. The site of the symbiosis may be the rumen, as in many herbivores, or the intestine, as in termites and woodroaches. The animal provides raw materials—frequently cellulosic substances that it ingests but itself cannot utilize—to the protozoa or bacteria it shelters, and the microorganisms in turn enzymatically degrade the ingested substrates to yield metabolic wastes that are absorbed and utilized by the animal. Without these microbial products the macro-symbiont would starve even though it was ingesting a wealth of forage, wood, or other presumably nutrient materials. The role of the microbial community is revealed experimentally by artificially depriving the animals of their microscopic partners; under such conditions wood-eating insects, for example, show severe nutrient-deficiency symptoms, they are weak and listless, and their ovaries may be altered. Not only protozoa and bacteria participate, but also genera of yeasts that live in a harmonious alliance with insects (Koch, 1963).

Symbiosis is likewise evident in animals that live on the blood of vertebrates. The heterotrophs borne by these blood-sucking metazoa are apparently essential for the digestion of the blood that is consumed. Leeches, ticks, lice, and bedbugs in this way contribute some good to the symbionts they shelter while doing evil to the species on which they feed.

Symbioses involving metazoa do not require that the microorgan-ism reside within the body of its associate, as in the *endosymbioses* cited above. Sometimes the site of interaction is the exterior surface of the animal. As a case in point, a few oceanic fish are colonized by luminescent bacteria that seem able to derive organic nutrients and support from their hosts. The light generated by the bacteria serves the fish in good stead in the darkness of the ocean depths.

Beetles, ants, and other insects participate in intriguing relation-ships in which the animals actually farm fungi in areas specifically de-signed for that purpose. A few insects burrow into the wood of trees, and the tunnels thus created become lined with a mycelium network that is a source of food for the larvae. By contrast, leaf-cutting ants of the genus *Atta* fragment leaves, pile up the pieces, and then domes-ticate fungi on the litter heaps. The fungi are cultivated with care, and ultimately they are eaten by the ants and their larvae. Because abandonment of the microorganisms by the ants is a prelude to a microbial invasion of the fungus growth and a rapid demise of the fungal farm, the microorganisms must benefit as well. Clearly, the ants manipulate the environment in a manner to favor the particular food fungus and exclude unwanted species.

CHARACTERISTICS OF MICROBIAL SYMBIOSIS

Functional relationships tie symbionts together ecologically and physiologically as well as frequently morphologically, and many associations differ biochemically from either of the two components.

The symbionts commonly function together as an autonomous entity, rather than as discrete organisms, and at times the two not only occupy the same habitats but they are disseminated as a unit. The union may result in a morphological alteration such that the resulting structural entity seems in almost every respect a distinct organism; this is well exemplified by the lichen.

Several traits are characteristic of many symbionts. Among the more obvious of these are the following five.

(a) Invasiveness. A microbial *endosymbiont*, that is, an individual residing within a microorganism, higher plant, or animal, must have the ability to join together with its host at the proper site and time. This entails a capacity for active invasion on the part of the endosymbiont, as is evident in the *Rhizobium*-legume relationship, or it may merely involve a passive incorporation, as when endosymbionts are ingested.

(b) Susceptibility to invasion. The host must be receptive to penetration by its partner, a receptiveness that may last for but a short time span.

(c) Effectiveness. Both partners must be symbiotically competent. Effectiveness or competence is probably best assessed by determining the degree of benefit accorded by one symbiont to the other, but this measurement is often difficult to make. In the *Rhizobium*-legume relationship, however, effectiveness is readily determined from the quantity of N_2 fixed. Ineffectiveness is well known in this bacterium-plant interaction, a deficiency attributable to discrete legume or microbial determinants, and the result is a nodule teeming with bacteria unable to catalyze N_2 fixation.

(d) Resistance to destruction by the symbiotic partner. The endosymbiont must be able to withstand destructive mechanisms of the host in which it resides. For example, the algae dwelling inside protozoan cells must somehow avoid being inactivated by digestive enzymes of the animal. Nevertheless endosymbionts are digested under certain conditions, as with the *Chlorella* contained inside *Paramecium bursaria* (Parker, 1926).

(e) Regulation. In most symbioses one organism governs the population density of the second. Such regulatory mechanisms are quite potent, for the cell density or mass of endosymbionts dwelling within animals or plants does not generally exceed a fixed range of values.

As yet the biochemical or morphological bases of invasiveness, susceptibility, effectiveness, resistance to destruction, and regulation have not been established, and there are similarly no known ways of predicting the invasiveness, susceptibility, or effectiveness of symbionts on the basis of laboratory tests performed on one of the organisms alone.

Only the *Rhizobium*-legume partnership has been extensively investigated with a view to securing mutants defective in symbiotic performance. These studies have revealed that bacterial invasiveness, host susceptibility, and effectiveness in the acquisition of N_2 are all mutable properties. Thus mutant bacterial strains and plant varieties derived from normally symbiosing parents have been obtained, and the new populations do not nodulate or, after nodulation, they fail to assimilate N_2. The plants have lost their susceptibility to invasion, the bacteria cannot infect to an extent sufficient to induce typical nodular growths, or a substance crucial for N_2 metabolism is no longer synthesized. The effectiveness of the rhizobia in participating in symbiotic N_2 fixation can be attenuated with ease, and indeed their ability to invade legumes may disappear entirely when the bacteria are repeatedly transferred on media containing amino acids. Comparable investigations have not been conducted with other associations, but it is likely that losses in symbiotic competence may be induced in the responsible organisms there, too.

The degree of reliance of one organism on its companion differs widely among the microorganisms involved in symbioses. At one extreme are the *facultative symbionts* that develop in nature either in a free-living state or in a close tie with a second species. *Rhizobium* is present in fields free of legumes, and plants capable of being nodulated develop well, provided that ample inorganic nitrogen is available, though their roots are devoid of nodules. Algae incorporated into lichens may be free-living, and certain of the animal species that bear algal associates can live without the microorganisms. The facultative symbionts multiply in nature in the free form, possibly not as well as when in symbiotic union, but still they grow. The following also proliferate in artificial conditions without their symbionts: *Paramecium bursaria* free of its zoochlorellae, the zoochlorellae without *P. bursaria,* fungal components of lichens, a variety of marine invertebrates lacking their specific algae, orchids devoid of their usual mycorrhizal fungi, *Nostoc* characteristic of the root nodules of cycads, and the flagellate *Crithidia oncopelti* without its bacterial endosymbiont. By contrast, though selected symbionts are indeed culturable in the laboratory in the absence of their associates, they are rare or unknown in nature in the free form, their ecological success requiring the partnership. These are *ecologically obligate symbionts.* Many flagellates and bacteria of the rumen are in this category.

At the opposite extreme are the true *obligate symbionts,* which not only have no free existence in nature but also cannot grow independently in vitro. *Endogone* and other mycorrhizal fungi require roots for active life, and some marine invertebrates seem to have an absolute need for their algal inclusions. Most and possibly all heterotrophic symbionts in the N_2-fixing nodules of nonleguminous roots are obligate symbionts, too; the occasional reports that the nodule organisms, such as those in *Alnus, Hippophae,* or *Ceanothus,* have been grown in culture must be regarded with skepticism pending independent verification. Yet, because infection by these mycorrhizal fungi and nodule inhabitants does occur, they must undoubtedly survive apart from their hosts.

Specificity is an unfailing and outstanding attribute of microbial symbionts. Specificity refers to the selectivity of the populations for the species with which they can couple. Inasmuch as the symbiotic state implies a condition wherein one organism constitutes an important or essential part of the environment of another, the extent of specificity reflects the kinds of living environments that permit the active development of a symbiont. Those groups exhibiting a high degree of specificity can only symbiose with a limited spectrum of compatible associates that have the exceptional biochemical or structural traits peculiarly suited to their partners. Low specificity is indicative of excessive symbiotic promiscuity and of an organism that can enter into union with a large number of strains, species, or genera because its requirements for symbiotic life are easily satisified.

Many fungi form mycorrhizae in conjunction with a multitude of plant species, and several plants accept a variety of mycorrhizal fungi. This low degree of specificity is well exemplified by *Cenococcum graniforme,* a fungus found in the ectotrophic mycorrhizae on the roots of numerous species of forest trees (Harley, 1959). Although it is true that obligate symbionts commonly exhibit a high degree of selectivity, this does not apply to some of the mycorrhizal fungi, which are strikingly promiscuous. The existing evidence also indicates that a single algal group can serve as phycobionts in a wide array of lichen types, and what appears to be the same alga is present in different species and sometimes genera of lichens. Similarly, a single fungus can apparently symbiose with dissimilar species of algae (Ahmadjian, 1965; Haynes, 1964).

As a rule, however, symbiotic specificity is marked, but the breadth of the range depends on the microbial group. Among nonlegumes bearing root nodules, presumably induced by actinomycetes, the same obligately symbiotic microorganism will infect representatives of the three genera of the family Elaeagnaceae (*Elaeagnus, Hippophae,* and *Shepherdia*) (Moore, 1964). Nevertheless, except for the Elaeagnaceae, the endophytes of one genus will not infect plants of another genus though able to nodulate other species of the host's genus. Still greater specialization is evident among those microsymbionts that are restricted to individual species of the genus *Coriaria*. An entirely analogous situation is evident in *Rhizobium,* some strains of which nodulate members of two or more legume genera, others are restricted to a collection of species within a single genus, and a few, so far as is known, are limited to a very narrow range of plant species. Extreme selectivity also characterizes occasional mycorrhizal fungi, like *Boletus elegans,* which seems to live on the roots only of *Larix* (Harley, 1959). Though numerous *Chlorella* strains infect *Paramecium bursaria,* nevertheless it is the chlorellae that are the widespread algal symbionts, and, conversely, the ciliate entering into a beneficial relationship with these algae is typically this paramecium. In rare instances a microorganism may express the full complement of symbiotic activities with but a few partners, as if these alone provide the appropriate microenvironment, but it reveals a portion of the processes necessary for a functional symbiosis with a larger array of associates; thus many *Rhizobium* strains infect, nodulate, and fix N_2 with a small assemblage of species of Leguminosae though they are capable of inducing non-N_2-fixing nodules on a large assortment of legumes. Nodules of the latter sort do not have the correct internal conditions to permit the bacterium to engage in N_2 assimilation.

Countless words but pitifully few experiments have been devoted to the biochemical or cytological bases for symbiotic specificity, and little information thus exists to account for the restriction of one symbiont to individual strains, species, or genera of associates. The only symbiosis that has received even a modest amount of study is that concerning *Rhizobium*; it has been suggested that an extracellular polysaccharide synthesized by these bacteria determines their specificities, but the experimental evidence is far from convincing.

The origin of the symbiotic habit has long fascinated biologists. As a rule, a species profiting from a neighbor's action will tend to be found in nature adjacent to its benefactor. Such loose relationships may then be subjected to selective pressures that improve the fitness of the association. Once an interaction having a degree of benefit— possibly unilaterally helpful at first—is established, one or both of the partners could become physiologically or morphologically modified, and the resulting alteration may further increase the fitness of the symbiosis or lead to the discarding of traits necessary for an independent but not for a dependent way of life. Thus the ability to synthesize a given growth factor is a requisite for multiplication in environments devoid of the compound but not for replication in a microenvironment wherein a neighbor generates and excretes the substance in question; the loss of the enzymes necessary for the synthesis of the metabolite can be balanced by the appearance of a new function, one that may be useful to the partnership and hence would be favored by natural selection. Irreversible loss of a trait essential for independent life or for competitiveness makes the species totally reliant on its neighbor: it is now an obligate symbiont. It is generally acknowledged that processes of this general sort have occurred in the evolution of a variety of obligate symbionts.

The evolving symbiotic relationship is subjected to natural selection much as is a free-living population, and should the interaction have an appreciable survival value, it will be selected as a distinct and functional entity. Mutually beneficial effects that allow for greater fitness are probably also enhanced under selection pressure. Undoubtedly, mutually useful associations have arisen in the past and continue to appear, but only those with a peculiar advantage will be preserved.

With a rise in the number of factors responsible for the uni- or bilateral dependency, the reliance of one species on the second for growth or for the carrying out of a particular reaction increases. It is likely, in addition, that the greater the degree of uniqueness or the more selective is any of these factors, the greater will be the extent of dependency, the degree of specificity, or both (Dubos and Kessler, 1963). Few of these factors, be they physiological activities or structural features present in the symbiont but lacking in its free-living counterparts, have been identified as yet. Because axenic culture of some symbionts is difficult whereas their free-living counterparts grow readily and are widely distributed, the strains participating in the symbiosis surely differ from their free-living sisters.

Overt structural or physiological characteristics of a symbiosis may be attributable to a trait of one of the associates, or they may arise only as a direct consequence of the interaction. The color of aquatic invertebrates, for example, results simply from the algae contained within their cells or tissues, and the high cobalt demand of nodulated legumes seems to arise by virtue of the bacterial requirement for cobalt (Evans and Kliewer, 1964). Similarly, though the endosymbiont probably has not been cultured, the cobalt needed for N_2 fixation by nodulated *Alnus, Myrica,* and *Casuarina* (Bond, 1967) probably reflects a requirement of the filamentous symbiont for this element. On the other hand, nodules and mycorrhizae are morphological units that owe their existence exclusively to the interplay between macro- and microsymbionts, and N_2 fixation occurs only when plants and microorganisms coexist. Morphological contrasts between the free-living and the symbiotic state of an organism are well known, too, as with the *Nostoc* cells that in culture but not in the lichen have a thick gelatinous sheath, with *Rhizobium* that in the nodule but not in conventional media shows its bizarre forms, or with the algae harbored by the turbellarian *Convoluta roscoffensis* that possess neither the flagella nor the cell walls of their free-living counterparts. Invertebrate hosts of algae likewise are anatomically different from the uninfected animals (Yonge, 1957).

For the association to be perpetuated, one symbiont must invade or be passed on to the cells or tissues of the second generation. This transfer is essential for the obligate symbiont, inasmuch as it is unable to live actively as an independent individual, but it must be accomplished for the facultative species as well if it is to participate. Contact between potential interactants is frequently casual and fortuitous, involving an accidental meeting as one organism moves, grows, or is passively transported to the second, and occasionally the contact results from the feeding habits of an animal participant. Certain transmissions probably are facilitated by a chemotactic response on the part of the microsymbiont. Transfer of microbial symbionts from one plant generation to the next may occur because the microorganism is seed-borne; the bacterium causing leaf nodule formation is transferred in this way from parent to new seedlings (Schaede, 1962). Animals transmit their microbial partners in numerous ways. Some algae residing in invertebrate tissues are passed to the next generation via the eggs, while with *Paramecium bursaria* the daughter cells gain their chlorellae from the parent paramecium as it divides. The female of certain insect species deposits a portion of its intestinal microflora adjacent to the newly laid eggs so that the larvae, after hatching, are able to suck up the microorganisms they need to survive (Gustafsson, 1968); this is illustrated in Fig. 11.5.

Definitive studies of symbiosis frequently require that the relationship be established under controlled conditions in vitro. This is usually readily accomplished by (a) combining the two microbial populations, as with *Paramecium bursaria* and its chlorellae; (b) inoculating a culture of the microsymbiont onto or into a suitable host, as with *Rhizobium, Klebsiella,* or fungi on the appropriate legume, *Psychotria,* or mycorrhiza-forming plant; or (c) applying to the host a tissue suspension bearing the nonculturable obligate microsymbionts, as with the endosymbiont of nodulating nonlegumes. The ease of estab-

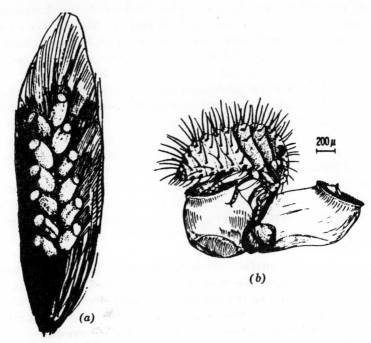

Figure 11.5 (*a*) Bacteria-rich coccoons between the eggs laid by female *Coptosoma scutellatum* on a leaf. (*b*) Newly hatched larva of the insect inserts its proboscis into a coccoon teeming with bacteria to suck out its contents (Buchner, 1965).

lishing the relationships cited is in marked contrast with the difficulty of initiating others. This difficulty is nowhere as apparent as with the lichens. The facility of growing the algal and fungal components separately led to the belief that synthesis of a lichen in the laboratory would be simple: the two populations need only be placed together. This assumption was soon discovered to be incorrect. Many attempts have been made to reconstruct a lichen in vitro, and these trials have generally been total failures, or the results have been inconclusive or subject to serious question. However, the first steps in synthesis of a lichen from the two components have been achieved, and structures analogous to those in naturally occurring lichens have been obtained (Ahmadjian, 1962).

The initiation, functioning, and continued existence of many, and possibly all, symbioses are dependent on conditions prevailing in the habitat. Given circumstances prevent the onset of the association, suppress its normal functioning, or favor its dissolution. Occasionally environmental factors favor the shift from a mutually beneficial interaction to a situation in which one species parasitizes its neighbor. A biochemical dependency is typical of both symbiosis and parasitism, the lines of demarcation between these two-membered interactions often being extremely fine, and the shift from a form of metabolic reliance providing mutual benefit to one resulting in a pathological condition is apparently neither difficult nor infrequent. Evidence for inhibition of the initiation of a symbiosis by environmental factors is found in the suppression of mycorrhiza synthesis in rich soils, though mycorrhizae are abundant in land of low fertility status (Harley, 1959), and in the retardation of infection and nodule formation by *Rhizobium* and by the root nodule heterotrophs of nonlegumes in the presence of high concentrations of inorganic nitrogen (Fahraeus and Ljunggren, 1967; Gardner and Bond, 1957). Destruction of an established symbiosis by environmental factors is illustrated by those lichen partnerships which break down when nutrients promoting independent development of the symbionts become available (Tobler, 1925). Aquatic animals harboring algae exemplify how symbiosis reverts to parasitism; for example, *Chlorella* is digested by the *Paramecium* cells when there is no light and the protozoan is starved (Pringsheim, 1928). Destruction and digestion of algal endosymbionts are typical of many invertebrates, and a condition of mutual benefit may be replaced by parasitism in the intestine of mammals, too, the microflora changing to induce a marked deleterious influence on the animal with which it previously lived in complete harmony.

EFFECTS OF SYMBIONTS ON EACH OTHER

The abundance and widespread distribution of symbiotic alliances attest to their enormous ecological value to the participants. The prominence or very survival of many populations can be credited entirely to their ability to couple in jointly beneficial enterprises. The exceptional physiological or morphological innovations that are the bases for symbioses permit the two companions to occupy sites where one or both would not grow at all or where the prominence of the separate populations would otherwise be far less. By joining forces with a second species, many microorganisms avoid the keen biological stresses imposed by their immediate neighbors. Such an escape is necessary for species incapable of surviving the rigors of biological interactions in heterogeneous communities, as in soil or water. Thus *Rhizobium* is shielded in the nodule, where it exists in essentially pure culture, and selected fungi unable to compete with heterotrophs that more rapidly metabolize plant remains or organic constituents of soil may escape to the hospitable confines of the mycorrhizal root. The advantage gained by other symbionts is attributable to their acquisition of a novel biochemical or structural feature providing them with a fitness trait that is useful in natural selection or in tolerating environmental extremes, such as the ability of lichens to withstand prolonged drought and temperature stresses that eliminate a high proportion of free-living organisms.

Symbionts are not altruistic. They are not affiliated with their partner for the latter's well-being but rather because they themselves obtain some useful return. This benefit frequently results from a nutritional interchange, and exchanges of nutrients are probably essential parts of the interplay between many symbionts. The microorganism residing within the tissues of a higher animal or plant unquestionably must be able to obtain its requisite nutrients from these tissues, but it also must find the microenvironment wherein it resides completely satisfactory for life. In certain symbioses the benefit to the microorganism seems to rest not on a nutritional foundation, at least not significantly so, but on some other feature of the interaction, such as a structure protecting against water loss or the lethal action of sunlight or a means by which one partner detoxifies the environment or utilizes wastes of the second.

That the gain to the symbiont is of greater than marginal consequence can be established in a number of ways. When artificially deprived of their intestinal flagellates, termites and woodroaches die of starvation because they cannot, without the protozoa, use the wood that they ingest. Lice devoid of microbial symbionts are weak, behave in a listless manner, and exhibit symptoms of starvation. Sterile larvae of certain insects normally colonized by microorganisms do not develop and soon die (Gustafsson, 1968; Koch, 1963). Algae are essential for the life of the marine animal *Convoluta roscoffensis*, and, conversely, zooxanthellae residing within protozoa and marine invertebrates depend on their hosts for existence in nature (Yonge, 1957). Without its zoochlorellae, *Paramecium bursaria* starves to death unless sustained with organic nutrients (Pringsheim, 1928). The crucial role of microorganisms for the development of their plant companions is widely known; for example, non-nodulated legumes are chlorotic and fare poorly in nitrogen-deficient soils, and nodule-free *Psychotria* is stunted and frequently does not survive for long (Humm, 1944). Mycorrhizae, too, are either important or essential for plants bearing them; the growth of some plants is poor and the development of others is soon arrested without the appropriate fungus. Furthermore the seeds of numerous orchid species will not germinate unless penetrated by the proper fungal associate.

What specific advantages are acquired by the symbionts, and what is the contribution of each organism to the association? In only a few symbiotic interactions is there a clear understanding of the basis for the mutual dependence, while in other relationships plausible hypotheses have been advanced though little experimental evidence has been gathered in support of the proposed hypotheses. The following are the major known or postulated means by which one symbiont affects its immediate neighbor. Some of these will be considered again in a later chapter.

(a) Increasing the growth rate. This is an obvious and possibly one of the most widespread effects. Such an influence is noted, for example, in paramecia infected by their specific algae (Karakashian, 1963) and in the greater growth of alga-bearing corals in sunlight than in darkness (Goreau, 1961). Sometimes the benefit is less evident in growth rate than in the population density, size, or mass of the organism.

(b) Enhancing metabolic activity. The stimulation may be expressed in terms of an increased rate of respiration, as postulated for the influence of nodule hemoglobin on *Rhizobium* (Burris and Wilson, 1952), or in terms of an unspecified increase in metabolic activity or efficiency, as proposed for corals and related coelenterates (Yonge, 1957).

(c) Providing available carbon as a result of photosynthetic CO_2 assimilation. Many investigators hold that the algal symbionts of lichens, protozoa, and aquatic invertebrates as well as plants bearing nodules or mycorrhizae form organic compounds during photosynthesis in amounts sufficient not only to satisfy their own demands but also to provide an appreciable quantity to serve as carbon sources for their heterotrophic companions (Fig. 11.6). The organic compounds may be transferred during active growth of the higher plant or, where algae are involved, either during the active phase of proliferation or following death of the alga and its digestion by the animal. The mere ability of paramecia, which are obligate heterotrophs, to multiply in inorganic solutions in the light when they harbor chlorellae demonstrates the focal role in the symbiosis of organic compounds synthesized by the photoautotrophs. Algae isolated from *Paramecium, Spongilla,* and *Chlorohydra* do indeed excrete large proportions of the

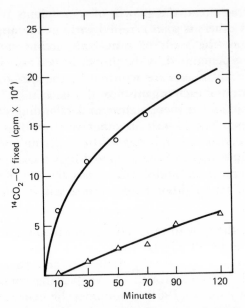

Figure 11.6 Fixation of $^{14}CO_2$ by the algal component (○) and its transfer to the medulla (△) portion of a lichen (Smith and Drew, 1965).

carbon fixed in photosynthesis (Muscatine et al., 1967), and animal tissues, at least of *Chlorohydra,* assimilate a sizable quantity of the CO_2 carbon acquired by their endosymbionts (Roffman and Lenhoff, 1969). By using $^{14}CO_2$, the transfer from photoautotroph to heterotroph of organic molecules generated in photosynthesis has been shown in lichens (Smith and Drew, 1965) and in mycorrhizal roots of pine seedlings (Melin and Nilsson, 1957). By contrast, animals like the giant clam *Tridacna* appear to obtain much of their nourishment by directly digesting the included algae.

(d) Converting an unavailable food ingested by the microorganisms' associate to an available form. Members of the microbial community of the rumen possess enzymes catalyzing the decomposition of cellulose and other polysaccharides, polymers that cannot be digested by the ruminant itself, to simpler molecules that both microorganisms and animal can either further metabolize or absorb and utilize directly. Flagellated protozoa dwelling in the alimentary tract of wood-eating termites perform an analogous function in converting cellulose, which makes up a high proportion of the termite's food, to simpler molecules without which the termite would starve; the host does not produce cellulose-degrading enzymes and hence cannot by itself digest this polysaccharide. The animal, in effect, lives on the wastes of microbial carbohydrate fermentation. Similar actions may be performed by the microbial communities in the digestive tracts of the horse, pig, and herbivorous insects.

(e) Supplying a higher plant with the organic nutrients it can neither obtain nor synthesize. It is widely believed that a principal function of the mycorrhizal fungi living together with orchids is to provide the heterotrophic plant with sugars and other metabolites. The mycelium apparently secures or forms these compounds as the hyphae attack the organic matter or dying tissues in the habitat, and the compounds are then transferred to the orchid or become available when the hyphae are digested by the macrosymbiont.

(f) Generating the CO_2 necessary for a photosynthetic partner. It is generally assumed that algae in symbiosis with protozoa and metazoan invertebrates get at least part of the CO_2 they require for proliferation from the animal's respiration. The importance to the algae of this respiratory CO_2 is not easy to assess, because the surrounding water normally contains an abundant supply, but the concentration of CO_2 in fresh or sea water may not be adequate to meet the high demand in animal tissues teeming with the autotrophs.

(g) Producing the O_2 required by an aerobic symbiont. During photosynthesis endosymbiotic algae generate O_2, an essential nutrient for the aquatic host. The algae of reef-forming corals, for example, liberate copious amounts of O_2 during the daylight hours. However, O_2 deficiency is often uncommon in habitats supporting these animals, so that this release may possibly not aid the animal. The O_2 evolved, on the other hand, could conceivably oxygenate tissues not receiving an adequate supply of the gas or contribute to the aeration of O_2-poor waters. Nevertheless, rigorous proof of a beneficial value or a dependence of metazoa on algal-produced O_2 does not yet exist.

(h) Removing O_2. The ruminant, together with its associated microflora, provides the anaerobic conditions essential for the cellulose decomposers it harbors.

(i) Assimilating N_2. No higher plant, unless it is invaded by a microbial symbiont, utilizes N_2. Infrequent claims have been made for the ability of non-nodulated plants to assimilate N_2, but the data are equivocal, the experimental methods are questionable, or independent confirmation is lacking. By metabolizing nitrogen present in the inexhaustible atmospheric reserve, the nodular tissue makes the plant independent of the small available supply in the soil, a process that may account in part for the prominence of legumes in nature and the pioneering role of some of the nodule-bearing nonlegumes. Nodulated plants surely must gain an advantage in interspecific competition in land where nitrogen is limiting, and they do fare well where little fixed nitrogen is available. It has long been taken for granted that the microbial cell is the site of N_2 fixation within the root nodule and that the N_2 fixed by the microsymbiont is rapidly made available to the plant tissue. That the N_2 assimilated is rapidly transported from the nodule to the rest of the plant is definite, but whether the N_2-metabolizing enzymes are of microbial or plant origin is largely unsure for those root nodules induced by heterotrophs on nonlegumes; it remains unresolved, therefore, whether the microorganism benefits these particular plants by fixing N_2 or by promoting formation of the root structure in which the plant's N_2-metabolizing enzymes are synthesized. On the other hand, fixation in legume nodules is clearly attributable to the bacteroids and not to the plant cells (Koch et al., 1967), so that here a role for the bacteria is well defined. Bacterial symbionts of *Psychotria* leaf nodules, *Anabaena* of cycad root nodules, and *Nostoc* of lichens make use of N_2 in axenic culture, and presumably they bring about the same reaction when located within the plants or lichens (Bettelheim et al., 1968; Douin, 1953; Henriksson, 1951). Blue-green algae, like *Anabaena* and *Nostoc,* characteristically liberate into their surroundings a high percentage of the nitrogen they obtain from N_2, and these nitrogenous compounds are probably transferred to and used by the allied organisms. Fixation of N_2 has likewise been demonstrated in the fern *Azolla* and the liverwort *Blasia,* both of which contain cavities inhabited by strains of blue-green algae, species of either *Anabaena* or *Nostoc* (Bond, 1959).

(j) Providing growth factors. Heterotrophic microsymbionts commonly require growth factors. For instance, *Rhizobium* strains generally need one or more B vitamins, some lichen fungi are dependent on an external source of biotin and thiamine, and diverse mycorrhizal fungi must be supplied with assorted B vitamins and amino acids. *Paramecium bursaria,* when deprived of its chlorellae, has a highly complex nutrition. Inasmuch as the symbiotic alliances exhibit no such requirements, so far as tested, it is likely that the source of the growth factors is the photosynthetic member, and in fact it is well known that lichen algae and higher plants excrete vitamins in axenic culture. Heterotrophs, too, may generate growth factors for their symbionts; this is exemplified by those microorganisms that provide

Table 11.3

The Vitamin Content of the Rumen of Steers Receiving Various Diets [a]

	Vitamin Content (µg/g dry material)			
	Hay and Concentrates		NaOH-Treated Straw + Casein	
Vitamin	Diet	Rumen Contents	Diet	Rumen Contents
Thiamine	5.0	3.0	0	1.8
Riboflavin	9.0	13.	1.0	12.
Nicotinic acid	32.	60.	2.1	52.
Pantothenic acid	19.	28.	1.2	18.
B_6	2.5	2.5	0.25	2.4
Biotin	0.12	0.22	0.004	0.17
Folic acid	0.25	2.3	0.08	1.0
B_{12}	0	6.5	0	8.3

[a] From Kon and Porter (1953).

their insect partners with essential B vitamins, carnitine, and possibly choline and sterols (Koch, 1963). Bacteria residing in herbivores have also been suggested to aid their companions by synthesizing vitamins and additional essential compounds present in insufficient supply in the food the animal consumes (Table 11.3).

(k) Supplying inorganic nutrients. A remarkable feature of the ectotrophic mycorrhiza is its greater rate of nutrient absorption than uninfected roots. Although part of this effect may be attributable to the fact that mycorrhizae have a greater surface area than uninfected roots, the rate is still higher even considering equivalent surface areas. Therefore the fungus specifically helps the plant by increasing its ion uptake, an influence that is particularly noticeable when the nutrient supply is low. The mycorrhizae are quite active in increasing the uptake particularly of phosphorus, but also of nitrogen and potassium. In addition, mycorrhizal plants can more readily obtain inorganic nutrients from relatively unavailable sources than can nonmycorrhizal plants (Murdoch et al., 1967). Similarly, algae contained within the animal body probably get inorganic nutrients, especially ammonium and phosphate ions, from the host's cells; these are probably end products of the metabolism of the animal cell (Droop, 1963). Algae are known, for example, to be able to assimilate all the phosphate that corals liberate. It has also been proposed that the lichen mycobiont, by virtue of its capacity to release inorganic nutrients bound in rocks or organic complexes, benefits the phycobiont by supplying it with inorganic ions.

(l) Utilizing metabolic wastes. In the process of satisfying their own nutritional needs, certain endosymbionts destroy wastes of their partners, products that could be toxic to the producing organism. This is of value in that one species detoxifies the internal environment of the second. Algae are believed to benefit their invertebrate partners by removing CO_2, ammonium, phosphate, and other wastes, acting in this way as biological waste disposal systems. In a striking illustration, alga-free Convoluta accumulates uric acid deposits, but these disappear once infection is accomplished (Droop, 1963).

(m) Shielding against deleterious abiotic factors in the environment. Fungal hyphae may protect the alga of the lichen from high light intensities and from desiccation; if true, this might account in part for the survival of algae in habitats where they would not persist alone. An organism that harbors another in its tissues or cells, as do the alga-bearing protozoa and metazoa, shields the individuals it contains from extreme conditions and abrupt changes in the surroundings.

(n) Protecting against parasites. Many symbionts synthesize compounds in vitro that are toxic to selected organisms, and these could conceivably benefit the partner. Thus mycorrhizae produced by Cenococcum graniforme and needles of conifers carrying such mycorrhizae apparently contain an antibiotic synthesized by the fungus (Krywolap et al., 1964), and mycorrhizal but not nonmycorrhizal roots of shortleaf pine are resistant to infection by the pathogenic fungus Phytophthora cinnamomi (Marx and Davey, 1967).

(o) Producing luminescence. Bioluminescent microorganisms colonizing deep-sea fish may be useful to the fish by luring organisms on which the fish feed or by attracting potential mates for the animals.

(p) Inducing a resting stage. The termite Termopsis angusticollis secretes a substance causing one of its symbiotic flagellates, Trichonympha, to form resistant cysts that aid the survival of the protozoan (Croll, 1966).

References

REVIEWS

Ahmadjian, V. 1967a. *The Lichen Symbiosis.* Blaisdell, Waltham, Mass.

Ball, G. H. 1969. Organisms living on and in protozoa. *In* T.-T. Chen, Ed., *Research in Protozoology,* Vol. 3. Pergamon Press, Oxford. pp. 565–718.

Bond, G. 1967. Fixation of nitrogen by higher plants other than legumes. *Ann. Rev. Plant Physiol.,* **18,** 107–126.

Buchner, P. 1965. *Endosymbiosis of Animals with Plant Microorganisms.* Interscience Publishers, New York.

Droop, M. R. 1963. Algae and invertebrates in symbiosis. *In* P. S. Nutman and B. Mosse, Eds., *Symbiotic Associations.* Cambridge University Press, London. pp. 171–199.

Dubos, R., and Kessler, A. 1963. Integrative and disintegrative factors in symbiotic associations. *In* P. S. Nutman and B. Mosse, Eds., *Symbiotic Associations.* Cambridge University Press, London. pp. 1–11.

Hale, M. E. 1967. *The Biology of Lichens.* Arnold, London.

Harley, J. L. 1959. *The Biology of Mycorrhiza.* Hill, London.

Harley, J. L. 1968. Mycorrhiza. *In* G. C. Ainsworth and A. S. Sussman, Eds., *The Fungi,* Vol. 3. Academic Press, New York. pp. 139–178.

McLaughlin, J. J. A., and Zahl, P. A. 1966. Endozoic algae. *In* S. M. Henry, Ed., *Symbiosis,* Vol. 1. Academic Press, New York. pp. 257–297.

Stewart, W. D. P. 1966. *Nitrogen Fixation in Plants.* Athlone Press, University of London.

OTHER LITERATURE CITED

Ahmadjian, V. 1962. *Am. J. Bot.,* **49,** 277–283.

Ahmadjian, V. 1965. *Ann. Rev. Microbiol.,* **19,** 1–20.

Ahmadjian, V. 1967b. *Phycologia,* **6,** 127–160.

Allen, E. K., and Allen, O. N. 1961. *In Recent Advances in Botany*, Vol. 1. University of Toronto Press, Toronto. pp. 585–588.

Allen, E. K., and Allen, O. N. 1965. *In* C. M. Gilmour and O. N. Allen, Eds., *Microbiology and Soil Fertility*. Oregon State University Press, Corvallis. pp. 77–106.

Becking, J. H. 1965. *Plant Soil*, **23**, 213–226.

Bettelheim, K. A., Gordon, J. F., and Taylor, J. 1968. *J. Gen. Microbiol.*, **54**, 177–184.

Bond, G. 1959. British Association: The Advancement of Science, **15**, 382–386.

Bond, G., and Scott, G. D. 1955. *Ann. Bot.*, **19**, 67–77.

Boullard, B., 1968. *Les Mycorrhizes*. Masson, Paris.

Burris, R. H., and Wilson, P. W. 1952. *Biochem. J.*, **51**, 90–96.

Centifanto, Y. M., and Silver, W. S. 1964. *J. Bacteriol.*, **88**, 776–781.

Croll, N. A. 1966. *Ecology of Parasites*. Heinemann, London.

Douin, R. 1953. *Compt. Rend.*, **236**, 956–958.

Evans, H. J., and Kliewer, M. 1964. *Ann. N.Y. Acad. Sci.*, **112**, 735–755.

Fahraeus, G., and Ljunggren, H. 1967. *In* T. R. G. Gray and D. Parkinson, Eds., *The Ecology of Soil Bacteria*. Liverpool University Press, Liverpool. pp. 396–421.

Gardner, I. C., and Bond, G. 1957. *Can. J. Bot.*, **35**, 305–314.

Goreau, T. F. 1961. *In* H. M. Lenhoff and W. F. Loomis, Eds., *The Biology of Hydra*. University of Miami Press, Miami. pp. 269–282.

Gustafsson, B. E. 1968. *In* D. H. Calloway, Ed., *Human Ecology in Space Flight III*. New York Academy of Sciences, New York. pp. 119–159.

Halldal, P. 1968. *Biol. Bull.*, **134**, 411–424.

Haynes, F. N. 1964. *Viewpoints Biol.*, **3**, 64–115.

Henriksson, E. 1951. *Physiol. Plantarum*, **4**, 542–545.

Humm, H. J. 1944. *J. N.Y. Bot. Garden*, **45**, 193–199.

Johnson, T. W., and Sparrow, F. K. 1961. *Fungi in Oceans and Estuaries*. Cramer, Weinheim, Germany.

Karakashian, S. J. 1963. *Physiol. Zool.*, **36**, 52–68.

Karakashian, S. J., and Karakashian, M. W. 1965. *Evolution*, **19**, 368–377.

Karakashian, S. J., Karakashian, M. W., and Rudzinska, M. A. 1968. *J. Protozool.*, **15**, 113–128.

Koch, A. 1963. *In* N. E. Gibbons, Ed., *Recent Progress in Microbiology*. University of Toronto Press, Toronto. pp. 151–161.

Koch, B., Evans, H. J., and Russell, S. 1967. *Plant Physiol.*, **42**, 466–468.

Kon, S. K., and Porter, J. W. G. 1953. *Proc. Nutr. Soc.*, **12**, XII.

Krywolap, G. N., Grand, L. F., and Casida, L. E. 1964. *Can. J. Microbiol.*, **10**, 323–328.

Marx, D. H., and Davey, C. B. 1967. *Nature*, **213**, 1139.

Melin, E., and Nilsson, H. 1957. *Svensk. Bot. Tidskr.*, **51**, 166–186.

Moore, A. W. 1964. *Can. J. Bot.*, **42**, 952–955.

Murdoch, C. L., Jackobs, J. A., and Gerdemann, J. W. 1967. *Plant Soil*, **27**, 329–334.

Muscatine, L., Karakashian, S. J., and Karakashian, M. W. 1967. *Comp. Biochem. Physiol.*, **20**, 1–12.

Parker, R. C. 1926. *J. Exptl. Zool.*, **46**, 1–12.

Pringsheim, E. G. 1928. *Arch. Protistenk.*, **64**, 289–418.

Roffman, B., and Lenhoff, H. M. 1969. *Nature*, **221**, 381–382.

Schaede, R. 1962. *Die pflanzlichen Symbiosen*. Fischer, Stuttgart.

Smith, D. C., and Drew, E. A. 1965. *New Phytologist*, **64**, 195–200.

Sund, G. C. 1968. *J. Protozool.*, **15**, 605–607.

Tobler, F. 1925. *Biologie der Flechten*. Borntraeger, Berlin.

Wollum, A. G., Youngberg, C. T., and Gilmour, C. M. 1966. *Proc. Soil Sci. Soc. Am.*, **30**, 463–467.

Yonge, C. M. 1957. *In* C. W. Hedgpeth, Ed., *Treatise on Marine Ecology and Paleoecology*, Vol. 1. National Research Council, Washington, D.C. pp. 429–442.

12

Competition

Numerous kinds of organisms frequently arrive during the colonization of previously unpopulated areas; yet, owing to the interactions initiated soon after growth commences at the site, few of the arrivals survive. Some of the potential pioneers are eliminated as a direct consequence of activities of their neighbors. Each cell, filament, and population must obtain from its surroundings needed organic compounds, inorganic nutrients, water, and, for some species at least, O_2, CO_2, and light. These very resources, however, are not in unlimited supply, and it is because of the insufficiency of biologically critical environmental resources that one of the major interspecific interactions, competition, has a profound impact on succession, natural selection, and the composition of microbial communities.

In axenic cultures the density of cells or mass of filaments usually rises until the supply of a component in the medium is exhausted. The cell density is proportional to the capacity of the medium to provide essential nutrients for the growth of that species. The situation is different when the species of interest is in rivalry with another. Its potential population density is not attained, because the second species uses many of the essential factors and depletes the supply of substances necessary for life.

Competition refers to the struggle between organisms for an essential resource—nutrients, water, light, space—that is present in the environment in an amount insufficient to meet the biological demand. The organisms are in rivalry for the same factor, and the supply or availability of this factor is too low to support populations as dense as would be found were the populations alone in the microcosm. In the climax, competition exists for whatever resource limits the density of the residents; that is, the resource for which the combined demands or requirements of the inhabitants exceed the immediate sup-

282

ply. The competitors do each other no direct harm in the sense of one cell feeding on its rival or producing toxins or enzymes that inhibit it; rather, the adverse influences arise indirectly through the struggle for mutual needs. In effect, two species are in competition only when the supply of a common requirement is too low to satisfy both. In *interspecific competition* the interaction involves individuals of different species; in *intraspecific competition* the struggle for the limited store is among individuals of the same species. Both are potent forces in natural selection.

A multitude of interspecific and intraspecific interactions have been described as a result of investigations of microorganisms in vitro and in vivo. In many instances the data strongly suggest that competition accounts for the behavior of the organisms under study. Interspecific competition has thus been proposed to explain the suppression of pathogenic bacteria colonizing burned animal tissues or the intestinal tract of germfree animals, the inhibition of plant pathogens by saprobic fungi, the dying out of aliens introduced into the sea with land drainage, the dominance of food-poisoning *Staphylococcus* in occasional food products, and the inhibition of fungi by bacteria jointly developing on cellulosic substrates. Intraspecific competition has been postulated to be the basis for the overgrowth of virulent plant and animal pathogens in cultures in which avirulent mutants arise, the assumption of dominance by antibiotic-sensitive mutants appearing in populations of resistant strains when the antibiotic is no longer present, and the displacement of bacterial populations showing one set of colonial characteristics by mutants producing dissimilar colony types.

COMPETITIVE DISPLACEMENT

In nutrient-rich habitats competition is inoperative at low population densities. In a laboratory study of two species this would be evident from the fact that each organism had similar growth rates in axenic and in two-membered cultures. As the abundance of cells or filaments rises, a point is reached where the available supply of some factor is exactly equivalent to that necessary to meet the demands of the two populations, with no diminution in their rates of multiplication; thereafter the concentration or level of the factor is insufficient to allow for maximum growth rates, the demand exceeding the supply. At this stage each population will get less of the factor that has become limiting than it would have obtained were it developing alone. Both the competitors will suffer to an extent. Nevertheless two species are generally not equally effective as competitors, and one will obtain or assimilate more of the limiting resource than the other; the latter, therefore, is more severely affected as a consequence. If the nutrient reserve in the environment is not replenished, as occurs in batch cultures in the laboratory, the final population density of each interactant will be less, to a minor or major extent, in mixed than in axenic culture; on the other hand, should fresh nutrients constantly enter the habitat—as is common in water, soil, sewage treatment plants, the gastrointestinal tract, and living tissues—one species will actually displace its neighbor.

Hence competition comes into play when the community is heterogeneous and the population density is high relative to the supply of a limiting feature of the environment. The precise number of cells or mass of filaments when competition becomes evident is governed by the amount of the limiting component. The concentration of the factor that can be said to be insufficient depends on its identity and the extent of the biological requirement for it: a micronutrient may be in plentiful supply though it is present in parts per million concentrations whereas a macronutrient may be in deficient quantity though occurring in a concentration of several per cent.

In certain poor or harsh habitats competition is of little significance. Should local conditions be so extreme, as on rocks colonized by lichens, that none of the factors for which microorganisms compete can be said to restrict growth, there will be no rivalry. As a rule the ratio of population density to a potentially limiting environmental resource is low in poor habitats, but the initiation of a rivalry requires that the ratio be high. Extensive predation, too, may keep the density of prey species so low that they fail to compete (Crombie, 1947).

A number of investigators have studied interspecific competition in vitro employing two-member cultures. The behavior of each species is usually noted and compared with that observed when the organisms are grown axenically. Where one organism has a highly distinct competitive advantage over the other because of its far greater growth rate under the prevailing conditions, no harm seems to come to it; when the competitive abilities of the two populations are not so markedly different, mutual harm is discernible. A typical instance involving *Escherichia coli* and *Staphylococcus aureus* is shown in Fig. 12.1a. The former bacterium produces no toxin active against the latter, but it is such a good competitor, presumably because of its shorter generation time, that its growth rate and final cell density

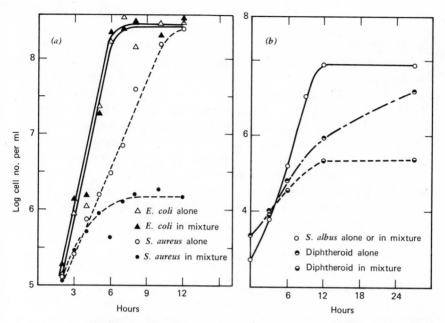

Figure 12.1 Competition between *Escherichia coli* and *Staphylococcus aureus* (a) and between *Staphylococcus albus* and a diphtheroid (b) (Oberhofer and Frazier, 1961; Annear, 1951).

are the same in the presence and absence of *S. aureus. S. aureus,* by contrast, fares poorly in mixed as compared with axenic culture. Figure 12.1*b* depicts a study of the same kind involving *Staphylococcus albus* and a diphtheroid. Typically the slowly growing organism continues to multiply so long as the limiting resource is still available to it; when the supply of this variable is exhausted, multiplication of the slow grower ceases. Similar experiments have been performed with species of the protozoan genera *Paramecium* and *Stylonychia* (Hairston and Kellerman, 1965; Gause, 1934). Of course, the fast growing organism stops multiplying when the limiting factor is used up, but this species will be present at a far higher population density.

Evidence for competition in natural ecosystems is difficult to obtain. In the climax community, furthermore, the operation of competition leads to a selection for better competitors and elimination of the less fit, so that much of the evidence has already been destroyed. Still the complete utilization of a given resource, organic nutrients for example, by a community is a strong indication that it is limiting, and should the addition of supplemental quantities of the factor in question bring about an increase in size or metabolic activity of one or more of the indigenous populations, a stronger case can be made for the operation of competition. Addition of increments of a non-limiting variable, as an inorganic nutrient, should not have an effect of this sort. Nevertheless, because of toxin production, parasitism, predation, nutrient interactions, and nonbiological reactions affected by the added materials, the results in vivo are frequently equivocal. In liquid media, conversely, as in the kinds of experiments cited above, direct evidence for competition can be gathered with little difficulty. Usually the level of the limiting factor in vitro is found to be low in the mixed culture, and addition of supplemental quantities relieves the suppression of one or both microorganisms; the chief concern in laboratory trials is to ensure that the detrimental effect recorded is the result of neither toxin production nor the direct feeding by the individuals of one population on cells of the other.

When nutrients enter a microenvironment containing two competitors whose cells are dying or are being washed out of that locale, one species may often completely eliminate and displace the second in time. The *competitive displacement principle,* sometimes termed the *competitive exclusion principle,* states that populations of two dissimilar organisms occupying the same area and having different growth rates will not long coexist, the one multiplying faster under the prevailing conditions displacing and causing the extinction of its neighbor. To illustrate by a simple laboratory test, if the pair are allowed to develop until they consume a limiting nutrient in the liquid medium employed, part of the solution is discarded, and the growth cycle is repeated until the limiting factor is used up again, and if the entire process is duplicated several more times, the faster-growing organism progressively makes up a higher proportion of the total number of cells and ultimately becomes the exclusive occupant of the microcosm. Conditions necessary for displacement include the following: (a) the populations must be physiologically dissimilar in some ecologically relevant way; (b) they must reside in the identical microenvironment; (c) they must require an identical resource; and (d) that resource must be limiting. The prominent physiological differences are those pertaining to an advantage, such as rate of replication, that one species has in obtaining the factor for which there is a struggle, and the organism possessing that beneficial trait in greater measure will tend to predominate. The better competitor is the more fit, and natural selection favors it. Should a variable like a growth factor, O_2, or a carbonaceous substrate restrict the development of one but not the other, or should the quantity of a resource that might be the basis for competition exceed the mutual demand, there will be no competitive displacement (DeBach, 1967; Hardin, 1960).

It is widely believed that competition is keenest between closely related strains and species. They tend to have more similar needs, nutrient requirements, biochemical capabilities, structural features, and tolerance ranges than dissimilar microorganisms. If this view is correct, then microorganisms with similar wants, performing analogous functions, and responding in like fashion to environmental stresses have a low probability of coexisting, and competition will tend to result in the elimination of the less fit rival. Although rigorous proof of this hypothesis is lacking, it is highly likely that populations coexisting in a microenvironment must be unalike in some ecologically significant way. The closely related microorganisms alluded to may be in identical taxonomic categories, but they need not; the essential issue is not systematic but rather ecological likeness, the role of the organisms in and their demands on the environment, rather than their position in handbooks of taxonomy. Still, because community heterogeneity rather than homogeneity is the rule in the decomposition of natural materials, such as plant remains, and inasmuch as countless genera contain only a single species in innumerable habitats, it seems that competition does indeed occur between related organisms with the extermination of the less fit.

AVOIDING COMPETITIVE DISPLACEMENT

Regardless of their spatial proximity, not all species resident in the same habitat are in competition. The limiting factor may not be identical for two populations, or they may occupy totally dissimilar ecological niches and hence are indifferent to each other's presence. Competition also may be reduced in intensity or prevented entirely if one or both of the interacting populations are subjected to attack by predators or parasites, an attack which can so drastically lower the number of cells that an essential environmental resource that might otherwise be in short supply remains plentiful. As the species that could potentially eliminate its neighbor is itself held in check by a predator or parasite, competitive displacement is effectively barred. The limiting factor, previously possibly a nutrient, is now predation or parasitism. Hence species occupying the same ecological niche can in fact coexist, even when competitive displacement might have been expected to run its course, when the better competitor is itself held in check by natural enemies.

An escape in time may also account for the coexistence of potential rivals. Species separated in time, as by the season of year favorable for their multiplication, will not compete directly for a variable in short supply. Furthermore, dissimilar populations may grow contemporaneously, but because properties of their habitat vary with time, one does not eliminate the other. Thus, for competitive displacement and the consequent selection to go to completion, the factor for which there is a struggle must not fluctuate during the course of time to an extent sufficient to upset the displacement; should the rivalry not be maintained and the resource not be kept limiting for an adequate period, the poorer competitor may persist. The properties of many habitats vary daily as well as seasonally, and the supply of the variable serving as the focal point for competition occasionally may, as a result, exceed the joint demand of the competitors so that the selection pressure is eased.

On the other hand, populations actively proliferating at the same time frequently avoid competition by a degree of spatial separation. The physical and chemical characteristics of macroenvironments vary in space as well as in time, and environmental heterogeneity is the rule rather than the exception in many regions. In some heterogeneous habitats, though organisms residing in the same microenvironment are subject to competitive displacement, the same will not take place among species that are spatially separate or displacement will be avoided when propagules of one species migrate to an adjacent microsite. Spatial separation, therefore, allows for the survival of populations having similar ecological requirements. It is probable, too, though proof is not at hand, that the greater the multiplicity of microhabitats, the greater is the number of rivals that coexist.

Competition among populations with similar needs may thus be avoided when each uses the same resource at a different time or place. Should the two organisms use somewhat dissimilar resources, such as carbon sources, the rivalry may also be minimized. Displacement may likewise be prevented when the poorer competitor is able to harm its rival directly, as by the formation of toxic metabolites. For these and undoubtedly other reasons, competitive displacement often does not go to completion.

In locales where rivals do coexist, the identities of the dominants will be governed by those physical and chemical factors affecting life in the region. Temperature, pH, O_2, moisture level, kind and amount of nutrients, and light intensity potentially have differential effects

Table 12.1

Environmental Factors Governing Success in Instances of Apparent Competition

Variable	Interacting Populations	Reference
Temperature	Staphylococcus—saprobic bacteria	Peterson et al. (1964)
Temperature	Skeletonema—Thalassiosira	Conover (1956)
Moisture	Fusarium—bacteria	Finstein and Alexander (1962)
pH	Pasteurella pestis: virulent and avirulent	Ogg et al. (1958)
O_2 level	P. pestis: virulent and avirulent	Delwiche et al. (1959)
CO_2	P. pestis: virulent and avirulent	Delwiche et al. (1959)

on the outcome of competition, and an organism winning out in one set of circumstances may be a minor member of the community or be totally displaced under dissimilar conditions. The impact of an environmental variable on the dominant species or strains in apparent instances of competition is shown in Table 12.1. For instance, at high temperature *Staphylococcus* is a good competitor with bacteria, whereas at low temperature it does not emerge as a significant member of the community (Peterson et al., 1964). In intraspecific interactions among *Pasteurella pestis* strains, avirulent mutants arising in a virulent culture outcompete the virulent population and displace it in the presence of O_2 but not under anaerobiosis (Delwiche et al., 1959).

COMPETITIVE ABILITY

Microorganisms vary enormously in their competitive abilities, and probably nearly every species is different in one or more physiological traits of consequence in competition, characteristics whose importance is governed by the resource serving as the focal point for the interaction. Traits associated with fitness in environments where there is little rivalry are often of marginal value in localities where there is a sharp struggle for a limited supply of an essential requisite, and competitive ability in the latter circumstances is a key facet of fitness.

The bases for the capacity of a species to compete effectively are largely unknown, but several features probably are involved. The following are proposed to be responsible for or contribute to the competitiveness of one or another organism.

(a) Growth rate. This may be assumed to be of prime importance. A species proliferating and making use of limiting nutrients or other resources rapidly would seem to have a distinct advantage over a slow grower.

(b) Tolerance to abiotic factors. The facility of growing at extremes to which the habitat is occasionally exposed—low temperatures in the spring, the heat of the summer, high light intensity, or low moisture levels—seems to be of major significance to occasional species.

(c) Tolerance to environmental fluctuations.

(d) Capacity to multiply at low concentrations of the limiting nutrient. Growth at low substrate levels is a property of a few species only. Some algae do notably well at extremely low nitrate and phosphate concentrations, and selected fungi multiply in liquids containing but a trace of organic nutrients.

(e) Efficiency in converting limiting nutrients into cellular constituents. The organism that can make a large quantity of cells or filamentous material per unit of nutrient assimilated may be favored.

(f) Requirement for growth factors. Where the concentration of vitamins or amino acids is minute, an auxotroph might be at a disadvantage by comparison with a prototroph.

(g) Ability to synthesize and store reserve substances and to use them when the food supply dwindles.

(h) Capacity to move from an area where the stock of the limiting resource is small to an adjacent, nondepleted microenvironment. Motility or the extension of filaments may be responsible.

It is not unexpected that high growth or colonization rates frequently determine the outcome of competition. The species multiplying more rapidly often can pre-empt a greater quantity of the limiting resource, especially if it is a nutrient, than adjacent populations. Even a very slight difference in growth rate among rivals could lead ultimately to displacement, given that nutrients are not exhausted and growth is not terminated in a manner that precludes one species from outgrowing the other. The facility for rapid proliferation is also advantageous during the explosive development common when simple organic nutrients suddenly are made available, as when organic materials are introduced into soil or water or when tissues or food products are exposed to colonization; in the initial burst of activity, bacteria with short generation times and fungi having spores that germinate readily and hyphae that spread rapidly are usually dominant among the pioneers. *Pseudomonas* and related gram-negative rods and terrestrial Phycomycetes are notorious in this regard. Most Phycomycetes are incapable of using the major carbonaceous substrates entering soil in the form of plant remains; yet their restriction to the small amounts of simple substrates present is compensated for by their having spores that germinate quickly and hyphae that develop readily, so that they can rapidly exploit the limited reserve of low-molecular-weight organic compounds and succeed in the initial phase of colonization (Garrett, 1956).

Studies in vitro have provided further information to reveal the benefits of rapid multiplication. For example, the ability of fungal hyphae to extend rapidly or of bacteria to replicate quickly is characteristic of good competitors in sterile soil inoculated with test cultures (Finstein and Alexander, 1962; Lindsey, 1965). Organisms possessing filaments that extend outward to colonize new substrates in advance of neighboring heterotrophs are well endowed for a pioneering role in succession, as suggested by work on model microcosms. Investigations of mutants arising in bacterial cultures also have shown that competition favors whichever has the shorter generation time, the wild type or the mutant, and should the mutant have the advantage, it outgrows and finally displaces the parent population (Atwood et al., 1951). The emergence of mutants with shorter generation times than the parents from which they originated seems to account for at least certain instances of the displacement of virulent pathogens by avirulent cells (Delwiche et al., 1959) and the natural selection for antibiotic-sensitive mutants in cultures of resistant bacteria, a selection that leads to the oft-observed loss of antibiotic resistance in cultures maintained in drug-free media (Herrmann and Steers, 1953). In continuous culture devices, like the chemostat, that permit the continuous growth of microorganisms in constant environmental conditions in the laboratory, there is also a favoring of bacteria with shorter duplication rates and the elimination of populations multiplying more slowly under the test conditions (Jannasch, 1968).

On the basis of the foregoing it should not be surprising that fast growers dominate in heterogeneous communities receiving readily available nutrients; these communities are characterized by a high degree of competition. The rhizosphere, for example, abounds with bacterial genera known for their vigorous proliferation, and heterotrophs multiplying slowly are relatively infrequent. The rapid growth permits more effective competition. Nevertheless species that invariably reproduce slowly do stand out in some densely inhabited natural ecosystems, so that it must be assumed that high growth rate is not the sole determinant of success; these dominants must have other qualities making them fit for their unique zones of habitation, such as the ability to use a highly resistant substrate that is decomposed so slowly that speed of colonization is of little consequence.

Tolerance of one or several abiotic factors may be of paramount significance in competition. Though a species grows over its full tolerance range in vitro, from the minimum to the maximum, in nature it is generally confined to a far narrower range of most of the ecologically important abiotic variables. Competition is responsible to a large extent for the narrower amplitude for existence of the species in nature as compared with axenic culture. Consider the case of two dissimilar competitors meeting along a gradient of intensities of an abiotic factor like temperature or pH. Each species is probably the better competitor at different portions of the gradient, so that competition favors the first or the second, depending on the intensity of the parameter in question. Microorganisms that seem, on the basis of their distribution in nature, to be most suited to one intensity may in fact have their optimum at a far different point in vitro. For instance, fungi as a group exhibit greater aptitude for the colonization of moderately acid habitats than do bacteria and actinomycetes; although this may result occasionally from the inability of the bacteria and actinomycetes to multiply at low pH, frequently the dominance of the fungi in regions of nonextreme acidity results from their greater competitiveness.

RESOURCES FOR WHICH MICROORGANISMS COMPETE

Any essential requisite is potentially the basis for a competitive interaction. The population must get from its habitat a source of energy, the elements necessary for the building of protoplasm, water, and, for the aerobes, O_2. Space must be adequate, too. Because the resources but not the inherent abilities of organisms to produce more of their own kind are limited, one or more of these resources become depleted as the resident populations multiply. Rivalries for carbonaceous substrates, nitrogen, phosphorus, O_2, light, water, and space have, for one ecosystem or another, been proposed as regulators of species composition of a community or of cell densities. Inorganic micronutrients, however, have yet to be seriously considered as the basis for a competitive interaction, for their supply usually exceeds the demand.

Energy sources are unquestionably limiting in many regions inhabited by heterotrophs, and the struggle for available carbonaceous substrates is probably a major microbial interaction in such ecosystems. It is important to emphasize the concept of available organic materials, because frequently much organic carbon is present, as in soil, but little of it is readily utilized (Alexander, 1961). Competition of this kind can be relieved merely by adding carbonaceous nutrients, a treatment leading to a burst of metabolic activity and a vast increase in the density of those populations that metabolize the specific compound introduced. It also appears that bacteria may be so effective in destroying simple organic compounds formed continuously in lake water that they can prevent heterotrophic proliferation of algae, despite the fact that the biomass of algae, many of which develop heterotrophically in axenic culture, far exceeds that of bacteria in the lake (Wright and Hobbie, 1966).

The abundance of predators or parasites is often regulated by the supply of their living food sources. As might be predicted, competition intensifies as the feeding causes a drop in the number of accessible prey or host individuals, and deliberate addition of the particular viable foods into the habitat generally increases the density of the feeder.

In regions where the carbon demand is satisfied, nitrogen competition may assume prominence. The immediate carbon demand is fulfilled and exceeded in waste waters or soils receiving appreciable amounts of readily oxidizable organic substrates, and the quantity of nitrogen needed for the large number of organisms that potentially could be supported might be greater than that present in the area. Moreover, nitrogen deficiencies are frequently induced or exaggerated by carbon amendments; the microorganisms proliferating at the expense of the carbon supplement assimilate a considerable amount of the inorganic nitrogen in the surroundings, and the level falls to a point where little or no ammonium or nitrate can be detected. Hence, where the heterotrophic need for carbon is fulfilled, a shortage of available nitrogen commonly develops, and a struggle for the little present in available form may be initiated. The ecological importance of nitrogen competition is disclosed in studies of the suppression of soil-borne plant pathogenic fungi like *Fusarium solani* (Snyder et al., 1959), *Ophiobolus graminis* (Garrett, 1956), and *Rhizoctonia solani* (Blair, 1943), a suppression that not only is promoted by additions to soil of organic materials that cause the microbial community to assimilate and tie up the inorganic nitrogen, but one that is nullified when supplemental nitrogenous materials are applied. The algae of aquatic ecosystems usually seem to be sated with the CO_2 they require for photoautotrophic life, and sometimes inorganic nitrogen regulates their biomass and abundance; however, the interspecific rivalry for this element among algae has not been explored.

Phosphate is rarely in short supply for terrestrial heterotrophs, but the concentration of this anion is often exceedingly low in water, particularly in view of the abundance of sunlight and CO_2 that the algae demand. Nitrogen deficiency may have a greater impact than that of phosphate in some waters, but often the nitrate level is high enough for photoautotrophs. Furthermore, N_2-fixing blue-green algae obtain their nitrogen, as well as carbon, from the atmosphere—or, more properly, from N_2 or CO_2 dissolved in the water—and therefore they often encounter a phosphorus insufficiency. Surprisingly, a thorough scrutiny of phosphorus competition among the algae has not been made, but evidence exists that bacteria do sometimes compete for inorganic phosphorus in lakes (Hayes and Phillips, 1958; Rigler, 1956).

A rivalry for low concentrations of growth factors is demonstrable in culture solutions. For example, *Phytophthora cryptogea* and *Arthrobacter* compete in vitro for the thiamine they both require (Fig. 12.2), and the suppression of *Staphylococcus aureus* by *Streptococcus diacetilactis* is attributable mainly to the assimilation of nicotinamide by the latter bacterium (Iandolo et al., 1965). Analogous effects have yet to be shown in nature, but it is likely that aquatic algae might occasionally compete for the B vitamins they need.

Competition for O_2 undoubtedly takes place among aerobic heterotrophs. Dissolved O_2 is generally exhausted when appreciable microbial growth occurs in nature, given a sudden influx or availability of organic materials—for example, in waters acquiring carbonaceous pollutants, soils receiving plant residues, plant tissues undergoing decay, and wounds penetrating the skin. In addition, the transfer of O_2 is slow to microhabitats within environments with a solid matrix, as in the floc of activated sludge, remote soil pores, or regions below the surfaces of heaps of organic refuse, and here, too, competition for O_2 is probably of consequence. In such circumstances aerobes will struggle

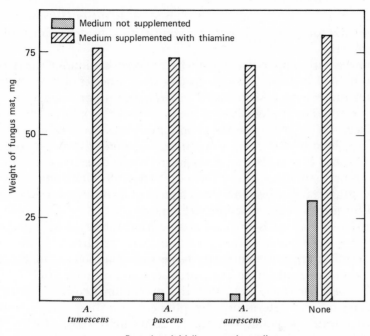

Figure 12.2 The effect of thiamine-requiring *Arthrobacter* spp. on growth of *Phytophthora cryptogea*. The bacteria were grown in a nutrient-poor medium containing thiamine. The cells were then removed by centrifugation, more nutrients were added, the supernatant was divided into two parts, and thiamine was added to one portion only. The fungus was then inoculated into the medium, and the weight of the fungus mat was determined after incubation (Erwin and Katznelson, 1961).

for the initially available O_2 as it is being used up, and for the O_2 that re-enters by diffusion from the air (Clark, 1965). Suppression of aerobes in dual-membered cultures in liquid media also has been attributed to competition for dissolved O_2.

It may be assumed that competition for light plays a role in aquatic algal communities as the phytoplankton becomes more dense and the intensity of light passing through the algal suspension progressively falls. Cells near the surface get their full share, underlying cells receive a smaller amount, and those at the lowest layer of the algal bloom are very likely to be light-limited. At the same time, evidence exists that CO_2 sometimes is inadequate in lake water to maintain optimum rates of photosynthesis at high algal densities (Wright, 1960); hence CO_2 too may be a basis for competition in dense phytoplankton growths.

Some question has been raised whether competition for water takes place in arid regions. Though species with a better water economy, greater capacity to absorb water, or drought-resistant stages are definitely favored in dry localities, it seems improbable that the decline or displacement of some species at low moisture levels results from a direct influence of one microorganism on another through competition. Instead it is likely that drought stress acts directly on the microorganisms themselves. Here the population densities and activities of the inhabitants are so low that chances of a rivalry for water, or indeed for any essential resource, are remote.

Unresolved as yet is whether competition for space is ecologically of consequence. Such a rivalry assumes that the physical dimensions of microsites are not sufficiently large to accommodate all the cells that otherwise could have been produced in a nutrient-rich habitat. A view once in vogue held that the maximum density of bacterial populations was determined by physical crowding, the organisms growing until they ran into space restrictions (Bail, 1929). This belief has fallen into disrepute, and it is now generally maintained that crowding does not govern the maximum number of bacteria possible in a liquid medium and that the population density of most species can be markedly increased by raising the nutrient concentration in laboratory media, provided that autoinhibitors are not produced. Thus in axenic cultures the number of bacteria may reach 10^{11} or more cells/ml of medium (Tyrrell et al., 1958), cell densities greater than found in nature. This suggests that competition for space among bacteria of aquatic environments probably does not occur. Furthermore, microscopic examination of environments with a solid matrix, such as soil, reveals the existence of uncolonized sites, so that there would seem to be little spatial restriction on the further development of bacteria, fungi, or actinomycetes, and hence a struggle for space here, too, is not particularly likely.

On the other hand, the belief that competition for space or lodging sites does happen is neither unreasonable nor totally ruled out. Minute localities are repeatedly enriched with particulate organic matter, and these particles often become coated with a protoplasmic mass. Fungal hyphae may struggle for micropores or channels in soil or for portals of entry through natural openings into plants, and the filaments of one species may thereby exclude those of another. The stones in trickling filter beds support a reasonably thick layer of cells, and here the nutrients continue to flow past the microbial mass and possible toxic products are washed away, circumstances that could allow for a competition for space within the cell mass. It is also not clear whether competition exists for lodging sites, as among the algae clinging to rocks in swiftly flowing streams, microorganisms adhering to intestinal walls, or cells attaching to surfaces of other structures over which a fluid is passed.

Nevertheless it still must be admitted that the significance of interspecific competition in natural communities is frequently unclear. Unequivocal evidence that a particular species is excluded from some environment because of competition is rare or lacking. The relevant experiments are difficult to perform. Yet many lines of evidence, albeit indirect, support the contention that competition is of profound importance not only in determining which species are present and which are excluded but also in governing the relative abundance of the indigenous organisms of natural ecosystems.

References

REVIEWS

Clark, F. E. 1965. The concept of competition in microbial ecology. *In* K. F. Baker and W. C. Snyder, Eds., *Ecology of Soil-Borne Plant Pathogens.* University of California Press, Berkeley. pp. 339–345.

Crombie, A. C. 1947. Interspecific competition. *J. Animal Ecol.,* **16,** 44–73.

DeBach, P. 1967. The competitive displacement and coexistence principles. *Ann. Rev. Entomol.,* **11,** 183–212.

Gause, G. F. 1934. *The Struggle for Existence.* Williams & Wilkins, Baltimore.

OTHER LITERATURE CITED

Alexander, M. 1961. *Introduction to Soil Microbiology.* Wiley, New York.

Annear, D. I. 1951. *Australian J. Exptl. Biol. Med. Sci.,* **29,** 93–99.

Atwood, K. C., Schneider, L. K., and Ryan, F. J. 1951. *Proc. Natl. Acad. Sci. U.S.,* **37,** 146–155.

Bail, O. 1929. *Z. Immunitaetsforsch.,* **60,** 1–22.

Blair, I. D. 1943. *Ann. Appl. Biol.,* **30,** 118–127.

Conover, S. A. M. 1956. *Bull. Bingham Oceanog. Collection,* **15,** 62–112.

Delwiche, E. A., Fukui, G. M., Andrews, A. W., and Surgalla, M. J. 1959. *J. Bacteriol.,* **77,** 355–360.

Erwin, D. C., and Katznelson, H. 1961. *Can. J. Microbiol.,* **7,** 945–950.

Finstein, M. S., and Alexander, M. 1962. *Soil Sci.,* **94,** 334–339.

Garrett, S. D. 1956. *Biology of Root-Infecting Fungi.* Cambridge University Press, London.

Hairston, N. G., and Kellerman, S. L. 1965. *Ecology,* **46,** 134–139.

Hardin, G. 1960. *Science,* **131,** 1292–1297.

Hayes, F. R., and Phillips, J. E. 1958. *Limnol. Oceanog.,* **3,** 459–475.

Herrmann, E. C., and Steers, E. 1953. *J. Bacteriol.*, 66, 397–403.

Iandolo, J. J., Clark, C. W., Bluhm, L., and Ordal, Z. J. 1965. *Appl. Microbiol.*, 13, 646–649.

Jannasch, H. W. 1968. *Appl. Microbiol.*, 16, 1616–1618.

Lindsey, D. L. 1965. *Phytopathology*, 55, 104–110.

Oberhofer, T. R., and Frazier, W. C. 1961. *J. Milk Food Technol.*, 24, 172–175.

Ogg, J. E., Friedman, S. B., Andrews, A. W., and Surgalla, M. J. 1958. *J. Bacteriol.*, 76, 185–191.

Peterson, A. C., Black, J. J., and Gunderson, M. F. 1964. *Appl. Microbiol.*, 12, 77–82.

Rigler, F. H. 1956. *Ecology*, 37, 550–562.

Snyder, W. C., Schroth, M. N., and Christou, T. 1959. *Phytopathology*, 49, 755–756.

Tyrrell, E. A., MacDonald, R. E., and Gerhardt, P. 1958. *J. Bacteriol.*, 75, 1–4.

Wright, J. C. 1960. *Limnol. Oceanog.*, 5, 356–361.

Wright, R. T., and Hobbie, J. E. 1966. *Ecology*, 47, 447–464.

13

Amensalism

Microbial communities supporting high densities of dissimilar populations are characterized by interactions detrimental to many of the inhabitants. Often prominent among these harmful interrelationships is amensalism, and the view that amensalism is ecologically consequential has gained favor from the innumerable reports that members of an enormous number of genera and species synthesize and excrete compounds in vitro injurious to the development of nearby populations. The responsible organisms are considered to condition their surroundings by excreting organic or inorganic metabolites deleterious to coinhabitant or transient populations, substances that may serve as chemical regulators of the community's composition.

Inhibitors generated as a direct result of microbial metabolism are widely considered to be protective devices or weapons in the struggle for existence. Many investigators have suggested that they are important in determining microbial distribution and population size and in governing succession. By modifying the environment in a manner unfavorable to other species, the toxin producer presumably improves its own chances for ecological success. It has also been suggested that toxin synthesizers are responsible for: (a) the elimination of certain pathogens that enter the throat, reach the skin, or come into contact with plant roots; (b) the control of soil-borne plant-pathogenic fungi that results when organic remains introduced into soil promote the development of enormous cell numbers; (c) the disappearance of aliens entering the sea from sewage outfalls or land runoff; and (d) the destruction of foreign bacteria reaching the rumen.

For the purposes of discussion, it is convenient to separate microbiologically synthesized toxins into three categories: inorganic substances, organics of low potency, and highly toxic organic compounds. Inorganic inhibitors such as hydrogen peroxide, ammonia, nitrite,

300

CO_2, O_2, and H_2S are implicated in biological changes in nature, and their importance, unfortunately, is often overlooked; the concentration needed for activity is in some instances large, but frequently only minute quantities are required. Organic inhibitors that must be present in high concentrations to be effective are also synthesized by a multitude of species, and the amounts generated exceed the levels required for potency; typical are simple fatty acids and ethanol. Finally, highly potent organic inhibitors are excreted by a diverse assemblage of heterotrophs and photoautotrophs. These are generally chemically complex metabolites, and to them is applied the term antibiotic.

A variety of methods have been employed to demonstrate amensalism in vivo. A common procedure is to plate dilutions of the natural specimen on a suitable agar medium and then look for colonies surrounded by clear zones in plates supporting many microbial types. The molten and cooled agar may be previously seeded with a test organism to determine its sensitivity to the toxins produced. Inhibitor formation by fungi is frequently first noted on a solid medium on which a susceptible organism fails to extend itself into a zone around the hyphae generating the harmful agent. The synthesis of inhibitors in two-membered cultures growing in liquid media is often suggested initially by the abrupt cessation of multiplication or the unexpected death of one of the two populations (Fig. 13.1). Other techniques are available to show the ubiquity and abundance of microorganisms capable of forming, in vitro at any rate, simple or complex substances that inhibit or kill microscopic forms of life.

The elaboration of an antimicrobial principle may be advantageous to the producer in that, by diminishing the growth rate or killing cells of the suscept, the injurious metabolite may reduce interspecific competition for nutrients, light, or physical sites for multiplication. Species that multiply slowly may not suffer from the handicap of low replication rates should they be active in releasing compounds harmful to their neighbors. During early phases of colonization or when large amounts of organic materials suddenly become available in an ecosystem, the toxins could serve to exclude many pioneers, with consequent gain to the producer of the poisonous substance. Amensalism may in these ways contribute to success in competition, but it is a phenomenon distinct from competition. Nevertheless, few microbial products are so potent or nonselective as to suppress all populations in the community, and the resistant species will thrive in areas where the metabolites are indeed generated. In this light, harmful products might be viewed as factors contributing to the

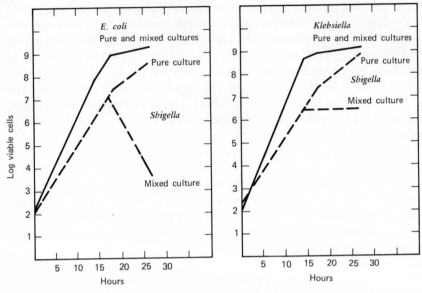

Figure 13.1 The inhibition of *Shigella flexneri* by two coliform bacteria. Bacteria were grown in axenic or two-membered cultures (Hentges, 1969).

makeup of the community rather than as the sole determinants responsible for the dominance of a particular species. Moreover, though amensalism occasionally has a widespread impact, many of the active compounds will have little more than a local effect, their zone of influence extending only for short distances away from the cell or filament responsible for the synthesis. Amensalism may also be responsible for the protection of a cell against invasion by nearby parasites or destruction by heterotrophs excreting enzymes that digest components of the microbial cell wall.

If it is assumed that the elaboration of a toxin by one population is ecologically advantageous, it follows that resistance to the substance by an adjacent population should also be beneficial. Vast differences do in fact exist in microbial tolerances. Such resistance is often attributable to the synthesis of detoxifying enzymes, the impermeability of the resistant cell to the offending compound, or the possession of metabolic pathways or cellular constituents unaffected by the poison. Whether the apparently advantageous attribute is toxin formation or resistance, its possession does not guarantee the ubiquity or abundance of an organism, for many other physiological or morphological traits contribute to ecological success. Furthermore, the producer of one toxin may be held in check because of its sensitivity to a compound excreted in the identical microenvironment by a neighbor, or it may itself suffer because the species eliminated by the inhibitor may have been just those that held down the competitors or parasites of the cells forming the injurious metabolite. It should also be borne in mind that suppression mediated by toxic principles can be evaded in a heterogeneous environment, the escape being accomplished by virtue of the availability of inhabitable microsites into which the compound does not penetrate.

TOXINS IN NATURE

Antimicrobial agents have been found in diverse habitats rich in microorganisms, and many of these substances presumably owe their genesis to the metabolism of microscopic residents of the region. As a rule the product is obtained in solution for laboratory study by extraction of samples of the environment in question, and then it is concentrated before a bioassay is attempted, so that the compound may have been collected from an enormous number of microenvironments and be concentrated far more highly than in the natural habitat. Such techniques generally do not reveal the existence of high local concentrations in vivo, and they give only a picture of the steady-state level of the inhibitor rather than the amount formed or decomposed in any unit of time. Nevertheless the results obtained are extremely informative and show the wide occurrence and variety of antimicrobial substances formed by microbiological processes.

As indicated in Table 13.1, substances harmful to bacteria, yeasts, algae, and fungal hyphae and spores have been observed in specimens obtained from widely different kinds of environments. Many organisms in addition to those listed are noted to be suppressed, too, in particular by substances derived from soil. Attempts to characterize the active principles, incomplete as the studies are, reveal that their chemical and antimicrobial properties are so dissimilar that no question remains that an array of compounds are involved in these several microbial habitats.

Because most antibiotics of chemotherapeutic value are derived from organisms isolated from soil, this environment has been of considerable interest in regard to its content of antimicrobial principles. Although it has been known for some time that soils contain water- or ether-soluble bacteriostatic and bactericidal components acting, for example, on *Azotobacter, Bacillus,* and *Rhizobium,* far greater

Table 13.1

Environments Containing Microbial Inhibitors

Environment	Susceptible Group	Test Species	Reference
Sea water	Bacteria	Staphylococcus	Saz et al. (1963)
	Fungal spores	Zygorhynchus	Borut and Johnson (1962)
Marine plankton	Yeasts	Rhodotorula	Buck and Meyers (1965)
Lake water	Algae	Scenedesmus	Zavarzina (1959)
Pond with Pandorina bloom	Algae	Chlorella	Rice (1954)
Soil	Bacteria	Staphylococcus	Klosowska (1958)
	Fungal spores	Trichothecium	Hessayon (1953)
	Fungal hyphae	Botrytis	Jefferys and Hemming (1953)
Flooded soil	Fungi	Native populations	Mitchell and Alexander (1962)
Peat	Bacteria	Native populations	Pochon and deBarjac (1952)
Rumen fluid	Bacteria	Escherichia	Wolin (1969)
Cheese	Bacteria	Clostridium	Grecz (1964)
	Fungi	Penicillium	Grecz (1964)
Saliva	Bacteria	Corynebacterium	Thompson and Shibuya (1946)
Potato leaves infected with Phytopthora	Viruses	Virus X	Hodgson et al. (1969)

attention has been given to the antifungal constituents. The toxins are found in soils from all corners of the globe, and they suppress fungal spore germination, hyphal development, or both, depending on the test species. Though it is possible that the inability of spores of certain fungal groups to germinate readily in soil results not from a toxin but from the lack of needed nutrients, an actual suppression must hold for those spores requiring no exogenous nutrients. These propagules germinate in distilled water; yet, when in contact with soil, no germ tubes emerge. Because of the inhibition, which affects different species to varying extents, the spores remain dormant for prolonged periods, but the suppression is overcome, for reasons still unclear, when suitable substrates are made available. The toxicity is abolished in large part when the soil is heated, but reappears upon subsequent colonization, suggesting the biological origin of the toxicity. Though the data are scant, the available evidence suggests that no one single compound but rather a complex of substances contributes to the antimicrobial action of natural soil (Lockwood, 1964; Park, 1967).

Bactericidal and fungistatic components are also found in sea water, but their identities and their contribution to the death of propagules derived from land or sewage remain the subject of continuing controversy. Part of the toxicity of the ocean and coastal waters is undoubtedly associated with their salinity and pH, but toxicity has also been attributed partly to microbial toxins (Kriss et al., 1967; Mitchell, 1968). Pond and canal waters bearing blooms of algae like *Anabaena, Microcystis, Oscillatoria, Pandorina,* and *Spirogyra* contain substances deleterious to other algae when tested in culture, but the evidence is not sufficiently good to conclude whether the active principles are made by the dominant photoautotroph of the bloom or its associated heterotrophs.

The tissues of the human mouth have been the subject of special scrutiny because of their remarkable resistance to infection, and the early belief that this resistance was linked with the antibacterial action of saliva has been confirmed experimentally. Growth of members of several bacterial genera is retarded by human saliva, and at least two substances seem to be involved, a simple and a high-molecular-weight compound. Some of the harmful effects may result from products of the salivary glands or scaled-off epithelial cells, but many researchers feel that metabolites excreted by bacteria resident in the mouth play a key role (Van Kesteren et al., 1942; Hoffman, 1966).

The manufacture of cheeses frequently relies on the metabolic changes induced by bacteria and fungi, and a high-quality product must be free of human pathogens and of significant numbers of cells that liberate compounds imparting "off" flavors. In addition to antimicrobials like nicin and diplococcin formed in milk by species of *Streptococcus,* components of the natural communities of certain aged cheeses make potent bactericides. *Brevibacterium linens* dwelling at the surface of ripened cheese is implicated in toxin biosynthesis, but yeasts and other bacteria may participate as well. Because these compounds are both effective and widespread, it seems likely that inhibitors elaborated by the indigenous microflora contribute to the population changes taking place during the curing of cheese (Grecz, 1964; Hirsch et al., 1952).

Extracts of the cellular material of a number of algae are antibacterial. Toxicity of this sort has been noted among the Bacillariophyta, Chrysophyta, Chlorophyta, and Pyrrophyta, and it is especially widespread among the seaweeds. The injurious cellular constituents are present among freshwater and marine algae and in samples taken from the tropical and temperate zones. Although results of studies on cell extracts imply a potential for a general or selective suppression of nearby bacteria, the significance of these toxins, or whether they are even released in aquatic environments, is almost completely unknown.

The excretion of injurious metabolites may also account for the overgrowth of autoinhibitor-forming bacteria by resistant mutants that appear. Autoinhibitors made by the parent population alter the selective conditions so that it is at a marked disadvantage and grows more slowly or dies out, while a mutant fit to proliferate in the modified circumstances gains the ascendancy by virtue of its tolerance to the metabolite of the wild type. The parental cells thus, by forming selectively toxic products, determine which of the spontaneous mutants that arise will be successful. Such inhibitor-mediated population changes have been reported for pathogens like *Salmonella typhimurium* and *Pasteurella tularensis* (Fig. 13.2), and the ultimate outcome with these bacteria is commonly a change in virulence of the culture (Page et al., 1951; Yaniv and Avi-Dor, 1953). Autoinhibition has also been postulated to account for the termination of massive blooms of aquatic algae (Lefevre, 1964) and for the replacement of one protozoan species by another.

Haematococcus pluvialis provides an interesting illustration of the apparent effect of microbial toxins in controlling the distribution of an organism. This green alga, although widely distributed, is almost invariably restricted to small, ephemeral pools of water as exemplified by rocky depressions, road ditches, and concrete basins. In nearly permanent ponds and pools, it is rare. Nutrients adequate for *H. pluvialis* are present in permanent bodies of fresh water, but it seems to be excluded by its sensitivity to toxins elaborated either by components of the established algal bloom or by heterotrophs accompanying the bloom algae. Areas where the temporary ponds arise, on the other hand, are exposed to prolonged desiccation and high temperatures, and *H. pluvialis* is beneficially endowed with resting spores that withstand extended periods of drought and elevated temperatures and are still able to germinate readily when environmental conditions become less extreme. Other potential colonizing algae, by contrast, generally either possess no known specialized resistant stage or their resting bodies germinate slowly. Hence the alga's susceptibil-

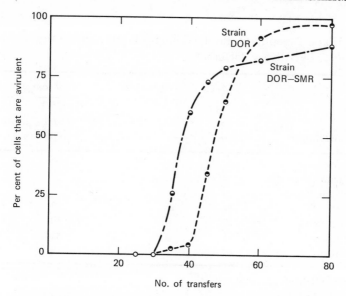

Figure 13.2 Appearance of avirulent cells on serial subculture of virulent strains of *Pasteurella tularensis* (Yaniv and Avi-Dor, 1953).

ity to toxins formed in permanent bodies of water is compensated for by its capacity to survive longer and to initiate growth sooner than potential competitors (Proctor, 1957).

INORGANIC INHIBITORS

A common mechanism of amensalism is acid production, and suppressions based on this kind of interaction are widely evident. The acids may be either inorganic or organic. Autotrophs altering the pH of their surroundings through excretion of inorganic acids drastically modify the composition of the community. Notorious instances are found among species of the chemoautotrophic genus *Thiobacillus,* which make copious quantities of sulfuric acid in oxygenated mine waters in contact with sulfide ores, in land drained free of sea water so that the previously submerged iron sulfides are exposed to air, or in soil treated with appreciable amounts of elemental sulfur. The pH falls precipitously, and the community of soils amended with sulfur or of streams receiving acid mine waters is largely decimated or is modified dramatically to one dominated by acid-tolerant

Table 13.2

Correlation between H_2O_2 Production and Inhibition of *Corynebacterium diphtheriae* by Salivary Bacteria [a]

Bacterial Group	No. of Strains Tested	Toxigenic Strains		Nontoxigenic Strains	
		Total No.	No. Producing H_2O_2	Total No.	No. Producing H_2O_2
Streptococci Producing green zones on blood agar	76	69	69	7	0
No such zones	10	0	0	10	0
Staphylococci	34	0	0	34	0
Diphtheroids	10	1	0	9	0

[a] From Thompson and Johnson (1951).

species. At the other extreme is amensalism due to the synthesis of metabolites, inorganic or organic, that cause an increase in alkalinity; this is illustrated in occasional rock pools bearing abundant growths of *Ulva*, an organism that may bring about a rise in alkalinity to the vicinity of pH 10, conditions so extreme that many algae are excluded (Atkins, 1922).

Innumerable groups of microorganisms make hydrogen peroxide, a relatively nonspecific but reasonably potent inhibitor of proliferation. Studies in vitro have established that the ability of one species to hold down another sometimes results from the generation of hydrogen peroxide. This product has also been proposed as a reason for the striking antibacterial action of saliva in vivo and as one of the metabolites governing the community composition of human saliva. On the basis of tests employing *Corynebacterium diphtheriae*, the view has been advanced that the antibacterial properties of saliva result specifically from the peroxide elaborated by individual groups of indigenous streptococci and not by other bacteria (Table 13.2). Interestingly, some salivary residents have the power to reverse the antibacterial potency of saliva, and these are the microorganisms containing catalase, a H_2O_2-decomposing enzyme. A few authorities have challenged the view that hydrogen peroxide plays a key role in the community makeup, arguing that enzymes like catalase and peroxidase will destroy the extracellular hydrogen peroxide so that harmful

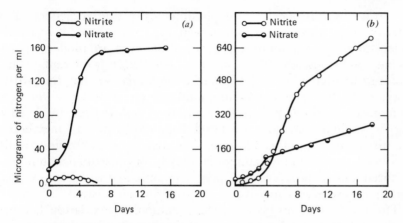

Figure 13.3 The effect of ammonium concentration on the accumulation of nitrite and nitrate in a slightly alkaline soil. Rate of ammonium addition: 125 μg (a) and 890 μg (b) per milliliter of soil solution (Stojanovic and Alexander, 1958).

levels are never attained. Nevertheless the steady-state concentration, representing the balance between microbial production and enzymatic destruction of peroxide, may be sufficiently high for the metabolite to be ecologically important (Kraus et al., 1957; Thompson and Johnson, 1951).

Another significant, injurious inorganic product of microbial metabolism is ammonia. Ammonia is liberated, occasionally in large amounts, in the decomposition of proteins and amino acids, the reduction of nitrate, and the hydrolysis of urea. Microorganisms living in proximity to a site from which ammonia is volatilized or in alkaline solutions where ammonia is formed are deleteriously affected, even though the nitrogen compound is a substrate for many of the organisms. Heterotrophs, chemoautotrophs like *Nitrobacter,* the phytoflagellate *Prymnesium,* and the diatom *Nitzschia,* among others, are all known to be harmed by minute quantities of ammonia. As a rule the effect becomes more pronounced with increasing pH, and the pH-controlled pattern of inhibition with several of the sensitive populations parallels the shift in the ammonia-ammonium equilibrium in a manner that makes it likely that ammonia rather than ammonium is the responsible toxicant. The outcome of such a toxicity is shown in Fig. 13.3. In aerated soils, as in aquatic ecosystems, ammonium nitrogen is oxidized to nitrite largely by *Nitrosomonas* and related chemoautotrophs, and the nitrite formed is further oxidized to nitrate by *Nitrobacter* and physiologically similar genera. The activity of the ammonium oxidizers, like *Nitrosomonas,* is relatively independent of ammonium concentration even in moderately alkaline conditions, but *Nitrobacter* is remarkably intolerant of ammonium at high pH, or presumably ammonia, so that the nitrite liberated by *Nitrosomonas* accumulates to high levels. Only when the ammonium oxidizers relieve the toxicity to *Nitrobacter,* by destroying the ammonium, does the latter begin to grow and metabolize the nitrite at an appreciable rate. The toxicity of microbiologically produced ammonium, or ammonia, may be of consequence in highly polluted waters, in bays or rock pools contaminated with bird excreta, and in soils receiving massive amounts of animal wastes.

The nitrite that builds up during ammonium oxidation because of the differential tolerances of *Nitrosomonas* and *Nitrobacter,* or physiologically related nitrogen oxidizers, itself could have a tremendous impact on the community inasmuch as nitrite is a potent poison. Such an influence is occasionally indicated when urea fertilizers are used in agriculture. As members of the soil community possessing the enzyme urease hydrolyze urea to ammonia, the pH climbs abruptly. The ammonia in turn not only is oxidized to nitrite but simultaneously checks the further microbiological oxidation of the nitrite. This urea-engendered nitrite accumulation apparently inhibits soil-borne fungi like pathogenic forms of *Fusarium oxysporum,* and it likewise results in a reduction in the severity of at least one plant disease caused by *F. oxysporum* (Sequeira, 1963).

Environmental changes associated with the biological consumption of O_2 regularly lead to the elimination of previous dominants. As obligate and facultative aerobes multiply in regions where the rate of diffusion of O_2 is inadequate to meet the demand, the concentration of dissolved O_2 declines until little or none is detectable. The E_h falls simultaneously, and the resulting decline of numerous indigenous populations may result from the absence of a major nutrient, O_2, as well as from an E_h too low to permit proliferation of aerobes; the precise impact of E_h on the community upset, however, has not been fully ascertained. Linked with the resulting anaerobiosis and the low potential is the appearance of other toxicants, H_2S and organic acids in particular, and hence an array of forces make anaerobiosis an effective means whereby one portion of the community does material harm to a second (Mitchell and Alexander, 1962; Meynell, 1963). Conversely, the O_2 liberated during algal photosynthesis, and possibly the high E_h characteristic of oxygenated waters, greatly diminishes the rate of biological processes dependent on the absence of O_2, like denitrification (Jannasch, 1960), or is responsible for the killing of obligate anaerobes.

Hydrogen sulfide accumulation is, as stated, a process typically associated with anaerobiosis. The sources of this potent destructive product are two, the sulfate reduced by *Desulfovibrio* or physiologically related anaerobes and the sulfur-containing amino acids attacked during proteolysis. Sulfide biogenesis is common in lake and marine sediments, flooded soils, ditches, pond mud, sewage digestion tanks, feces, and the intestine, and the sulfide formed inactivates countless protozoa in sewage digestion tanks and fungi in poorly aerated soil, as well as heterotrophs of the intestine.

Largely unrecognized by microbiologists is the fact that CO_2, a widespread product of organic matter decomposition and algal respiration, is also a differential inhibitor of microorganisms, and it is probably of ecological significance because of its nearly universal formation. Generally the CO_2 content of natural ecosystems rises as the O_2 level falls and as the chances for ready gas exchange with the overlying atmosphere diminish. It seems likely that the concentration of the toxicant in many habitats is far greater than the minimum required for an untoward biological effect. Sizable levels of CO_2 accumulate in the rapid initial phase of decomposition of plant remains or organic pollutants, in the vicinity of roots where both plant respiration and heterotrophic utilization of root excretions and sloughed-off tissues contribute to CO_2 evolution, and in bogs, food products, and necrotic animal and plant tissues. Tolerant communities probably are characteristic of these environments, but surprisingly little evidence exists to show that this is in fact true.

With scores of species growing in axenic culture, a slight increase in the partial pressure of CO_2 over the trace in the ambient atmosphere enhances vegetative development or one particular stage in the life cycle. The introduction of still more CO_2 into the culture vessel, however, uncovers the differential sensitivity of microorganisms to this metabolite. Such studies in vitro demonstrate that various heterotrophs are essentially indifferent to the ranges of CO_2 encountered commonly in nature, some are moderately sensitive, and others are markedly retarded by small amounts in the gas phase. Mycologists have paid particular attention to this problem, and their results indicate that mycelial extension, spore germination, and the formation of conidia, sporangia, and chlamydospores by fungi of diverse classes are retarded by CO_2 to different degrees. Investigations in vivo also strongly argue for the ecological importance of this product; for example, changes in the concentration of CO_2 in the soil air alter the relative abundance of fungi (Griffin, 1966) and the extent of infection of seedlings by specific pathogens (Louvet and Bulit, 1963). Furthermore, the distribution of strains of a fungus like *Rhizoctonia solani* appears to be regulated by its CO_2 tolerance; strains inhabiting aerial parts of plants are most sensitive to CO_2, those residing in underground habitats are most tolerant, and strains occurring near the soil surface exhibit an intermediate response (Durbin, 1959).

SIMPLE ORGANIC TOXINS

Simple organic molecules are generated in profusion during the metabolism of a natural community. Most of them are, at the concentrations they attain, innocuous, either because the compound is essentially nontoxic or because too little is formed. A small number, though never present in great quantities, are extremely potent and have an effect where and when a sufficient amount accumulates. Another group of microbial metabolites possess lesser toxicity, but they are excreted in such abundance and the steady-state concentration is maintained for so long that an ecological role cannot be disputed. Chief among this last category of organic products are the low-molecular-weight organic acids.

Investigations of two-membered cultures in vitro in which one member is inhibited by excretions of its associate disclose that quite frequently the effect is directly attributable to one or several organic acids. Almost invariably the organic acid toxicity—whether to fungi, algae, or bacteria—is great at low pH, with little or no suppression detectable under neutral or alkaline conditions (Fig. 13.4). The growth inhibition is not merely an acidity effect, inasmuch as only specific organic acids are functional. One or more of four volatile fatty acids are generally implicated in these model studies, namely, acetic, butyric, propionic, and formic acids; in some instances the active principle is acrylic or lactic acid.

Essentially identical observations have been made in investigations of natural ecosystems as well. The intestine of the mouse, for example, contains microbiologically formed antibacterial constituents sufficiently potent to prevent the establishment of *Salmonella typhimurium*. This protection against salmonellae is abolished when the intestinal microflora is upset by administration of antibiotics. Apparently it is the short-chain fatty acids, chiefly acetic and butyric, together with the low E_h and pH created by the gut community, that

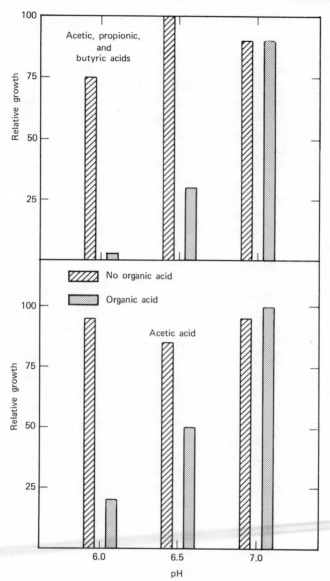

Figure 13.4 Inhibition of the growth of *Escherichia coli* by acetic acid or a mixture of acetic, propionic, and butyric acids at three pH levels (Wolin, 1969).

render impossible the establishment of *S. typhimurium*. Inhibition of this pathogen is overcome by antibiotic treatment, which decimates the microflora so that little of the appropriate organic compounds is produced, the acidity is not great enough, and the E_h is not sufficiently low (Meynell, 1963). The same three factors—volatile fatty acids, an acid reaction, and a low E_h—have been postulated to explain the inhibition of other bacterial pathogens in the intestine. The failure of *Escherichia coli* to grow in the rumen fluid of cows has likewise been ascribed to volatile fatty acids—acetic, propionic, and butyric—excreted by rumen heterotrophs (Wolin, 1969). *Salmonella* entering milk after pasteurization, although able to multiply readily during the manufacture of cheese made from the tainted milk, dies out when the cheese is cured, a decline that has been attributed to the volatile fatty acids, whose concentration increases as the cheese ages (Goepfert et al., 1968). Free fatty acids are also found in reasonable concentrations on the human skin. These acids, originating to a significant extent from the metabolism of the skin residents, have been assigned a role in keeping low the population density of the yeast *Pityrosporum ovale* and in eliminating *Streptococcus* (Marples, 1965).

In environments rich in sugars or polysaccharides but free of O_2, the lactic acid bacteria proliferate extensively and elaborate sufficient lactic acid to reduce drastically the density of nearly all species initially present, an activity so intense that the resulting material is preserved and protected from spoilage. The drop in pH induced by the lactic acid bacteria is a clear case of amensalism, one that is evident in the abrupt decline in bacterial numbers and the consequent long keeping quality of cabbage, cucumbers, and plant remains fermented for the preparation of sauerkraut, pickles, and silage, respectively. The acidity of the vagina of adult women and the resulting inability of acid-sensitive cells, including many human pathogens, to flourish is the result of the metabolism of the indigenous bacteria, chiefly strains of *Lactobacillus* as well as possibly other acid formers (Rosebury, 1962). Similarly, the acetic acid synthesized by *Acetobacter* from the ethanol formed in the fermentation of fruits kills much of the established community and prevents the establishment of new arrivals. The low pH per se may not be the sole basis for the detrimental effect, because the toxicity of lactic and acetic acids at low pH is greater, at least against certain microorganisms, than can be accounted for on the basis of acidity alone.

A particularly interesting instance of amensalism mediated by an organic acid has been observed in penguins feeding on *Euphausia superba*. The gastrointestinal contents of these penguins are almost devoid of bacteria and show toxic properties. The *E. superba* consumed by the penguins, however, themselves graze on marine phytoplankton that possesses antibacterial activity. The antimicrobial agent affecting the gastrointestinal microflora of penguins has been traced to an alga, *Phaeocystis,* that makes large amounts of acrylic acid. Although acrylic acid is toxic neither to *E. superba* nor to the penguins, the level in the penguin is high enough to have an effect on the animal's potential gastrointestinal inhabitants (Sieburth, 1959, 1960).

Alcohols synthesized microbiologically may play a role in amensalism as well. Thus populations growing on food might make enough alcohol to retard the development of spoilage organisms. No question would be raised in regard to the effectiveness of the ethanol formed by yeast in wine fermentation. The concentration of this metabolite is clearly great enough to affect not only the yeast manufacturing the alcohol but also anaerobic bacteria that could otherwise have developed at the expense of the organic nutrients still available.

ANTIBIOSIS

Microorganisms are noted for their capacity to form highly potent antimicrobial agents, the antibiotics. An *antibiotic* is a substance produced by one organism which, in low concentrations, kills or inhibits the growth of another organism. Sometimes the suscept stops growing but retains its viability, but occasionally the antibiotic has a lethal effect. Chemicals of this kind are synthesized in axenic cultures of bacteria, fungi, actinomycetes, and algae; yet not only are many species unable to make any but, indeed, activity is often limited to rare strains of a species. The antibiotics inhibit or kill bacteria, fungi, protozoa, algae, yeasts, actinomycetes, and viruses, in culture at any rate; yet all are selective to a greater or lesser extent, acting only on a fixed collection of organisms. The size of the assemblage may nevertheless be quite small or reasonably large. These poisons either are excreted by living cells as part of their normal metabolism or are on occasion released solely upon the autolysis of dying or dead cells.

When tested in culture, isolates from many habitats exhibit competence in synthesizing antibiotics or related substances, and obtaining such strains and showing their performance in vitro are easily accomplished (Fig. 13.5). Active populations have now been obtained from soil, lake bottom mud, rivers, lakes, estuaries, the open sea, plant roots, leaf surfaces, seed coats, compost piles, sewage disposal systems,

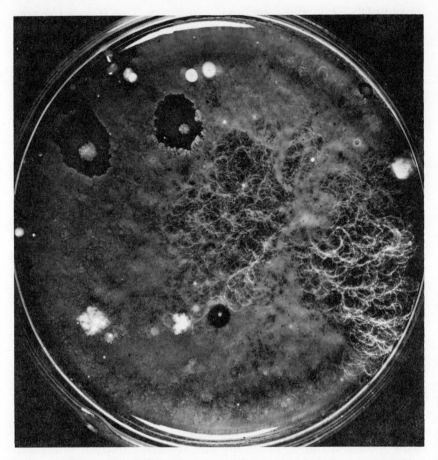

Figure 13.5   Microorganisms growing on an agar plate prepared from a soil dilution. The clear zones surrounding the colonies result from the suppression of nearby organisms by antibiotics produced by cells in the active colonies (courtesy of A. Kelner).

and the mouth, skin, and superficial infections of man and animals, so that an extensive list has been compiled of genera with the potential for harming their neighbors through antibiosis. The percentage of the isolates showing activity in vitro varies with the habitat, the time of year the sample is taken, and the assay organism employed for sensitivity testing. A high percentage of isolates may be toxic to one assay organism whereas a few at best may be deleterious to a second.

Bacteriocins represent a class of compounds behaving like the antibiotics. They are protein or protein-like substances that act as highly specific antibiotics in that they have a very narrow antibacterial spectrum. These inhibitors are elaborated only by selected bacterial strains, and they kill solely strains of the same or of closely related species. Bacteriocin biosynthesis, as well as sensitivity, is found in *Bacillus, Bordetella, Diplococcus, Escherichia, Pasteurella, Pseudomonas, Salmonella,* and *Shigella.* Metabolites with a similarly narrow antimicrobial spectrum likewise may be released by ciliates (Bradley, 1967).

Granted that cells and filaments having the potential for antibiotic biosynthesis are both ubiquitous and abundant, nevertheless tests in vitro showing their frequency or the effectiveness of their products shed little light on the ecological issue of whether or not these potent molecules appear and function in nature. Toxicity in laboratory media is so striking that it is easy to be led astray in extrapolating to natural ecosystems, but there is no necessary correlation between the abundance of cells capable of making antibiotics and the actual generation of these compounds in vivo. The existence of a potential for some biochemical process should never be taken as proof that the transformation does indeed occur. Many antibiotics are generated exclusively in rich media or their biosynthesis occurs solely in a restricted range of circumstances, and the concentration or kinds of nutrients necessary and the particular set of conditions may rarely if ever be found in a given ecosystem. To this date, in spite of the wealth of knowledge about antibiosis in model laboratory systems and the unequivocal success of microbiologically produced antibiotics in chemotherapy, little direct evidence exists that the substances are of consequence in vivo. The question concerned with the ecological role of antibiosis is not one that can be resolved in axenic or two-membered cultures pampered in the laboratory, but rather only in nature.

Nevertheless the importance of antibiotics to the well-being of the populations that produce them and their ecological role have prompted considerable debate. Seven major arguments against their importance have been advanced. (a) Antibiotics chemically identical with those elaborated in culture have not been found in many natural ecosystems. (b) The demise of aliens deliberately introduced into a heterogeneous community is not linked with the build-up of toxic compounds active against the intruder. (c) Populations that excrete potent antibiotics do not necessarily dominate, in spite of their presumed advantage, and they are frequently minor components of a community. (d) Cells of antibiotic-resistant populations are not apparently more abundant than sensitive individuals in the well-studied habitats, though they might be expected to be more prevalent if toxins were implicated in the regulation of community composition. (e) Antibiotics are inactivated or destroyed readily in many environments. Detoxication of added chemicals results from adsorption onto colloids, microbial degradation, nonbiological decomposition, or the formation of complexes with tissue constituents, body fluids, or inanimate materials. Microbial enzymes catalyze the destruction of a large number of antibiotics, and some of these compounds even may serve as carbon sources for heterotrophs. (f) Nutrients may be insufficient in quantity or of the wrong type to promote antibiotic biosynthesis, an argument of particular relevance in a milieu with a deficient nutrient supply. The formation of numerous antibiotics requires specific substrates and a fixed range of conditions. Though the organisms may multiply when the specific circumstances are not satisfied, the particular product in question never appears. (g) The effectiveness of a soluble inhibitor might well be dissipated as it diffused through an aquatic habitat. The concentration can be high in the confines of a flask, but the compounds could be diluted below their effective concentration in bodies of water. Proponents of the view that antibiosis is little else than a laboratory curiosity in effect believe that antibiotic producers have been isolated, tamed, and indulged for the purpose of exploiting a specific biochemical potential, much as rare and generally unproductive animal or plant types were domesticated by early man and pampered in such a manner that they now yield sizable quantities of a commodity of value to mankind.

Advocates of antibiosis as an ecological fact, by contrast, have strong arguments for placing these harmful metabolites in the forefront of the struggle for existence. A widely held assumption is that toxins account for the success of one population in becoming established or gaining the upper hand over a second in colonizing a substrate when no differences in their growth rates or competitiveness are apparent, although factors unknown to the investigator, but yet not antibiosis, could underlie the differences. Still, a property of so many and of so diverse an assemblage of species derived from such an assortment of ecosystems must confer a benefit on the possessor. It is argued by advocates of antibiosis that were these compounds inconsequential in nature, the capacity to synthesize them would have been lost in time; but since the activity has been preserved, it probably provides some selective advantage. Even if production is short-lived because suitable conditions are not maintained for long, the formation of an antibiotic may have appreciable value while the producer is growing, and the resulting compound may be beneficial despite its instability, biodegradability, or susceptibility to adsorption or complexing, the persistence in an active state being long enough to confer an advantage in interspecific associations. Readily inactivated inhibitors may still have an influence if a reasonably high steady-state level is maintained between the rates of excretion and detoxication. Moreover, granted that countless macroenvironments are too impoverished to support consequential toxin formation, minute local areas still may be well supplied with organic substrates for heterotrophs, or nutrients may occasionally become available in habitats generally poor in organic materials. In addition, not a few ecosystems harboring microorganisms are notoriously rich in carbonaceous substrates, and these could well be sites of antibiotic biosynthesis. For photosynthetic organisms, the paucity of oxidizable organic matter is not a deterrent, and carbon limitation is a poor argument against antibiosis by algae flourishing in freshwater or marine habitats. The fact that the antibiotic former is not dominant, it is argued further, does not negate or minimize the role of these metabolites in the well-being of the producer; it itself may be held in check for other reasons, as by its susceptibility to parasitism, inability to compete effectively with resistant populations, or the low concentration of organic substrates it can utilize. Granted also that dilution may be of consequence in the aquatic milieu, two populations still are commonly in such close proximity that the excretions of one can have a pronounced physiological effect on the other. Finally, as indicated in Table 13.1, selected environments do indeed contain microbial inhibitors; though some of the suppressing agents may be inorganic and others may be organic compounds active only when in high concentrations, and hence not considered antibiotics, indirect and sometimes direct evidence supports the opinion that at least a few are antibiotics.

The phenomenon of antibiosis is regularly viewed from the vantage point of the total ecosystem, but it is likely that in many habitats the microenvironment is the locus of antibiotic biogenesis. These minute, discontinuous loci could become colonized by toxin producers, and because the organic matter level is abundant and other circumstances are favorable, locally high concentrations may accumulate there. In the microhabitat of the active population, the species composition might well be regulated by the presence of antibiotics; yet extracts made from large samples of the given macroenvironment may not disclose the presence of these locally important inhibitors. Not only may the nutrients be of the proper kind and amount for the antimicrobial substances to be elaborated, but biological and chemical inactivation may be too slow to prevent significant accumulation. Microecosystems wherein antibiosis could be a potent force are possibly to be found in soil containing plant remains, on the surfaces of plankton undergoing decay, along roots excreting simple organic molecules, or in regions where fresh substrates are available for colonization intermittently. It is at the microenvironmental level, too, where bacteriocins could be prominent in allowing one population to inhibit and displace closely related bacteria.

A striking illustration of how antibiotics may be related to microbial survival is provided by studies of *Cephalosporium gramineum,* a pathogen of wheat. From the time one wheat crop dies until it is replaced the following year with new plants, this fungus must persist or maintain itself. As it forms no resting structure, the parasite is restricted to a hyphal existence; yet it is a poor competitor in the very host tissues where it previously gained a foothold and where it must now persist. Interestingly, fresh isolates excrete in vitro a wide-spectrum antifungal substance identical chromatographically with a toxin found in infested wheat tissue that had been naturally colonized. The hypothesis that the ability to make this antibiotic is ecologically advantageous to *C. gramineum* is supported by two observations: (a) cultures maintained in the laboratory soon lose this synthetic capacity, that is, selection preserves the antibiotic formers in nature whereas the non-toxin producers spontaneously arising during growth of the fungus are eliminated in vivo; and (b) toxin-synthesizing populations of *C. gramineum* are better able to exclude fungi endeavoring to colonize dead tissues than are inactive strains (Bruehl et al., 1969).

The importance of antibiosis in soil has been the object of considerable scrutiny. Though the existence of fungistatic and bacteriostatic principles in this environment is beyond doubt, their effectiveness, origin, and chemical identities remain uncertain. The toxicity revealed in laboratory tests possibly represents a response to a concentration of antibiotics made in a multitude of microenvironments, compounds that may be of the kinds that are neither quickly destroyed enzymatically nor tenaciously bound to clay colloids, or of the types that are produced as rapidly as they are inactivated. In addition, evidence exists, although often it is not overly rigorous, that distinct antibiotics can be elaborated in soils inoculated with specific fungi, bacteria, or actinomycetes (Table 13.3). The toxins sometimes only appear when the soils are amended with organic nutrients, a situation that may be akin, on a macrocosmic scale, to that of innum-

Table 13.3

Antibiotics in Soil and Related Environments Following Inoculation with Toxin-Forming Cultures

Antibiotic Producer Used as Inoculum	Environment	Antibiotic Detected	Reference
Bacillus subtilis	Soil	Bulbiformin	Vasudeva et al. (1963)
Bacillus subtilis	Seeds in soil	Bulbiformin	Singh et al. (1965)
Cephalosporium gramineum	Straw buried in soil	Antifungal	Bruehl et al. (1969)
Penicillium frequentans	Seeds in soil	Frequentin	Wright (1956)
Penicillium purpurogenum	Soil	Antibacterial	Mirchink and Greshnykh (1961)
Streptomyces sp.	Corn rhizosphere	Antifungal	Rangaswami and Vidyasekaran (1963)
Trichoderma viride [a]	Seeds in soil	Gliotoxin	Wright (1956)
Trichoderma viride	Soil plus plant remains	Gliotoxin	Wright (1952)

[a] Indigenous rather than inoculated.

erable nutrient-rich microenvironments, and they also are synthesized on seed surfaces, in straw buried below ground, and in the rhizosphere. The finding that gliotoxin is a constituent of soil containing an indigenous rather than an inoculated population of *Trichoderma viride* is especially noteworthy. On the other hand, the frequently observed fungistatic and bacteriostatic components of soil could well be simple microbial products rather than potent, complex compounds of the sorts usually considered to be antibiotics. Still, in the last analysis, the likelihood is small that a clear picture will emerge until information is provided on the identity of these fungistatic and bacteriostatic metabolites and their distribution and concentration in soil microenvironments.

An organism susceptible to a toxin is considered to be harmed by virtue of its sensitivity; yet work on soil fungistasis has uncovered a likely benefit accruing to the suscept. Soils contain enormous numbers of viable fungal conidia, and since many develop into germ tubes with no exogenous nutrients, their failure to germinate in soil may well reflect the presence of inhibitory compounds. Yet were the spores to germinate, the emerging hyphae would encounter an environment deficient in the organic substrates obligatory for vegetative development, and death by starvation or lysis would ensue. The fungistatic restraint on germination of many species is overcome, fortunately and surprisingly, when nutrients become available or when a root exuding simple organic compounds extends into the vicinity of the conidia so that the spore dormancy is broken at the same time that food becomes accessible to the emerging mycelium. In this way the toxin in effect aids the suscept.

Occasional mycorrhiza-forming fungi synthesize antibiotics in culture, a finding that has been used to bolster the view that the fungus protects its plant symbiont from soil-borne pathogens. In fact, soil in an area containing physiologically active *Pinus densiflora* mycorrhizae is largely free of bacteria and actinomycetes, and the mycorrhizal symbiosis involving *Pinus* spp. and either *Tricholoma matsutake* or *Cenococcum graniforme* suppresses many microorganisms in vitro. Moreover, the antibiotic of the latter fungus is detectable not only in the mycelium but also in mycorrhizae, roots, and needles of pines, and soil containing *C. graniforme* sclerotia (Krywolap et al., 1964; Ohara, 1966).

Antimicrobial principles have been noted in other microbial habitats. For example, a discrete antibacterial toxin has been observed in marine waters and characterized as a large, thermolabile molecule (Saz et al., 1963). Certain coelenterates contain antimicrobial terpenoids apparently elaborated by the dinoflagellates teeming within the animal's body, the terpenoids possibly excluding alien microorganisms (Ciereszko, 1962). Antibiotic formation has also been recorded on the skin; in particular, the skin-inhabiting fungus *Trichophyton mentagrophytes* synthesizes penicillin as it spreads on the hedgehog, and *Staphylococcus aureus* isolates from infected skin areas are therefore penicillin-resistant, in contrast with the widespread penicillin intolerance of most naturally occurring strains of *S. aureus* (Smith and Marples, 1964). Foods supporting extensive microbial proliferation are potentially ideal sites for antibiosis, owing to the abundance of utilizable nutrients and the intense interspecific stresses. For example, antibiotics are known to be synthesized by *Streptococcus* in milk, and the dominance of lactic acid-forming *Streptococcus* strains in mixed dairy starter cultures occasionally is the result of antibiotic production by the dominants. The antimicrobial potency of aged Limburger, Liederkranz, Brick, Munster, and other well-ripened cheeses is likewise ascribed to antibiotics (Grecz, 1964; Lightbody and Meanwell, 1955).

Several authorities have implicated organic inhibitors in seasonal successions and in the typical sequence of population changes that takes place in the progression leading to the climax community. The possible role of organic or inorganic acids and other toxins in microbial successions has already been briefly considered in Chapter 4. Presumably the toxin former eliminates or prevents establishment of sensitive species, but it in turn is displaced by populations that not only are unaffected by the metabolites in question but also can compete effectively with the toxin producer. Pioneers may exhibit a low tolerance to biologically formed inhibitors, since little or none is present early in succession, but subsequent colonists must be able to withstand any still present from prior inhabitants of the site as well as those currently generated. A successor does not become established, it is argued, if it is sensitive to those of its predecessors' metabolites remaining in the area. Microbiologically formed antimicrobials thus have been assigned a focal role in the succession typical of the ripening of certain cheeses and in the seasonal succession of algal species in bodies of water (Grecz, 1964; Lefevre, 1964). When algae attain enormous cell densities, harmful substances are easy to demonstrate, and these compounds, whether they are of algal origin or come from the accompanying heterotrophs, might govern the identity of the successors. For example, toxin production has been proposed to explain why *Microcystis aeruginosa* is the dominant blue-green alga in selected ponds (Vance, 1965) and why the flagellate *Olisthodiscus luteus* achieves seasonal dominance in Narragansett Bay (Pratt, 1966).

However, it should be borne in mind that, more often than not, amensalism is at best just one of the controlling factors in succession, dominance, or the elimination of aliens. Antibiotics are probably very rarely the sole determinant, but they may be extremely significant in governing ecological success or failure. At the same time the impact of amensalism—and also of abiotic stresses, competitiveness, or freedom from parasites—depends on the organism, the time, and the particular ecosystem.

References

REVIEWS

Brian, P. W. 1960. Antagonistic and competitive mechanisms limiting survival and activity of fungi in soil. *In* D. Parkinson and J. S. Waid, Eds., *The Ecology of Soil Fungi*. Liverpool University Press, Liverpool. pp. 115–129.

Lefevre, M. 1964. Extracellular products of algae. *In* D. F. Jackson, Ed., *Algae and Man*. Plenum Press, New York. pp. 337–367.

Park, D. 1967. The importance of antibiotics and inhibiting substances. *In* A. Burges and F. Raw, Eds., *Soil Biology*. Academic Press, New York. pp. 435–447.

Sieburth, J. M. 1968. The influence of algal antibiosis on the ecology of marine microorganisms. *In* M. R. Droop and E. J. F. Wood, Eds., *Advances in Microbiology of the Sea*, Vol. 1. Academic Press, New York. pp. 63–94.

Waksman, S. A. 1961. The role of antibiotics in nature. *Perspectives Biol. Med.*, 4, 271–287.

OTHER LITERATURE CITED

Atkins, W. R. G. 1922. *J. Marine Biol. Assoc. U. K.*, 12, 789–791.

Borut, S., and Johnson, T. W. 1962. *Mycologia*, 54, 181–193.

Bradley, D. E. 1967. *Bacteriol. Rev.*, 31, 230–314.

Bruehl, G. W., Millar, R. L., and Cunfer, B. 1969. *Can. J. Plant Sci.*, 49, 235–246.

Buck, J. D., and Meyers, S. P. 1965. *Limnol. Oceanog.*, 10, 385–391.

Ciereszko, L. S. 1962. *Trans. N.Y. Acad. Sci.*, Ser. 2, 24, 502–503.

Durbin, R. D. 1959 *Am. J. Bot.*, 46, 22–25.

Goepfert, J. M., Olson, N. F., and Marth, E. H. 1968. *Appl. Microbiol.*, 16, 862–866.

Grecz, N. 1964. *In* N. Molin, Ed., *Microbial Inhibitors in Food*. Almqvist and Wiksell, Stockholm. pp. 307–320.

324

Griffin, D. M. 1966. *Trans. Brit. Mycol. Soc.*, **49**, 115–119.

Hentges, D. J. 1969. *J. Bacteriol.*, **97**, 513–517.

Hessayon, D. G. 1953. *Soil Sci.*, **75**, 395–404.

Hirsch, A., McClintock, M., and Macquot, G. 1952. *J. Dairy Res.*, **19**, 179–186.

Hodgson, W. A., Munro, J., Singh, R. P., and Wood, F. A. 1969. *Phytopathology*, **59**, 1334–1335.

Hoffman, H. 1966. *Advan. Appl. Microbiol.*, **8**, 195–251.

Jannasch, H. W. 1960. *J. Gen. Microbiol.*, **23**, 55–63.

Jefferys, E. G., and Hemming, H. G. 1953. *Nature*, **172**, 872–873.

Klosowska, T. 1958. *Acta Microbiol. Polon.*, **7**, 45–49.

Kraus, F. W., Nickerson, J. F., Perry, W. I., and Walker, A. P. 1957. *J. Bacteriol.*, **73**, 727–735.

Kriss, A. E., Mishustina, I. E., Mitskevich, N., and Zemtsova, E. V. 1967. *Microbial Populations of Oceans and Seas*. Arnold, London.

Krywolap, G. N., Grand, L. F., and Casida, L. E. 1964. *Can. J. Microbiol.*, **10**, 323–328.

Lightbody, L. G., and Meanwell, L. J. 1955. *J. Appl. Bacteriol.*, **18**, 53–65.

Lockwood, J. L. 1964. *Ann. Rev. Phytopathol.*, **2**, 341–362.

Louvet, J., and Bulit, J. 1963. *Ann. Inst. Pasteur*, **105**, 242–256.

Marples, M. J. 1965. *The Ecology of the Human Skin*. Thomas, Springfield, Ill.

Meynell, G. G. 1963. *Brit. J. Exptl. Pathol.*, **44**, 209–219.

Mirchink, T. G., and Greshnykh, K. P. 1961. *Mikrobiologiya*, **30**, 1045–1049.

Mitchell, R. 1968. *Water Res.*, **2**, 535–543.

Mitchell, R., and Alexander, M. 1962. *Soil Sci.*, **93**, 413–419.

Ohara, H. 1966. *Proc. Japan Acad.*, **42**, 503–506.

Page, L. A., Goodlow, R. J., and Braun, W. 1951. *J. Bacteriol.*, **62**, 639–647.

Pochon, J., and deBarjac, H. 1952. *Ann. Inst. Pasteur*, **83**, 196–199.

Pratt, D. M. 1966. *Limnol. Oceanog.*, **11**, 447–455.

Proctor, V. W. 1957. *Ecology*, **38**, 457–462.

Rangaswami, G., and Vidyasekaran, P. 1963. *Phytopathology*, **53**, 995–997.

Rice, T. R. 1954. *U.S. Fish Wildlife Serv. Fishery Bull.*, **54**, 227–245.

Rosebury, T. 1962. *Microorganisms Indigenous to Man*. McGraw-Hill, New York.

Saz, A. K., Watson, S., Brown, S. R., and Lowery, D. L. 1963. *Limnol. Oceanog.*, **8**, 63–67.

Sequeira, L. 1963. *Phytopathology*, **53**, 332–336.

Sieburth, J. M. 1959. *Limnol. Oceanog.*, **4**, 419–424.

Sieburth, J. M. 1960. *Science*, **132**, 676–677.

Singh, P., Vasudeva, R. S., and Bajaj, B. S. 1965. *Ann. Appl. Biol.*, **55**, 89–97.

Smith, J. M. B., and Marples, M. J. 1964. *Nature*, **201**, 844.

Stojanovic, B. J., and Alexander, M. 1958. *Soil Sci.,* **86,** 208–215.

Thompson, R., and Johnson, A. 1951. *J. Infect. Dis.,* **88,** 81–85.

Thompson, R., and Shibuya, M. 1946. *J. Bacteriol.,* **51,** 671–684.

Vance, B. D. 1965. *J. Phycol.,* **1,** 81–86.

Van Kesteren, M., Bibby, B. G., and Berry, G. P. 1942. *J. Bacteriol.,* **43,** 573–583.

Vasudeva, R. S., Singh, P., Sen Gupta, P. K., Mahmood, M., and Bajaj, B. S. 1963. *Ann. Appl. Biol.,* **51,** 415–423.

Wolin, M. J. 1969. *Appl. Microbiol.,* **17,** 83–87.

Wright, J. M. 1952. *Nature,* **170,** 673–674.

Wright, J. M. 1956. *Ann. Appl. Biol.,* **44,** 561–566.

Yaniv, H., and Avi-Dor, Y. 1953. *J. Bacteriol.,* **66,** 6–9.

Zavarzina, N. B. 1959. *Trud. Vsesoyuz. Gidrobiol. Obshch.,* **9,** 195–205.

14

Parasitism

A *parasite* is an organism that feeds on the cells, tissues, or body fluids of another and usually larger organism, the *host,* which is commonly injured in the process. The parasite is dependent, to a lesser or greater extent, on the host at whose expense it is maintained, and the former remains in intimate physical and metabolic contact with the latter for a considerable part or for the entire life of the parasitic individual. No major group of plants, animals, or microorganisms is wholly free of attack by microbial parasites, although in some instances few are known. Diverse taxonomic categories of microorganisms contain parasitic representatives, moreover, this way of life characterizing selected fungi, bacteria, protozoa, viruses, and actinomycetes, and the structural or biochemical modifications associated with the shift to a parasitic mode of nutrition among these groups are numerous.

A high percentage of parasites reside in a unique environment, that created by the living organism, and a study of the interactions of these parasites with their hosts is thus an actual extension of ecology in the sense that the interactions involve a relationship between cells of a population and their metabolically active surroundings as provided by the host. The host individual, or some appropriate segment of it, is the immediate habitat, serving as the particular microenvironment that directly affects the parasite in much the same way as the inanimate habitat influences and sustains free-living populations. By entering into parasitism, an organism foregoes the difficulties of interspecific competition and relinquishes to the host many of the problems of obtaining food and coping with the external environment. The host is the food source and, for those microorganisms resident within the tissues or cells of another species, the regulator of crucial factors affecting the well-being of the parasite.

327

The dependent organism contributes little, if anything, for the benefit of its host; rather it has a detrimental role that ranges from modest to devastating. As it grows and exploits a portion of the host's constituents or metabolic activities, the microorganism has an adverse effect on the multiplication, health, or functioning of the associated individual. Though little direct harm is discernible in occasional instances, innumerable parasites regulate the abundance of host individuals, and at times they may reduce the host population to a point approaching extinction.

Lines demarcating parasitism from related kinds of two-membered interactions are difficult to draw. At one side there is a gradual transition between exclusively free-living species and those having a parasitic existence. The border between commensalism and parasitism is frequently vague, too, but according to the usage adopted here the two types of partnerships can be differentiated because no harm comes to the associate in the former interaction while in the latter one of the two companions is harmed. Nevertheless a given two-membered association in nature may not be sufficiently well characterized to fit neatly into the category of either commensalism or parasitism. Microbial predation and parasitism are likewise often hard to distinguish clearly, as both entail a benefit to one and generally some harm to individuals of another species. The parasitic and predatory habits are sometimes easily differentiable, to be sure, but occasional dependent microorganisms possess features of both sorts of relationships. However, numerous parasites are in prolonged contact with and feed continuously on their hosts, they are generally smaller than the individuals they exploit, and many attack a remarkably narrow range of organisms; by contrast, the contact period is generally short in predation inasmuch as the victim is commonly killed and digested rapidly, the predator usually lives free of and is larger than its prey, and it exhibits less specificity in its food habits.

Species living independently as well as by parasitism are termed *facultative parasites*. *Obligate parasites*, conversely, must live on the cells, tissue components, or body fluids of a viable individual for part or all of their life cycles. The so-called *endoparasites* live inside an organism, as illustrated by bacteriophages harbored within the bacterial cell, microorganisms residing within fungi and algae, or heterotrophs inhabiting blood, the alimentary tract, or a multitude of plant and animal tissues. *Ectoparasites*, on the other hand, are localized on the external surfaces of their host; *Bdellovibrio* attached to the outside of gram-negative bacteria (Fig. 14.1) or protozoa inhabiting the skin or gills of vertebrates are considered to be ectoparasites.

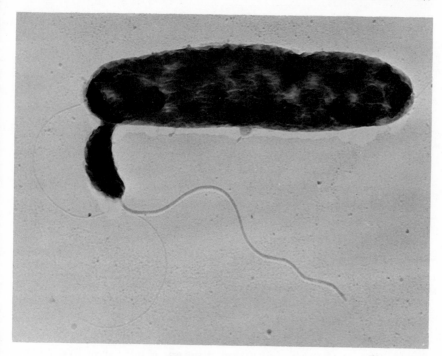

Figure 14.1 Attachment of *Bdellovibrio bacteriovorus* to a bacterial cell (courtesy of H. Stolp).

The present chapter is devoted almost entirely to *intermicrobial parasitism,* wherein both host and parasite are microorganisms. Some of the observations, however, apply likewise to systems involving a higher plant or animal as host, but additional attention to relationships involving higher organisms is given in a later chapter. In the realm of intermicrobial parasitism, little difficulty is encountered in distinguishing between the *balanced,* or nondestructive, and the *destructive* types of parasitism. In the former, the feeding is prolonged and the host survives; in the latter, the parasites multiply and soon bring about the death and destruction of their host.

SOME CHARACTERISTICS OF PARASITES

Certain parasites, notably among the protozoa and fungi, possess structural attributes distinguishing them from their free-living counterparts. Not a few ciliates and flagellates have morphological features permitting their attachment to the metazoa they dwell within or upon. In some protozoa the change from independent to parasitic feeding habits is paralleled by the appearance of a new sort of digestive apparatus, and occasional protozoa have complicated life cycles associated with their transmission from infected to healthy animals. Fungi may develop specialized hyphae permitting their attachment to hosts, or they sometimes produce morphological peculiarities facilitating penetration of plant cell walls, invasion of underlying cellular material, or feeding on susceptible tissue constituents.

An invariable trait of parasitic microorganisms is the restricted range of species they can exploit. *Host specificity*, as previously stated in Chapter 5, describes the range of hosts suitable for a given parasite. Each has a particular group of acceptable hosts; for example, selected bacteria or fungi may be invaded, but not algae, protozoa, or higher animals. As free-living forms are capable of inhabiting only a fixed set of environments, so it is with the parasites; among the latter, because the habitat is a viable individual, the net effect of environmental selectivity is expressed in terms of well-defined host ranges. The breadth of the range varies markedly, a few parasites being highly specific and attacking solely occasional strains of a single species or members of related species of an individual genus, others invading a wide array of hosts representing different genera, families, orders, and even classes. Still, the host range in nature is often narrower than that observed in the laboratory, presumably because conditions in vivo are not conducive for penetration of cells of all potential hosts, although were invasion to occur, replication of the parasite probably would proceed.

Specificity is generally believed to result from a high degree of specialization such that an organism's versatility in habitat selection, living on different host species in the case of parasitism, is quite restricted. The factors involved and the precise mechanism underlying this selectivity have so far eluded definition, but speculation has not been hindered by a lack of experimental data. A very narrow host range, and hence a high degree of specialization, is typical of viruses and various fungal parasites of plants, protozoa, and other fungi. The rust fungi affecting cereal crops infect only a few varieties of the host species in nature. Protozoa causing major diseases of man, domesticated animals, and wildlife also frequently are capable of infecting but a single or closely related species. The selectivity associated with a narrow host range is often compounded by a striking tissue specificity, as with the predilection of *Neisseria meningitidis* for the spinal fluid and meninges of the brain or of *Corynebacterium diphtheriae*

Table 14.1

Host Specificity of Parasitic Microorganisms

Parasite	Host	Reference
Parasitic Fungi: Narrow Range		
Calcarisporium parasiticum	Physalospora	Barnett (1963)
Dispira simplex	Chaetomium	Benjamin (1961)
Gonatobotryum fuscum	Ceratocystis	Barnett (1968)
Rhizophydium sphaerocarpum	Spirogyra	Barr and Hickman (1967)
Wide Range		
Bacteriophage	Pasteurella, Shigella	Lazarus and Gunnison (1947)
Bdellovibrio bacteriovorus	Escherichia, Pseudomonas, Salmonella	Shilo and Bruff (1965)
Blue-green algal virus	Lyngbya, Phormidium, Plectonema	Safferman (1968)
Entamoeba histolytica	Man, dog, rabbit	Dogiel (1964)
Phymatotrichum omnivorum	2000 dicotyledonous species	Blank (1953)
Piptocephalis xenophilia	Ascomycetes, Fungi Imperfecti, Phycomycetes	Dobbs and English (1954)
Rhizoctonia solani	Plants, Mucor, Pythium	Butler (1957)

for the mucous membranes of man's upper respiratory tract. Representative microorganisms capable of invading and doing harm to a rather large as well as a small number of host species are listed in Table 14.1. In addition to the listed groups having wide ranges, species of *Mycobacterium, Pasteurella,* and *Streptococcus* are known to cause disease in dissimilar animals, including man.

Host ranges may on occasion be broadened in the laboratory. Thus bacteriophages can be induced experimentally to develop in previously resistant bacteria and bring about their lysis, and fungi, too, sometimes acquire the capacity to invade new hosts. A similar widening of range may and probably does take place in nature as well.

The reproductive ability of most parasites is enormous. This is necessary, particularly for those unable to grow in nature as saprobes, because only a minute proportion of the propagules produced during proliferation in or on the host survive the journey to an uninfected individual. The restriction of the microorganism to one or a few host species magnifies the problem of successful dissemination. An immense number of the propagules become nonviable before conditions adequate for renewed growth are encountered, and this high mortality rate must be compensated for by a high reproductive capacity, lest the species disappear; this is true of many human-pathogenic bacteria, viruses, and plant-pathogenic fungi disseminated by aerial transport. The risk is minimized somewhat in species able to produce endospores, cysts, chlamydospores, sclerotia, or some other type of resting stage that resists inactivation until an uninfected host is again encountered. Several parasites improve their chances of successfully establishing themselves because they are attracted, albeit over a short distance, to their host or to particular tissues where their renewed development is possible; for example, germ tubes emerging from the spores of various fungi extend toward the hyphae of nearby host fungi, and infective stages of several protozoan species move in a manner improving their chances of locating a host.

OBLIGATE AND HYPERPARASITISM

For its continued existence the obligate parasite must be intimately associated with and grow at the expense of a suitable host. Some of these host-dependent microorganisms have a free-living phase and rely on the metabolic activities of another individual at just selected stages in their life cycles; others grow and replicate only if they constantly have access to the metabolic machinery of a second individual. For both these categories a key ecological factor is the continuous functioning of the physiological processes of the species with which they are linked, and they cannot be cultured apart, at least not by any known technique, from the viable individuals or cells providing them with sustenance. A few can be propagated in tissue culture, but never in cell-free media or on nonliving substrates (Yarwood, 1956). The state of absolute dependence is, for some of these parasites at least, probably more apparent than real, reflecting a lack of understanding of the details of their highly complex nutrition. Yet, though it is likely that several ultimately will be cultured, possibly only after new techniques have been devised, nevertheless even these organisms, and indeed many facultative parasites too, must be considered ecologically obligate parasites.

Obligate parasites are spread among an assortment of taxonomic groups, a scattering so wide that it seems probable that absolute host dependency must have evolved independently in the various categories of bacteria, protozoa, Phycomycetes, Ascomycetes, and Basidiomycetes. The most abundant obligate parasites include the following: (a) viruses having bacteria, actinomycetes, blue-green algae, higher plants, or animals as hosts; (b) the rickettsiae that reproduce only in animal cells; (c) *Mycobacterium leprae,* the leprosy bacillus; (d) bacteria living within the cells of protozoa (Drozanski, 1963) and bacteria apparently developing at the expense of certain algae; (e) *Bdellovibrio bacteriovorus,* a parasite of bacteria that generally cannot be grown in cell-free media, although free-living mutants have been obtained (Shilo and Bruff, 1965); (f) a number of protozoa pathogenic for man and animals, such as species of *Plasmodium* and *Toxoplasma;* (g) fungi like those Zoopagales that penetrate and destroy protozoan cells, or like the chytrid, *Rhizophydium planktonicum* (Canter and Lund, 1948), which causes epidemics in algae; (h) fungi parasitizing unrelated fungi; and (i) a surprisingly large number of plant-parasitic fungi, including the powdery mildew, downy mildew, and smut fungi causing diseases of major economic importance (Yarwood, 1956). A high proportion of the obligate parasites live within the organism with which they are associated, essentially in an intracellular habitat, as in the case of viruses, rickettsiae, *M. leprae,* and assorted chytrids, but a portion of the fungal structure or a phase of the life cycle may be extracellular.

Parasites, no exceptions to the nearly universal proneness of living things to attack, are clearly liable to infection themselves.

> *Great fleas have little fleas*
> *Upon their backs to bite 'em,*
> *And little fleas have lesser fleas,*
> *And so, ad infinitum.* [de Morgan, 1872]

Organisms that invade and do harm to parasites are termed *hyperparasites;* in essence, one individual attacks a second that is itself parasitic on a third. For example, despite their small size, protozoan parasites serve as hosts for bacteria, fungi, and dissimilar protozoa. An illustration of hyperparasitism is presented in Fig. 14.2. Parasitic ciliates, flagellates, and amebae all suffer in this way as hosts for microscopic forms of life. Pathogenic fungi are also set upon as they develop on susceptible plants in nature, both bacteria and fungi apparently being able to live on the cellular constituents of these pathogens. Especially widespread are the fungi that invade and kill morphological forms of the rust and powdery mildew fungi harmful to

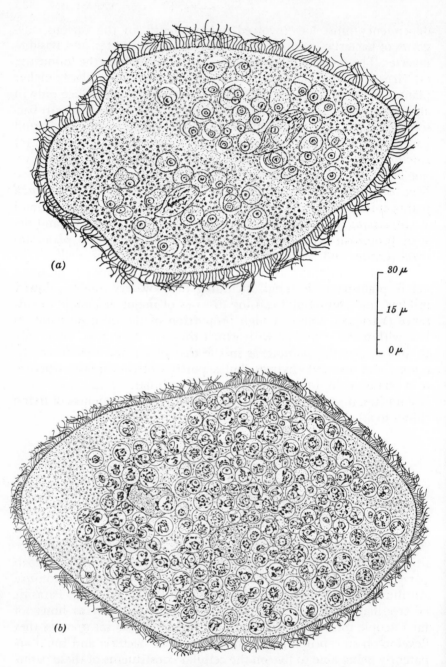

(a)

30 μ

15 μ

0 μ

(b)

Figure 14.2 *Entamoeba* hyperparasitic in the cells of *Zelleriella,* a ciliate living in the rectum of toads. Both vegetative stages (*a*) and cysts (*b*) of the ameba are evident. Cilia are diagrammatic (Stabler and Chen, 1936).

Table 14.2

Examples of Microbial Hyperparasitism

Hyperparasite	Host for Hyperparasite	Host for Primary Parasite	Reference
Allantosoma	Ciliates	Horse	Dogiel (1964)
Bacteriophages	Pseudomonas tabaci	Tobacco	Fulton (1950)
Chytridium parasiticum	Chytridium suburceolatum	Rhizidium richmondense	Willoughby (1956)
Chytrids	Ciliates	Cattle	Noble and Noble (1964)
Darluca filum	Puccinia graminis	Bluegrass	Bean (1968)
Ectrogella bessey	Olpidiopsis schenkiana	Spirogyra	Sparrow and Ellison (1949)
Xanthomonas	Puccinia graminis	Wheat	Klement and Kiraly (1957)

economic crops; some of these hyperparasites, like *Darluca filum*, can be particularly common on rust species. From Table 14.2 it is evident that viruses, protozoa such as *Allantosoma*, bacteria of the genus *Xanthomonas*, and fungi like *Chytridium*, *Darluca*, and *Ectrogella* can hyperparasitize protozoa, bacteria, and fungi like *Olpidiopsis* and *Puccinia*. Hyperparasites in turn are not immune to infection so that, considering hyperparasitism as secondary parasitism, tertiary parasitism does occur; thus bacteriophages infect a strain of *Xanthomonas* which is a hyperparasite of *Puccinia graminis* var. *tritici*, itself pathogenic to wheat (Klement and Kiraly, 1957); similarly, the fungus *Sphaerita* invades amebae that are infective for *Opalina*, a protozoan parasite of frogs (Dogiel, 1964).

PARASITES OF MAJOR MICROBIAL GROUPS

An astonishing array of protozoa are subjected to endo- or ecto-parasitic fungi, bacteria, and dissimilar protozoa. Frequently the parasites are destructive, and the host is killed and its structural integrity is lost soon after infection has begun. Some endoparasites, however, are balanced and remain within the protozoan cell for its entire existence and for the life of innumerable generations of daughter cells. At times protozoa containing these balanced parasites exhibit no morphological signs of infection, but occasionally hypertrophy or similar structural disturbances are readily detected.

Considerable attention has been given to the fungi feeding on protozoa. Some of them might be better, and often are, considered predators rather than parasites, but they shall be considered here together with other microorganisms living at the expense of unicellular animals. Among the frequent attackers of protozoa are a number of chytrids that infect *Amoeba, Entamoeba,* and several genera of ciliates; the chytrids *Sphaerita, Nucleophaga, Olpidium,* and *Rhizophlyctis* all can live in this way. The so-called predacious fungi, commonly members of Zoopagales or Moniliales, capture viable amebae or testaceous rhizopods and feed on either the moribund or the nonviable cells; these fungi are quite abundant in decomposing plant materials, dung, and soil (Duddington, 1968).

Protozoa also can be infected by coccus- or rod-shaped bacteria. The physiology or taxonomic placement of the bacteria is ill defined, in large part because most of the bacteria have yet to be grown apart from their hosts. The bacteria are usually endoparasites multiplying within the animal and reaching massive numbers in their intracellular habitat. These individuals, clearly differing from the bacteria that normally serve as food sources for the protozoa, are not destroyed by the digestive enzymes of their hosts, and many, but apparently not all, cause death of the protozoan. Some of the bacteria inhabit the outer surfaces of the protozoan cell, but the parasitic behavior of such organisms is poorly understood. Genera of flagellates, ciliates, and amebae are suitable hosts for the bacteria (Ball, 1969; Drozanski, 1963).

Intracellular bodies resembling microorganisms in several important ways have been observed in the cells of *Entamoeba,* flagellates like *Trichonympha,* Sporozoa, and a number of genera of ciliates. Some of these rod-shaped or coccoidal intracellular inclusions are widely held to be either bacterial endoparasites or symbionts of the protozoa; their identities and function probably must await axenic culture or at least additional critical studies. In the case of the intracellular bodies contained within *Paramecium,* however, careful attention and considerable effort have been devoted to an understanding of their behavior and function. In strains of this ciliate large numbers of such particles are evident, and if transferred to a sensitive population of paramecia, some will result in the death of the newly exposed cells. The paramecium particles show, as might be expected, specificity for the protozoa they are able to infect. It has been proposed that these intracellular entities represent bacteria or a form of life intermediate between viruses and bacteria. One kind of particle localized within the cells of *Paramecium aurelia* has indeed been cultured axenically, and it has been identified as a gram-negative bacterium (Sonneborn, 1959; Van Wagtendonk et al., 1963).

Specialized protozoa are known to live either within or upon the cells of unrelated protozoa, and these endo- and ectoparasites are seemingly widespread. In occasional populations all individuals are hosts to the invaders, and the individuals, once infected, may contain great numbers of small protozoa within their bodies. Sometimes the parasite will even encyst within the host's cell. Many of the small-celled species are hyperparasites of protozoa pathogenic for man or other animals, and some are capable of attacking the protozoan cyst stage. Parasites of these sorts occur in families of ciliates, amebae, and flagellates, while among the hosts are *Opalina, Paramecium, Stentor,* and *Trichodina.*

Soil-inhabiting, aquatic, wood-rotting, and plant-pathogenic fungi are subject to parasitism, an attack frequently leading to the loss of viability of the structure affected. Often the incitant of the harmful effect is another fungus, termed a *mycoparasite,* that may coil around the hyphae of its host or penetrate and ramify extensively within the filaments (Fig. 14.3). On occasion the structure of the invader is well concealed and cannot be distinguished from the protoplasmic constituents of its host for a considerable period of time. Some mycoparasites also produce external reproductive structures as they feed on their host's cellular constituents. The extent of damage done by the fungi ranges from no demonstrable effect to a disruption of the cellular contents that invariably has fatal consequences. Mycoparasitism is not restricted to rare taxonomic groups of hosts or parasites, and long lists of species serving as attackers or suscepts have been prepared. Among the organisms able to develop ectoparasitically or to invade portions or essentially the entire mycelium of suitable hosts are representative Phycomycetes, Ascomycetes, Fungi Imperfecti, and Basidiomycetes. Not only are vegetative hyphae assailed, for conidia, oospores, zoospores, oogonia, sporangia, chlamydospores, and sclerotia are prone to infection (Barnett, 1963). Individual Basidiomycetes, Ascomycetes, and Fungi Imperfecti are also capable of penetrating lichens and causing mild or serious damage to the lichen thallus (Watson, 1948). Certain yeasts are likewise susceptible to destructive invasion; for example, *Candida* is invaded by a bacterium that grows within its cells and ultimately brings about their rupture (Akiba and Iwata, 1954).

Algae, too, are set upon by an assortment of microorganisms indigenous to lakes, offshore waters, and the open seas of widely separated regions. Diverse fungi are pathogenic not only for the abundant but

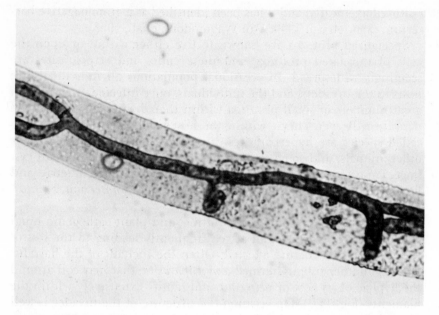

Figure 14.3 *Rhizoctonia solani* growing within *Gilbertella persicaria* (courtesy of E. E. Butler).

also for the uncommon algae, and those photoautotrophs that become prominent only sporadically are regularly faced with parasites that arise at almost the same time. Representative freshwater and marine Phycomycetes, Ascomycetes, and Fungi Imperfecti are thus known to do harm, but unquestionably it is the chytrid fungi, many of which can attack a variety of algal species, that stand out. Most uniflagellate chytrids are wholly or partially ectoparasitic, whereas the biflagellates tend to be internal parasites of planktonic algae. The injurious chytrids include species of *Rhizophydium, Chytridium, Zygorhizidium, Septosperma,* and *Chytriomyces,* and the algal hosts for these and other fungi include a heterogeneous collection of freshwater and marine red, green, and blue-green algae as well as diatoms. Bacteria, some of which have yet to be cultured axenically, also seem to live as ectoparasites of algae, but details of their parasitic behavior are far from clear. The algal group known as dinoflagellates contains genera that similarly assail suitable algal hosts; for instance, strains of *Gonyaulax* and *Paulseniella* will invade diatoms as well as other dinoflagellates. The infection produced by ecto- or endoparasites of algae is under many circumstances quite mild, but not a few of the invaders

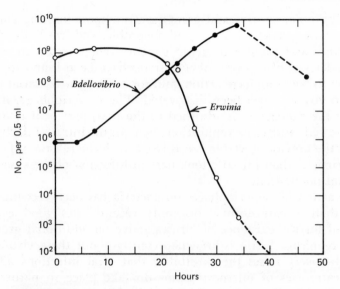

Figure 14.4 Development of *Bdellovibrio bacteriovorus* on *Erwinia amylovora* (Stolp and Starr, 1963).

cause appreciable structural or metabolic injury and decimate the host population.

Bacteria are parasitized by two kinds of organisms, vibrios and viruses. *Bdellovibrio bacteriovorus*, a vibrio-shaped bacterium about 0.3 μ in diameter, is obligately parasitic and requires viable cells to support its growth. These minute organisms are ubiquitous in soil, water, and sewage, and one strain or another is able, in vitro at any rate, to lyse species of *Aerobacter, Erwinia, Escherichia, Hyphomicrobium, Lactobacillus, Proteus, Pseudomonas, Rhodospirillum, Serratia,* and *Streptococcus*. The infection process is initiated by attachment of *B. bacteriovorus* to the outside of a bacterium. Shortly thereafter the morphology of the host undergoes a change, and the vibrios are seen to be lodged between the cell wall and protoplast of the host. The vibrio fragments into daughter cells, and soon thereafter the bacterial host cell undergoes lysis (Seidler and Starr, 1969). Figure 14.4 depicts changes in the population of the vibrio when it is feeding on a suitable bacterium.

Viruses having bacterial hosts are called *bacteriophages*. Some also act on actinomycetes, and though these are sometimes termed actinophages, they are generally categorized together with the bacteriophages. Viruses are known to infect a sizable array of bacteria, including gram-positive and gram-negative, spore-forming and non-spore-forming, strictly aerobic and anaerobic, and free-living as well as animal- and plant-pathogenic genera. When tested in culture, species derived from scores of ecosystems have been found to be susceptible to bacteriophage action, and for only a few bacterial groups have no viruses as yet been discovered. The bacteriophage, initially outside the bacterium, is adsorbed to the host, part of it is injected into the cell, and virus replication then begins intracellularly. Ultimately the previously viable cell is lysed, with the release of far more viral particles than initially took part in infection of the susceptible bacterial population.

The action of bacteriophages on bacteria has been recognized for more than a half-century, but only recently has good evidence emerged for the existence of viruses active on additional groups of microorganisms. Little is known presently about these viruses, but their discovery raises the possibility that viral infections affecting other categories of microorganisms do take place in nature. The newly discovered viruses, all of which show the anticipated host specificity, parasitize strains of *Penicillium* (Banks et al., 1968), the commercial mushroom *Agaricus bisporus* (Hollings et al., 1963), and blue-green algae like *Lyngbya, Plectonema,* and *Phormidium* (Safferman, 1968). Viruses of the latter kind are quite common in waste stabilization ponds supporting algal communities. Evidence also exists for a virus-like lytic agent attacking green algae (Tikhonenko and Zavarzina, 1966).

HOST-PARASITE RELATIONSHIP

Microorganisms differ markedly in their susceptibility to parasites and in the influence the invaders exert on them. Such differences are governed by inherent properties of the cells under attack, the virulence of the particular parasite, and prevailing environmental conditions. If the host is viewed as an environment for the invader and the host-parasite relationship as essentially an interaction between a small organism and its habitat, then invasiveness or virulence is essentially a measure of the parasite's colonizing ability, and the resistance of the host may be deemed to reflect the suitability of the environment for the invader. This resistance is determined by constituents and metabolic responses of the exposed individual, and though little is known about the precise details, defense mechanisms are present in cells of many microorganisms to aid them in warding off or limiting the extent of damage inflicted by a potential or actual parasite. The prospective invader may fail because it is unable to become attached to or penetrate the cell surface or, if attachment or entry has been accomplished, to grow in or upon the host. Furthermore, the microbial host may, in response to invasion, erect physical barriers or produce antimicrobial agents that limit the spread or entirely eliminate the parasite, and evidence exists, albeit frequently indirect, that mechanical or biochemical factors are responsible for the differences between resistant and susceptible species or between resistant and susceptible portions of a single filamentous organism. In mycoparasitism, for example, the invading fungus may be checked or walled off by physical barriers, as in the case of *Rhizoctonia solani* hyphae developing within *Mucor recurvus* (Butler, 1957). Occasional Phycomycetes construct septa across their filaments and thus seal off and prevent the further linear extension of the parasite's hyphae. In other fungi the older hyphae are resistant, possibly because of the melanin they contain, and dark or melanin-containing mycelium and spores are often less liable to invasion than similar but light-colored structures.

Alterations in the external or internal environment of the host—be it a microorganism, plant, man, or animal—modify its susceptibility either slightly or dramatically, and the outcome of the host-parasite relationship in intermicrobial parasitism or in human, animal, or plant disease is frequently regulated wholly by the prevailing environmental circumstances. A host may be in contact with countless of its microbial enemies yet show no signs of harm because ambient conditions favor, for reasons usually unclear, increased host resistance or diminished microbial virulence. At given periods virulence may be pronounced; at other times it is inconsequential because the habitat of the parasite or the external habitat to which the host is exposed has altered in some manner.

Among the variables that promote or prevent the invasion or extensive multiplication by parasites of animals, plants, or microorganisms are host nutrition, temperature, pH, age of host, light intensity, and stress conditions. Nutrient deficiencies or excesses in the host have been found to affect the reproduction of viruses, bacteria, protozoa, and fungi both in vitro and in vivo. Sometimes the difference between success and failure of a parasite is governed merely by the presence of adequate amounts of a single growth factor, but generally the precise basis for the difference is unexplained. Similarly, animals, plants, or microorganisms of one age or exposed at one temperature are not deleteriously affected by their parasites, whereas individuals of a dissimilar age, at times older, at time younger, and those exposed at higher or lower temperatures become significantly diseased. Light intensity and pH also influence the susceptibility of plants and microorganisms to parasitism. Stress conditions, including injury to tissues and host debilitation, likewise modify the resistance of algae to fungi, bacteria to bacteriophages, plants to bacteria, and humans to many pathogenic agents. Nevertheless little information is at hand to explain, in physiological terms, the variations in resistance and to allow for the definition of biochemical mechanisms underlying the modified susceptibility induced by the particular environmental change.

Strains of a single parasite species vary enormously in the harm they can do, under even the most favorable circumstances. One strain may be extremely potent while a second is wholly innocuous. Similarly, the first tested species of a genus may affect one host, the next species of the genus may harm a second but not the first host, while a third species may exist only in the free-living state. The same parasite, moreover, may have entirely different impacts on unrelated host species; at one extreme little evil may be perpetrated on the associate and the relationship may border on or indeed be commensalism, while at the opposite extreme the disruption in the host's function or structure might be so severe that death results. Where there is injury or an impairment in function, it may arise because the parasite assimilates metabolites essential for the well-being of the host, enzymatically destroys cellular or tissue components, causes abnormal and deleterious growths, forms toxic products, or otherwise disturbs the metabolism of the infected individual in such a way that it can no longer survive or its ability to compete with disease-free individuals in the same community is significantly reduced.

Numerous dependent microorganisms behave as destructive parasites, devastating populations of susceptible species in the immediate environs. Thus bacteriophages, *Bdellovibrio,* and viruses of blue-green algae kill and lyse their respective hosts, and the infective agent is then transmitted to healthy individuals. Some bacterial endoparasites and the Zoopagales group of fungi likewise overwhelm amebae, and many terrestrial and aquatic fungi destroy much of the protoplasm of their fungal and algal hosts. Mass mortalities are also common to terrestrial and aquatic vertebrates, invertebrates, and seed-bearing plants, especially when the potential host population is introduced into a new region or when a parasite is transported to a locality where it was previously unknown. Parasites have thus decimated populations of cereals bred for resistance to pre-existing strains of rust fungi or stands of chestnut trees in the United States, which were suddenly exposed to the agent of chestnut blight, *Endothia parasitica,* a fungus that presumably originated in China. Epidemics with catastrophic consequences also took their toll among the Indians of the New World and among the Polynesians, who had no contact with small-pox, tuberculosis, and measles until the white explorer arrived, and the available historical evidence suggests that the resulting epidemics in some instances wiped out 90% or more of the previously unex-posed populations.

To maintain themselves, destructive parasites, if obligate, must move from the individuals they killed to healthy organisms, in the meantime losing vast numbers of propagules that fail to survive the journey. Devastating disease outbreaks that lead to mass destruction result in unquestioned harm to the susceptible species, but the para-site suffers simultaneously inasmuch as its food source becomes scarce or is eliminated from that ecosystem. In the latter instance the para-site may be eradicated, too, only reappearing, if at all, when chance carries its propagules into a susceptible host population that has survived and proliferated. Hence the life of the destructive parasite in a particular microenvironment, if not in the larger ecosystem, is tenuous at best.

The utter devastation potentially brought about by destructive parasites probably occurs infrequently in nature, whether the depop-ulation is of the sort familiar in human or veterinary medicine or in crop husbandry, or is of the kind involving microbial hosts. The rarity of these disasters may be attributable to (a) the emergence of genotypes not susceptible to the destructive parasite, as when bacteria or actinomycetes mutate to become resistant to the bacteriophages in their immediate vicinity; (b) the existence of a genetically hetero-geneous population of the potential host, the individuals susceptible to any given parasite or pathogen making up only a small proportion of the total; (c) biological restrictions preventing an unbridled in-crease in the parasite density—it may be a poor competitor, subject to amensalism, or be itself set upon by a hyperparasite when it be-comes overly abundant; (d) physical restrictions, as in soil or other environments composed of innumerable, spatially separate micro-habitats permitting the multiplication of a suscept near but not immediately adjacent to its microscopic enemies; (e) poor transmis-sibility of the parasite; and (f) sensitivity of the dependent microor-ganism to abiotic stresses. Were it not for factors of these sorts, both the obligate parasite and its host species would be eliminated.

A goodly number, possibly the majority, of host species persist be-cause they live in harmony with their endo- or ectoparasitic enemies rather than being destroyed by them. These balanced parasites derive nutrients from their associates but do not exterminate them; they are thus in a sense better attuned to a host-dependent mode of life than are the destructive groups, in that the food source is not eliminated but continues to replicate and generate fresh metabolites for the invader to feed upon. The greedy or the complete exploiter of the parasitic niche is the one that finds additional hazards in perpetuating its own kind, because overexploitation is linked with an uncertain future; the nonglutton guarantees for itself a luxurious repast and a comfortable, viable habitat insofar as the associate is able to provide them. The obligate parasite in particular is a metabolic pauper, and risky indeed is the existence of such an organism were it so imprudent as to destroy the population that sustains it. Balanced parasites are thus quite prevalent, and overt and fatal diseases, especially those affecting higher animals and plants, are likely to be infrequent in natural communities.

Balanced parasitism is readily noted in intermicrobial parasitism and in interactions between micro- and macroorganisms. Numerous instances are known, for example, of balanced mycoparasitism and of reasonably harmonious relationships between bacteria and protozoa. In addition, the *temperate* bacteriophage enters into a balanced association with bacteria; the virus is borne intracellularly and is transferred to succeeding generations of daughter cells without lysing or destroying the resulting new populations. Only when the virus is brought in contact with cells for which it is virulent does massive lysis proceed, and the presence of the viral particle is thereby made evident. This phenomenon, *lysogeny,* is apparently widespread, and lysogenic bacteria are generally believed to be a major reservoir for these viruses in nature (Adams, 1959). A degree of balance likewise characterizes the relationships between protozoa of the genus *Trypanosoma* and subterranean mammals and between *Mycobacterium tuberculosis* or *Plasmodium* and man.

The state of destructive parasitism is inherently unstable, but it is maintained inasmuch as there are limits to the catastrophic decline in abundance of the host species. The instability is beautifully illustrated in the game of tag played by cereal crops and races of the rust fungus *Puccinia,* a game promoted and prolonged under the aegis of man. As the rust fungus assumes prominence in fields of cultivated wheat, agriculturalists develop and introduce new crop varieties that are resistant to the prevailing race of the pathogen. Few of the *Puccinia* propagules can grow at the expense of the disease-resistant variety, but still an extremely small number are able to reproduce because they are unaffected by whatever traits determine resistance in the wheat population. These fungi are favored by natural selection and proliferate to an extent that the new wheat variety, previously deemed resistant, is now clearly rust-susceptible. To right the situation and allow for production of healthy grain plants, agriculturalists are compelled to introduce yet another resistant variety, and a new round in the tag game is initiated. Figure 14.5 shows the change in rust susceptibility of cereal varieties in North America and illustrates the reason for the need to replace repeatedly the prevalent varieties of wheat and oats employed in agricultural practice.

The continued existence of the obligate parasite, obligate in nature if not in fact, and its host reflects the success of both in the struggle for existence. Several of the factors underlying the inability of the destructive parasite to eliminate its host have already been considered, but the rust-grain example discussed above exemplifies quite clearly how destructive parasites are probably frequently protected from self-elimination. Both host and parasite populations are, to a lesser or greater extent, genetically heterogeneous. During the increase in parasite density, the abundance of suscepts declines, but a small number of resistant host individuals—resistant because of the inherent inhomogeneity of the population or the appearance of mutants—remain. These have a selective advantage over their susceptible neighbors, and hence they multiply. At the same time the number of parasites falls as a result of the overexploitation of the sensitive host individuals and the now inadequate food supply, but either the innate genetic diversity of this microbial population or the appearance of mutants sets the stage for the proliferation of microorganisms able to attack hosts possessing the modified genotype. Such a *genetic feedback* mechanism of population regulation functions to damp the extreme fluctuations in density of the interacting organisms. Alternatively, the density of the susceptible host may rise because of the scarcity of its natural enemies and the greater fitness, in the absence of parasitism, of the sensitive than of the resistant individuals. The parasite subsequently gains the ascendancy again. Cycles of this sort are undoubtedly repeated ad infinitum and lead to a tenuous stability based on repeated fluctuations.

Another damping mechanism is evident in virus-animal relationships such as that of the myxomatosis virus. After the virus was first introduced into Australia to control the hordes of European rabbits that had become established, not only did the mortality rate attributable to the viral disease decline coincidentally with the appearance of resistant rabbits, but attenuated strains of the virus began to replace the virulent strain. The latter replacement presumably can be ascribed to the short life of rabbits invaded by the virulent virus,

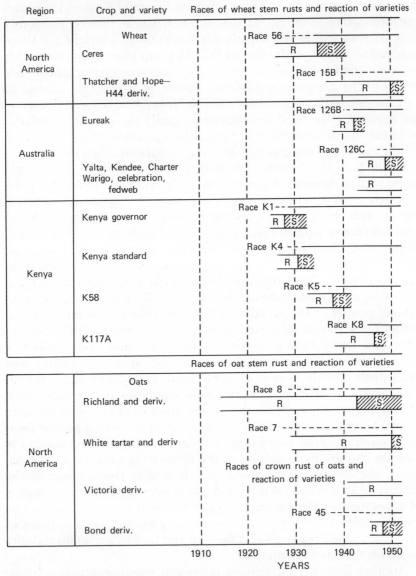

Figure 14.5 Relationship between the abundance of rust races in North America and the rust resistance of varieties of wheat and oats (Johnson, 1953). The relative abundance of the fungal race is indicated by dotted (scarce) or solid lines (abundant). Plant resistance: R—resistant; S—susceptible.

whereas those infected with the less virulent forms survived longer and thereby provided a more durable supply of the virus, which could only be transferred to healthy rabbits by means of a mosquito vector (Pimentel, 1968). Virulent parasites transmitted by an animate vector that feeds on diseased, viable individuals may often be selected against, and their less harmful counterparts favored, by the diminishing opportunities for the vector to find a living host bearing the virulent agent as the disease takes its toll. Hence, by a variety of mechanisms, severe parasitic attack is infrequent, localized, or damped by appropriate ecological changes.

BENEFITS OF PARASITISM

The precise benefits acquired by the parasite as a result of its unique mode of existence are ill defined or poorly understood. The facultative organisms, alternating as they do between a free-living way of life and one based on exploitation, are not metabolically dependent on their hosts but benefit from their ability, on occasion, to forego the rigors of independence for the comparative ease of parasitism, thereby obtaining nutrients unavailable to those of their neighbors that are incapable of infecting living cells. The facultative or obligate parasites may utilize either simple host-synthesized compounds that participate in dynamic physiological processes or complex structural constituents of the cell or tissue; where the food source is wholly host-created, the parasite need not have the enzymes necessary to convert complex molecules to the simple compounds that are assimilated by the cell. These enzymes must be present in many free-living heterotrophs. Indeed, inasmuch as certain enzymes are apparently not essential for growth in the intracellular habitat, they have been lost in the course of evolution to parasitism, a loss that is reflected not only in the absence of the appropriate biochemical activity but also frequently in the appearance of highly complex nutritional patterns. Presumably the parasite making enzymes that catalyze the synthesis of growth factors already present in its intracellular microhabitat is often at a selective disadvantage by comparison with its nutritionally less versatile equivalents and is, in time, displaced by them; such parasites rely on their hosts for one or more essential substances, factors that can be provided in culture media for many organisms but cannot be supplied in the case of the obligate parasite —because, for example, the substances may be highly labile and lose activity when the host dies or when attempts are made to extract the crucial metabolites.

The obligate parasite is generally conceded to obtain the metabolites needed for its growth almost constantly and, as a rule, in ample amounts. It can multiply only when a host is available to generate the vital substances or conditions it requires, and it is these deficiencies that govern its host dependency. Occasionally the development of the parasite is limited because it needs more of a factor than the host supplies or because it is localized in or on the host at a site where the metabolite concentration is low; for example, multiplication of the obligate mycoparasite, *Gonatobotryum fuscum,* is restricted because its host fungus, a species of *Graphium,* does not provide enough of an essential vitamin (Barnett, 1968). The number of factors responsible for the nutritional or metabolic dependency is largely unknown, but one may expect that often only a single physiological deficiency is involved. On the other hand, there is no reason to doubt that some relationships are far more intricate and entail several biochemical substances that the parasite is unable to make for itself. At the extreme of dependency are the viruses, which neither release biochemically useful energy nor catalyze the reactions necessary for synthesis of the nucleic acids and proteins required for viral replication, and they therefore rely on the energy of the host cell and divert part of its biochemical activity to make viral constituents.

The host may provide the microorganisms it supports not only with nutrients but also with a multitude of other factors that contribute to the maintenance of a relatively constant microenvironment. The osmotic pressure and pH encountered by the parasite are not those of the habitat at large but rather those of the microhabitat inside or on the surface of a living cell, and, by governing the intensity of these variables, the host may benefit its dependents. Furthermore, the intracellular parasite may be shielded from harmful chemicals that affect the associated species, and endoparasites of warm-blooded animals exist in a region of constant temperature that is not shared by most taxonomically related organisms. Moreover, many endoparasites are sheltered so effectively that they cannot be located by potential hyperparasites or predators.

ECOLOGICAL EFFECTS OF
INTERMICROBIAL PARASITISM

Parasites are integral components of innumerable communities, and they undoubtedly exert a degree of control over many indigenous populations as well as possibly preventing the establishment of some invading species. The outcome of interspecific competition may be modified, moreover, because of an attack on one of the competitors, or parasitism may allow two competitors to coexist in a locality where one might be expected to displace the other. In addition, a naturally occurring epidemic may upset the original balance in a community so that a dominant species becomes of minor significance or dies out, while an uninfected species assumes prominence. Because they are probably held in check at fixed times of the year in selected ecosystems and run rampant at other periods, parasites may alter the season when an organism, such as an aquatic alga, reaches its maximum density. Unfortunately, the ecological importance of most kinds of intermicrobial parasitism is unsure, inasmuch as the bulk of the research, which is extensive, has been concerned almost solely with in vitro questions. Nevertheless, sufficient evidence is available to permit a general outline of the impact of parasitism on a few groups of microbial hosts to be sketched.

Epidemics of immense proportions regularly occur in populations of freshwater and marine algae. The common culprits are fungi, frequently Phycomycetes and particularly the chytrids, and the frequency and severity of these outbreaks suggest that they have a significant impact in a variety of aquatic ecosystems. Chytrid infestations of freshwater algae are not rarities, the devastation of populations of susceptible hosts having been recorded frequently and in widely separated areas of the world. For reasons as yet obscure, susceptible algae commonly remain largely free of infection for much of the year, but then suddenly the fungi begin to multiply rapidly and appear everywhere to bring about appreciable algal destruction. A typical instance of chytrid infection is that caused by *Rhizophydium planktonicum,* a fungus that actively attacks the lake-inhabiting diatom *Asterionella formosa* (Canter and Lund, 1948). Figure 14.6 reveals one such relationship. Fungus-caused epidemics have likewise been noted in communities of marine algae, and fungi like *Podochytrium, Ectrogella,* and *Olpidium* have been implicated in the destruction of lake or marine diatoms. Phycomycetes are known to do serious damage to seaweeds, too. Fungi may also destroy large numbers and materially threaten populations of red algae, like *Porphyra,* that are cultured commercially for food (Arasaki et al., 1960).

Mycoparasites are ubiquitous in soil, but their precise role in suppressing their fungal hosts remains nebulous. Infection of sensitive organisms has been observed under natural conditions (Boosalis, 1956; Drechsler, 1943), but the capacity of fungi to grow away from their attackers and the frequent inability of the parasite to find its

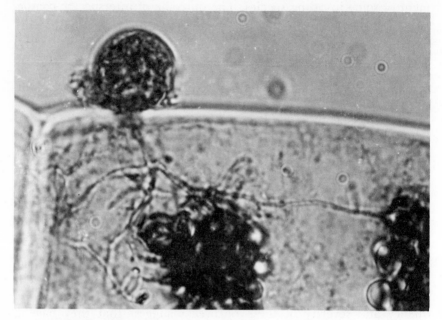

Figure 14.6 *Rhizophydium sphaerocarpum* parasitic on *Spirogyra* sp. (Barr and Hickman, 1967).

host in an environment as physically heterogeneous as soil may limit the impact of mycoparasites on the subterranean community. Still mycoparasites can contribute to the eradication of resting bodies formed by soil fungi, structures that usually are not overly numerous and cannot escape from the invader; thus the dormant sclerotia of *Sclerotinia trifoliorum,* though potentially able to retain their viability in nature for many years, may be slowly inactivated by *Coniothyrium minitans* (Tribe, 1957). It has also been proposed that the colonization of newly cut logs or stumps by wood-rotting Basidiomycetes is facilitated because they first invade established fungi (Griffeth and Barnett, 1967).

Cultivated mushrooms also have their enemies, and the economic importance of the edible mushroom has prompted a modest amount of research. In addition to fungi affecting the underground portions of these Basidiomycetes, the aboveground parts are invaded by species of *Mycogone, Sporodinia, Trichoderma,* and *Verticillium,* occasionally with the destruction of nearly the entire mushroom crop (DeVay, 1956). Virus-induced disorders also create difficulties in the commercial production of mushrooms (Hollings et al., 1963).

Although protozoa are subject to a variety of invaders, the precise impact of these parasites has not been elucidated. The intestinal tracts of man, domesticated animals, and wildlife support innumerable protozoa, many bearing clear signs of microbial infection, and fungi feeding on soil protozoa are likewise common. The high incidence of infection suggests that these natural enemies act to regulate the qualitative or quantitative composition of the protozoan community, but investigations in vivo designed to answer such ecological questions have been remarkably rare.

The abundance, ubiquity, and wide range of genera destroyed in culture media by bacteriophages and *Bdellovibrio* have prompted the view that such parasites are of consequence in aquatic environments, sewage, or soil, possibly in the regulation of population size. Data to support this contention are meager, at best, though the hypothesis is surely attractive. In specialized circumstances it is not implausible that the abrupt decline of a dense bacterial population is brought about by viral infection, and an activity of this sort has been considered to be responsible for the death of *Shigella dysenteriae* (Gispen and Gan, 1950) and for the rapid disappearance of *Vibrio comma* introduced into river water by fecal contamination. The mere fact that an obligate parasite, like the bacteriophage, exists in polluted water is a priori evidence that susceptible individuals are present, but whether this parasitism governs host abundance or frequency or, indeed, whether the parasite acts destructively or in a balanced fashion—as a temperate bacteriophage, for example—is still uncertain (Chambers and Clarke, 1966). Though attributed occasionally to bacteriophages, the bactericidal action of sea water on *Escherichia coli* cells entering the marine environment from sewage outfalls and land drainage appears instead to be the result of a complex of forces, and viruses alone cannot account for elimination of this bacterium (Carlucci et al., 1961). In addition, the high incidence of blue-green algal viruses in waste stabilization ponds has been taken as a basis for the argument that these viruses affect the algal community of the ponds in some manner, and they have also been assumed to contribute to the death of blue-green blooms in other aquatic habitats.

Bacteriophages have an unquestioned impact on microbial processes of direct concern to man, as in the industrial production of streptomycin effected by *Streptomyces griseus*, the mycelium of which can be invaded during its growth in commercial fermentation tanks, or in the suppression of lactic acid formation in dairy starter cultures containing acid-generating *Streptococcus* strains. Bacteriophages lytic to species of *Rhizobium*, the bacterial genus containing representatives fixing N_2 in symbiosis with leguminous plants, are found on legume roots, in nodules, and in soils supporting nodulated plants, and it was therefore natural to assume that these viruses significantly affect *Rhizobium*, especially since legumes like alfalfa or clover sometimes show poor vigor if continuously cropped on the same land. It was argued that the virus is present in the nodule and, by parasitizing the rhizobia, reduces markedly the N_2 fixation catalyzed in that root structure. However, the evidence that the bacteriophage is the underlying cause of the lack of vigor of alfalfa or clover is highly dubious, and its alleged role in reducing symbiotic N_2 fixation awaits clarification (Anderson, 1957).

Parasites may be important in nature not only by weakening or destroying individuals they invade but also because they can modify the host's genotype as well as distinct phenotypic characters of ecological relevance. Clearly the parasite must be of the balanced type. Most information in regard to changes of this kind is derived from studies of bacteria infected by temperate bacteriophages able to transmit genetic traits of the cell from which the virus was released to a second cell that is penetrated by the viral particle. In this process of transduction, briefly considered in a previous chapter, the characteristics transmitted may alter the nutrition or physiology of the infected cell, or, as demonstrated with *Staphylococcus* (Blair and Carr, 1961), they may lead to modifications in the recipient so that it acquires resistance to antibiotics to which it was formerly sensitive.

Hyperparasitism, a well-characterized laboratory phenomenon, is not simply an in vitro curiosity. Early investigators of bacterial viruses were greatly interested in using bacteriophages for the treatment of human disease, as biological controls for pathogenic agents, but their dreams of success came to naught (d'Herelle, 1926). However, hyperparasitism by viruses is of enormous importance in diphtheria, a disease of man caused by *Corynebacterium diphtheriae*. This aerobic bacterium is commonly localized in a lesion in the throat or nasopharynx, where it synthesizes a highly potent toxin that is transmitted by the bloodstream to remote organs in which extensive damage is done. Hyperparasitism enters the picture because only *C. diphtheriae* strains lysogenic for unique bacteriophages elaborate the toxin and hence are pathogenic, and only when a nonlysogenic cell becomes infected by the appropriate virus does it acquire toxin-producing ability (Freeman, 1951; Groman, 1955). The hyperparasite thus has a striking effect not so much on the host wherein it lives, namely the bacterium, but rather on man. Toxin-forming strains may even have an advantage in natural selection over nontoxigenic *C. diphtheriae* populations in that a virus which is a balanced parasite of the former often is destructive to the latter, an advantage that may contribute to the increasing dominance of toxin producers during some diphtheria outbreaks (Mouton, 1960). In addition to *C. diphtheriae,* infection of strains of *Streptococcus* and *Staphylococcus* with temperate bacteriophages derived from toxin-producing bacteria of the same genus may result in the newly infected cell acquiring the capacity to make toxin.

Hyperparasites are prominent during numerous plant disease outbreaks, the pathogen being infested as it develops on its host. Fungi like *Cephalothecium, Cladosporium, Coniothyrium, Darluca,* and *Tuberculina,* to cite but a few, are common hyperparasites in nature, and a few of these fungi appear to suppress disease in the field. For example, when *Darluca filum* is successfully established in bluegrass pustules induced by the pathogen *Puccinia graminis* f. sp. *poae,* further development of the bluegrass disease ceases. In nearby fields free of *D. filum,* by contrast, disease caused by strains of *P. graminis* progresses unabated (Bean, 1968). Similarly, large numbers of bacteriophages are present on the leaves of tobacco plants naturally infested with *Pseudomonas tabaci,* the cause of wildfire disease (Fulton, 1950). Plant pathologists have endeavored to exploit hyperparasitism by applying appropriate fungi or bacteriophages to plants or soil, but practical crop protection by such techniques has not reached the stage of reality or of widespread use.

LYSIS

Heterotrophs lysing cells and filaments of their neighbors are both widespread and abundant, and they have been found on leaves and in soil, the rhizosphere, and compost heaps as well as in the sea. They are particularly numerous and active in soil, especially microorganisms digesting fungal mycelium, and it is these *mycolytic* groups that have been of greatest ecological interest. The density of mycolytic bacteria, for example, has been reported to range from 10^4 to $10^8/g$ of soil. *Bacteriolytic* myxobacteria can also be quite plentiful, notably in fields receiving manure and compost, their populations occasionally exceeding 10^5 propagules/g. Experimental observations have revealed that indigenous lytic groups become still more common when hyphae or cells of fungi or bacteria are deliberately introduced into soil or sea water; furthermore, the frequency of lytic cells and hyphae might also be expected to rise when organic nutrients supporting heterotrophic development are introduced into an area.

As shown in Table 14.3, the ability to produce lytic enzymes is a characteristic of many taxonomic categories. Genera of true bacteria like *Aeromonas, Bacillus, Flavobacterium, Pseudomonas,* and *Staphylococcus,* myxobacteria such as *Chondrococcus, Myxococcus,* and *Sorangium,* actinomycetes like *Micromonospora* and *Streptomyces,* and the fungus *Chalaropsis* all contain bacteriolytic strains. The bacterial genera susceptible to such digestion are numerous, the ones presented in the table being only a partial listing. It is evident from Table 14.3 that a genus resistant to one lytic population may be destroyed by another. In addition, as indicated in the table for *Arthrobacter,* strains of a single species or species of a genus vary markedly in their susceptibility to lysis. A wide assortment of fungi coming from all classes are also subject to this kind of parasitism, and the organisms responsible, in vitro at least, for this activity include representative bacteria, actinomycetes, and fungi. The few yeasts so far investigated are easily destroyed in this manner, too, and though the information is still scanty, blue-green and green algae are subject to lysis induced by bacteria and fungi.

Considerable effort has been directed at establishing the causes of lysis. The walls of fungi, bacteria, actinomycetes, and algae contain components that confer rigidity on the cell. The wall-localized macromolecules responsible for this rigidity maintain the unique shape of the individual and permit growth of the salt- and metabolite-rich cell in a hypotonic environment. The portion of the cell underlying the wall, the protoplast, is osmotically sensitive on account of its high osmotic pressure, and a substance synthesized by a nearby organism that affects the rigidity or brings about destruction of the backbone component of the wall ultimately leads to rupture of the cell membrane and an explosive disruption of the protoplast as the high internal osmotic pressure is no longer restrained by the stiff outer structure. Thus, as a rule, extensive digestion of the wall usually means death for unicellular microorganisms, though for fungi and possibly other filamentous groups, parts of the hyphae may remain intact or new filaments may begin to grow at the same time as portions of the pre-existing structure are undergoing lysis.

A number of macromolecules are present in microbial cell walls, some conferring rigidity, others being closely linked with the constituent serving as the wall's backbone. The identity of the polymer varies among the major taxonomic groups of microorganisms, but the major ones are (a) cellulose, laminarin (a β-1,3-glucan), and related glucose-containing polysaccharides in many fungi, yeasts, and algae; (b) a peptidoglycan (sometimes called murein, mucopeptide, or gly-

Table 14.3

Microorganisms Producing Lytic Enzymes and Species Susceptible to These Enzymes

Lytic Organism	Susceptible to Lysis	Resistant to Lytic Species	Reference
		Bacteriolytic Organisms	
Aeromonas	Bacillus, Clostridium	Pseudomonas, Salmonella	Coles and Gilbo (1967)
Chalaropsis	Corynebacterium, Streptococcus	Mycobacterium, Proteus	Hash (1963)
Flavobacterium	Pediococcus, Staphylococcus	Micrococcus	Sakaguchi et al. (1966)
Myxobacterium	Arthrobacter, Micrococcus	Arthrobacter, Escherichia	Ensign and Wolfe (1965)
Sorangium	Bacillus, Sarcina	Rhizobium, Xanthomonas	Gillespie and Cook (1965)
Streptomyces	Corynebacterium, Bacillus	Streptococcus, Sarcina	Mori et al. (1960)
		Mycolytic Organisms	
Agarbacterium	Achlya, Pythium	—	Mitchell and Wirsen (1968)
Bacillus	Alternaria, Penicillium	Pythium, Saccharomyces	Mitchell and Alexander (1963)
Pseudomonas	Fusarium	Rhizoctonia	Potgieter and Alexander (1966)
Streptomyces	Aspergillus, Sclerotium	Cladosporium, Rhizoctonia	Bloomfield and Alexander (1967)
Streptomyces	Mucor, Penicillium	Alternaria, Helminthosporium	Aguirre et al. (1963)
Verticillium	Hemileia, Puccinia	—	Garcia Acha et al. (1965)

355

copeptide) in true bacteria and certain blue-green algae; (c) chitin and other N-acetylglucosamine-containing polysaccharides in the mycelium of numerous fungi; (d) mannans and xylans in some algae; and (e) as yet poorly characterized polysaccharides in diverse genera of algae and fungi. Obscure as yet is the chemistry of the surface structures of endospores, cysts, chlamydospores, and other resistant stages in the life cycles of countless species.

Lysis may be effected by one of two general means: the cell may be disrupted by activities of individuals of another population or the cell may destroy its own surface components. The former is known as *heterolysis,* the latter as *autolysis.* In enzymatic heterolysis the cellular integrity is abolished by extracellular, cell wall-degrading enzymes synthesized by a neighboring population, and microbiologically formed enzymes hydrolyzing all the macromolecules mentioned above have been described and characterized. These enzymes digest a key structural linkage in the wall-localized polymer, destroying its rigidity and leading to disruption of the organism's structure and hence to death of the cell. One species may cause lysis of a second in other ways, too, as by modifying the cells so that autolysis is initiated; for example, selected antibiotics cause autolysis of bacterial cells or fungal mycelium, and a microorganism excreting toxins of this sort, if indeed they are formed in nature, may induce nearby cells to digest their own surface components and bring about their own demise. The mode of action of a few of these antibiotics is understood, their function being to interfere with cell wall biosynthesis. Lysis may also be brought about by substances that digest the cytoplasmic membrane or affect it in some other way.

Innumerable toxic agents promote autolysis. Some are remarkably simple in structure; for example, common microbial products like ammonia or acetic acid are responsible for pH-dependent autolysis of the phytoflagellate *Prymnesium parvum* (Shilo and Shilo, 1962). Many of these toxins are active at extremely low concentrations and have a complex structure, suggesting that they are antibiotics. Few of the autolysis-inducing compounds observed in microbial cultures in the laboratory have been adequately characterized, but among those that have been identified are antibiotics of great practical importance. As a case in point, penicillin is a selective inhibitor of the synthesis of wall components necessary to maintain the morphological integrity and viability of penicillin-sensitive bacteria. The susceptible population must be actively multiplying in order for penicillin to be effective, and the bacterial cells, although dividing, no longer can make the rigid component of the wall so that the osmotic pressure of the newly generated protoplasts causes them to burst. The mode of action of antibacterial antibiotics like bacitracin, Vancomycin, and D-cycloserine likewise appears to involve inhibition of cell wall biosynthesis (Strominger and Ghuysen, 1967). Other antibacterial agents influence the cytoplasmic membrane rather than the walls. Among the characterized antifungal antibiotics inducing autolysis are amphotericin B, cycloheximide, Filipin, and nystatin.

Starvation and O_2 deficiency may predispose a cell to autolysis. An inadequate supply of nutrients, especially of energy-yielding substrates, favors autolysis of fungi and some bacteria in culture media, and similar microbiologically produced deficiencies undoubtedly occur in natural habitats during competition.

ENZYMATIC HETEROLYSIS

Chitinase, β-1,3-glucanase, and β-1,4-glucanase are the enzymes responsible for the lysis of those fungi containing chitin, laminarin, or cellulose as the structural backbones of the hyphal walls. Laminarin is a β-1,3- and cellulose a β-1,4-glucan, and the catalysts responsible for the hydrolysis of these polysaccharides are therefore designated β-1,3- and β-1,4-glucanases. The mycelial walls of many but far from all Phycomycetes, Ascomycetes, Fungi Imperfecti, and Basidiomycetes are composed of chitin, a glucan, or both. Microorganisms excreting chitinase and glucanases are widespread and abundant, notably in soil, and the hydrolytic reactions catalyzed by these extracellular enzymes liberate oligosaccharides and sugars from the wall polysaccharides. At the same time the ruptured protoplast releases an assortment of potential nutrients that, together with the wall-derived degradation products, support both the lytic organisms and immediately adjacent cells. Different or additional polysaccharide-hydrolyzing enzymes are needed for the lysis of yeasts and those fungi that contain dissimilar polysaccharides as structural backbones of the cell walls.

Enzymatic heterolysis of bacteria is readily detected in vitro by noting the clearing when a lytic organism is inoculated into a medium containing a turbid cell suspension or when the responsible enzymes are incubated with the cells, but evidence for their lysis in nature is far less easily collected than with the fungi, because of the smaller size of the bacterium. Bacteriolytic enzymes, like those that are mycolytic, generally act by hydrolyzing a wall macromolecule conferring rigidity on the cell, solubilizing or markedly degrading the polymer, and thereby bringing about lysis and death of the organism. Because the chemical compositions of bacterial and fungal walls are entirely dissimilar, however, different enzymes and, almost always, totally different populations are responsible for bacteriolysis and for mycolysis. The building blocks of the peptidoglycan in the walls of simple rods and coccoidal bacteria are acetamido sugars and a few amino acids, and this enormous molecule is composed of glycan strands, peptide units, and peptide bridges. Its degradation can be effected by three categories of enzymes: glycosidases or carbohydrases acting on the wall polysaccharides, endopeptidases breaking the peptide cross-linkages, and enzymes cleaving the points where the polysaccharides and peptides are joined (Strominger and Ghuysen, 1967).

Myxobacteria are considered to be predators, but their predation is characteristically associated with the elaboration of extracellular lytic enzymes enabling them to disrupt the bacteria, and presumably the fungi, on which they feed. The constituents released from the prey as a consequence of its lysis are then utilized as nutrient sources. Ubiquitous soil myxobacteria such as *Archangium, Chondrococcus, Myxococcus,* and *Sorangium* thus live at the expense of *Aerobacter, Arthrobacter, Bacillus, Micrococcus, Pseudomonas,* and other gram-negative and gram-positive bacteria, although selected strains of these edible genera are often partially or totally resistant to lysis by individual populations of myxobacteria.

Lysis of fungi and bacteria has been well explored in vitro and, albeit to a lesser extent, in vivo, but little attention has been given to lysis of algae and protozoa. The chemical composition of walls of selected blue-green algae is remarkably similar to that of bacteria, and they are destroyed by the lytic enzyme, lysozyme. Furthermore, a *Cytophaga* isolated from a waste stabilization pond lyses species of green and blue-green algae (Stewart and Brown, 1969), and *Fomes* and other fungi can have an identical effect on *Chlorella* (Warren and Winstead, 1965). The largely ignored protozoa seem to be susceptible, too, as indicated by the fact that *Acanthamoeba* is attacked by a *Pseudomonas*-like bacterium (Drozanski, 1956).

LYSIS IN NATURE

Regardless of the precise mechanism involved, a lytic population benefits by eliminating prospective competitors or by attacking suscepts in such a manner that nutrients are obtained from the walls and protoplasts that are destroyed (Fig. 14.7). The same popu-

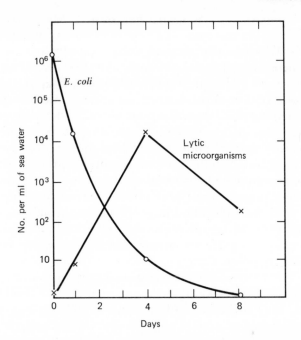

Figure 14.7 The death of *Escherichia coli* and the concomitant increase in lytic microorganisms in sea water (Mitchell and Yankofsky, 1969). By permission of the American Chemical Society.

lation simultaneously modifies the surrounding community to a lesser or greater extent, eliminating susceptible cells and filaments or lysis-sensitive stages of organisms that exist in more than one morphological form. Nevertheless, lysis is not an all-pervasive means of regulating community composition inasmuch as sensitive populations exist in proximity to cells capable of excreting substances promoting lysis, though possibly not in the same microhabitat. At the microenvironmental level, to be sure, it is not inconceivable that a population is decimated by enzymatic heterolysis or by toxin-induced autolysis.

No environment has been investigated as thoroughly with regard to lysis, at least of fungi, as soil, and results derived from these studies indicate that this kind of interaction is quite important in determining the morphological forms of fungi that persist in soil. Mycolysis is also widely regarded as a key factor governing the abundance and activity of fungi and limiting their spread in terrestrial habitats, but how significant it is in comparison with competition or amensalism has yet to be assessed. Mycolysis is readily noted by direct microscopic observation of soil specimens. For example, fungal hyphae are seen to ramify rapidly through organic materials incorporated below ground, but the mycelium that develops is observed to be quickly surrounded by bacteria and actinomycetes. Colonization of the filaments is accompanied by decomposition of the hyphal walls and digestion of the underlying protoplast. The hyphal tips of many parasitized soil Phycomycetes appear morphologically normal, but shortly behind the tips the filament is devoid of internal contents and is enveloped by bacteria. An appreciable percentage of the rod-shaped bacteria and actinomycetes occupying the fungal surface undoubtedly subsist on components of dead cells and are not the cause of the organism's demise, but since colonization of the mycelium and the decline in number of viable fungal propagules are accompanied by a sharp rise in the abundance of *Bacillus, Pseudomonas,* and *Streptomyces,* representatives of which are mycolytic, it seems likely that many of the heterotrophic colonists are responsible for and gain benefit from mycolysis (Lockwood, 1967; Mitchell and Alexander, 1963). Many genera of fungi are destroyed in soil in this manner. Most research has been centered on the destruction of innocuous subterranean inhabitants or plant pathogens, a large number of which produce hyphae readily destroyed by lysis, but animal and human pathogens entering soil are eliminated in like fashion. For instance, *Blastomyces dermatitidis* introduced into soil with saliva from a dog dying of blastomycosis soon dies out, and no viable cells can be isolated from the contaminated area. Such lysis may account for the very brief persistence of *B. dermatitidis* and other human- or animal-pathogenic fungi and yeasts and may explain why the disease agents cannot be detected several days following known instances of natural contamination (McDonough et al., 1965).

Scores of fungi, including species causing major plant diseases, survive below ground by virtue of the persistent spores they form. These structures do not develop into actively metabolizing hyphae unless organic nutrients are made available from dead plant remains or a root exuding low-molecular-weight compounds extends to the vicinity of the resting body. However, when germination is triggered by the appropriate organic compounds, the emergent germ tubes bear little chemical relationship to the dormant spores from which they arose, and they are remarkably prone to lysis. Their continued growth and extension require that their mycolytic enemies be evaded in some way, sometimes by successful penetration of host roots, occasionally by outgrowing the lytic organisms, at times by locating a friendly refuge or microhabitat (Chinn and Ledingham, 1961; Cook and Snyder, 1965).

The countless observations of mycolysis and the frequency of its occurrence have prompted debate on the exact mechanism functioning in soil. The organisms chiefly responsible, evidently strains of *Streptomyces, Bacillus,* and *Pseudomonas,* are typically synthesizers of potent extracellular lytic enzymes. Their abundance is easily demonstrable because, when soil suspensions are plated on agar supplemented with mycelium of test fungi, their colonies are surrounded by a clear zone arising from the digestion of the hyphal material. However, members of these very same genera, *Streptomyces* in particular, excrete antifungal antibiotics so that toxin-induced autolysis may be implicated as well as enzymatic heterolysis. The finding that nonviable hyphae devoid of enzymatic activity are lysed suggests the importance of enzymatic heterolysis, since autolysis is not possible in these structures. On the other hand, inasmuch as living hyphae are decomposed even when they are separated from soil by a membrane preventing the passage of enzymes, a role for autolysis is indicated, too (Lloyd and Lockwood, 1966; Mitchell and Alexander, 1962).

Bacteriolysis by myxobacteria is also not merely a laboratory curiosity. Terrestrial myxobacteria subsist largely on a bacterial diet, and their enzymes rapidly disrupt the cells of indigenous gram-negative and gram-positive species. The elimination of alien fungi and bacteria entering sea water is correlated with the emergence of bacteria capable of attacking cell walls of the aliens by means of extracellular enzymes, and the consequent lysis may be one of the critical factors contributing to the lack of survival in sea water of innumerable organisms of terrestrial or sewage origin (Mitchell and Wirsen, 1968; Mitchell et al., 1967). An analogous kind of destruction may be of consequence on plant surfaces, as exemplified by the *Verticillium* lysing rust spores inhabiting coffee leaves (Garcia Acha et al., 1965).

Susceptibility to lysis is, beyond question, of immense ecological relevance, but of tremendous importance, too—in ecology as in agriculture and human and veterinary medicine—is the resistance to this mode of parasitism. A large number of fungi, protozoa, and bacteria, including many causing disease of man, livestock, and agricultural crops, are innately refractory to heterolysis. In some instances this resistance is an attribute of the sole morphological form of the species, but more frequently only a single one of the several stages of the organism is immune from attack. Innate resistance to lysis is suggested by the prolonged viability in nature of a non-growing population, microscopic evidence showing little or no destruction of a particular structure, the inability to obtain isolates lysing a test culture, or the very slow growth of lytic organisms on the refractory forms. The susceptibility to any particular lytic population differs from one test species to another, and the extracellular enzymes elaborated by the parasite act, at best, on a limited spectrum of microorganisms; nevertheless the very existence of microbial groups or structures resistant to all neighbors implies that they have unique constituents, presumably in their walls or outer coverings, that prevent enzymatic destruction.

Endospores of *Bacillus* and *Clostridium,* cysts of protozoa, sclerotia, chlamydospores, and occasionally fungal hyphae and conidia retain their viability in soil for months and sometimes years without being lysed, although the community of which they are a part is large, physiologically heterogeneous, and metabolically active. The sensitivity to lysis of a variety of fungi is given in Table 14.4. Vegetative cells emerging from the resistant endospores, cysts, sclerotia, chlamydospores, or conidia, however, are usually prone to attack, though a few notable exceptions have been found, especially among the fungi. The capacity to form resistant bodies, and thereby survive the destruction of vegetative cells, maintains the species and confers on the possessor of this ability a selective advantage not held by its neighbors. Moreover, because the recalcitrant structures are made by plant pathogens as well as by *Bacillus anthracis* and *Clostridium,* which are agents of livestock and human disease, serious economic or public health problems may arise because of their durability in soil.

Few explanations have been offered for the remarkable persistence of these structures in nature. They must have a means of shielding themselves from the onslaught of their neighbors, a protection against enzymatic heterolysis or toxin-induced autolysis. In the case of *Mortierella parvispora,* a representative of a fungal genus not only abundant but also apparently often present in soil in the mycelial form, the resistance has been ascribed to a heteropolysaccharide in the walls that makes the organism refractory to enzymatic hydrolysis. The longevity of hyphae containing a heteropolysaccharide backbone is not unexpected, because a macromolecule containing several sugar units and possibly more than a single kind of linkage between the individual building blocks probably requires for its hydrolysis a number of different enzymes excreted by individuals of dissimilar

Table 14.4

Resistance and Susceptibility of Fungi to Lysis by Soil Microorganisms

| Fungus | Sensitivity of Morphological Stages | | Reference |
	Resistant	Susceptible	
Alternaria solani	Dark hyphae[a]	Light hyphae	Lockwood (1960)
Aspergillus phoenicis	Conidia[a]	Hyphae	Bloomfield and Alexander (1967)
Fusarium oxysporum	Chlamydospores	Conidia, hyphae	Sequeira (1962)
Glomerella cingulata	Chlamydospores	Hyphae	Lloyd and Lockwood (1966)
Helminthosporium sativum	Conidia	Germ tubes	Chinn and Ledingham (1961)
Helminthosporium victoriae	Conidiophores[a]	Hyphae	Lloyd and Lockwood (1966)
Rhizoctonia solani	Hyphae[a]	—	Lockwood (1960)
Sclerotium rolfsii	Sclerotia[a]	Hyphae	Bloomfield and Alexander (1967)
Thielaviopsis basicola	Normal chlamydospores[a]	Albino chlamydospores	Linderman and Toussoun (1966)

[a] Contain melanin or other dark pigments.

populations, each of which must be in the vicinity of the mycelium, and the likelihood is small of having organisms with the entire set of requisite enzymes immediately adjacent to the filaments in question (Pengra et al., 1969).

It is noteworthy that a high proportion of lysis-insensitive resting structures or vegetative cells of the fungi are dark in color (Table 14.4). They are assumed or have been shown to contain melanin or melanin-like substances as the pigment. The sensitive stages of these fungi, by contrast, are light in color and have little or no melanin. It is now known that the melanized surfaces protect the polysaccharide components of the wall from digestion, thereby warding off disaster. Several lines of evidence, in part considered in Chapter 8, point to the critical protective function of the surface-localized pigment: (a) microorganisms capable of utilizing melanin, melanized hyphae of *Rhizoctonia solani,* or melanin-containing conidia of *Aspergillus phoenicis* as sole carbon sources for growth have yet to be isolated;

(b) the melanized mycelium of *Aspergillus nidulans* is unaffected by enzymes lysing a nonpigmented mutant of the same fungus; (c) the resistance of *A. nidulans* hyphal walls to digestion increases as its content of this dark, polyaromatic constituent rises; (d) removal of the melanin-containing spicules from the surface of *A. phoenicis* conidia makes the previously resistant spores sensitive to polysaccharide-hydrolyzing enzymes participating in lysis; (e) microorganisms digesting walls of *Sclerotium rolfsii* hyphae are inactive toward its sclerotia, the sclerotia having a melanin outer layer overlying a coil of mycelial strands; and (f) melanin itself is not appreciably decomposed in soil, a necessary prerequisite if this polyaromatic compound is to protect microbial structures for long periods in nature (Bloomfield and Alexander, 1967; Kuo and Alexander, 1967). Nevertheless, numerous lysis-insensitive microorganisms are devoid of melanin, and hence other shielding devices must be present in these refractory structural elements.

References

REVIEWS

Adams, M. H. 1959. *Bacteriophages.* Interscience Publishers, New York.

Anderson, E. S. 1957. The relations of bacteriophages to bacterial ecology. *In* R. E. O. Williams and C. C. Spicer, Eds., *Microbial Ecology.* Cambridge University Press, London. pp. 189–217.

Ball, G. H. 1969. Organisms living on and in protozoa. *In* T.-T. Chen, Ed., *Research in Protozoology,* Vol. 3. Pergamon Press, Oxford. pp. 565–718.

Barnett, H. L. 1963. The nature of mycoparasitism by fungi. *Ann. Rev. Microbiol.,* **17**, 1–14.

Dogiel, V. A. 1964. *General Parasitology.* Oliver and Boyd, Edinburgh.

Duddington, C. L. 1968. Fungal parasites of invertebrates. Predacious fungi. *In* G. C. Ainsworth and A. S. Sussman, Eds., *The Fungi,* Vol. 3. Academic Press, New York. pp. 239–251.

Lockwood, J. L. 1967. The fungal environment of soil bacteria. *In* T. R. G. Gray and D. Parkinson, Eds., *The Ecology of Soil Bacteria.* Liverpool University Press, Liverpool. pp. 44–65.

Madelin, M. F. 1968. Fungi parasitic on other fungi and lichens. *In* G. C. Ainsworth and A. S. Sussman, Eds., *The Fungi,* Vol. 3. Academic Press, New York. pp. 253–269.

Stolp, H., and Starr, M. P. 1965. Bacteriolysis. *Ann. Rev. Microbiol.,* **19**, 79–104.

Strominger, J. L., and Ghuysen, J.-M. 1967. Mechanisms of enzymatic bacteriolysis. *Science,* **156**, 213–221.

Yarwood, C. E. 1956. Obligate parasitism. *Ann. Rev. Plant Physiol.,* **7**, 115–142.

OTHER LITERATURE CITED

Aguirre, M. J. R., Garcia-Acha, I., and Villanueva, J. R. 1963. *Experientia,* **19**, 82–83.

365

Akiba, T., and Iwata, K. 1954. *Japan. J. Exptl. Med.*, **24**, 159–166.

Arasaki, S., Inouye, A., and Kochi, Y. 1960. *Bull. Japan. Soc. Sci. Fisheries*, **26**, 1074–1081.

Banks, G. T., Buck, K. W., Chain, E. B., Himmelweit, F., Marks, J. E., Tyler, J. M., Hollings, M., Last, F. T., and Stone O. M. 1968. *Nature*, **218**, 542–545.

Barnett, H. L. 1968. *Mycologia*, **60**, 244–251.

Barr, D. J. S., and Hickman, C. J. 1967. *Can. J. Bot.*, **45**, 423–430.

Bean, G. A. 1968. *Phytopathology*, **58**, 252–253.

Benjamin, R. K. 1961. *Aliso*, **5**, 11–19.

Blair, J. E., and Carr, M. 1961. *J. Bacteriol.*, **82**, 984–993.

Blank, L. M. 1953. *In Plant Diseases: The Yearbook of Agriculture*. U.S. Department of Agriculture, Washington, D.C. pp. 298–301.

Bloomfield, B. J., and Alexander, M. 1967. *J. Bacteriol.*, **93**, 1276–1280.

Boosalis, M. G. 1956. *Phytopathology*, **46**, 473–478.

Butler, E. E. 1957. *Mycologia*, **49**, 354–373.

Canter, H. M., and Lund, J. W. G. 1948. *New Phytologist*, **47**, 238–261.

Carlucci, A. F., Scarpino, P. V., and Pramer, D. 1961. *Appl. Microbiol.*, **9**, 400–404.

Chambers, C. W., and Clarke, N. A. 1966. *Advan. Appl. Microbiol.*, **8**, 105–143.

Chinn, S. H. F., and Ledingham, R. J. 1961. *Can. J. Bot.*, **39**, 739–748.

Coles, N. W., and Gilbo, C. M. 1967. *J. Bacteriol.*, **93**, 1193–1194.

Cook, R. J., and Snyder, W. C. 1965. *Phytopathology*, **55**, 1021–1025.

de Morgan, A. 1872. *Budget of Paradoxes*. Longmans, Green, London.

DeVay, J. E. 1956. *Ann. Rev. Microbiol.*, **10**, 115–140.

d'Herelle, F. 1926. *The Bacteriophage and Its Behavior*. Williams & Wilkins, Baltimore

Dobbs, C. G., and English, M. P. 1954. *Trans. Brit. Mycol. Soc.*, **37**, 375–389.

Drechsler, C. 1943. *Phytopathology*, **33**, 227–233.

Drozanski, W. 1956. *Acta Microbiol. Polon.*, **5**, 315–317.

Drozanski, W. 1963. *Acta Microbiol. Polon.*, **12**, 9–23.

Ensign, J. C., and Wolfe, R. S. 1965. *J. Bacteriol.*, **90**, 395–402.

Freeman, V. J. 1951. *J. Bacteriol.*, **61**, 675–688.

Fulton, R. W. 1950. *Phytopathology*, **40**, 936–949.

Garcia Acha, I., Leal, J. A., and Villanueva, J. R. 1965. *Phytopathology*, **55**, 40–42.

Gillespie, D. C., and Cook, F. D. 1965. *Can. J. Microbiol.*, **11**, 109–118.

Gispen, R., and Gan, K. H. 1950. *Antonie van Leeuwenhoek J. Microbiol. Serol.*, **16**, 373–385.

Griffeth, N. T., and Barnett, H. L. 1967. *Mycologia*, **59**, 149–154.

Groman, N. B. 1955. *J. Bacteriol.*, **69**, 9–15.

Hash, J. H. 1963. *Arch. Biochem. Biophys.*, **102**, 379–388.

Hollings, M., Gandy, D. G., and Last, F. T. 1963. *Endeavour*, **22**, 112–117.

Johnson, T. 1953. *Biol. Rev.*, **28**, 105–157.

Klement, Z., and Kiraly, Z. 1957. *Nature*, **179**, 157–158.

Kuo, M.-J., and Alexander, M. 1967. *J. Bacteriol.*, **94**, 624–629.

Lazarus, A. S., and Gunnison, J. B. 1947. *J. Bacteriol.*, **53**, 705–714.

Linderman, R. G., and Toussoun, T. A. 1966. *Phytopathology*, **56**, 887.

Lloyd, A. B., and Lockwood, J. L. 1966. *Phytopathology*, **56**, 595–602.

Lockwood, J. L. 1960. *Phytopathology*, **50**, 787–789.

McDonough, E. S., Van Prooien, R., and Lewis, A. L. 1965. *Am. J. Epidemiol.*, **81**, 86–94.

Mitchell, R., and Alexander, M. 1962. *Proc. Soil Sci. Soc. Am.*, **26**, 556–558.

Mitchell, R., and Alexander, M. 1963. *Can. J. Microbiol.*, **9**, 169–177.

Mitchell, R., and Wirsen, C. 1968. *J. Gen. Microbiol.*, **52**, 335–345.

Mitchell, R., and Yankofsky, S. 1969. *Environ. Sci. Technol.*, **3**, 574–576.

Mitchell, R., Yankofsky, S., and Jannasch, H. W. 1967. *Nature*, **215**, 891–893.

Mori, Y., Kato, K., Matsubara, T., and Kotani, S. 1960. *Biken's J.*, **3**, 139–150.

Mouton, R. P. 1960. *Antonie van Leeuwenhoek J. Microbiol. Serol.*, **26**, 297–304.

Noble, E. R., and Noble, G. A. 1964. *Parasitology: The Biology of Animal Parasites.* Lea and Febiger, Philadelphia.

Pengra, R. M., Cole, M. A., and Alexander, M. 1969. *J. Bacteriol.*, **97**, 1056–1061.

Pimentel, D. 1968. *Science*, **159**, 1432–1437.

Potgieter, H. J., and Alexander, M. 1966. *J. Bacteriol.*, **91**, 1526–1532.

Safferman, R. S. 1968. *In* D. F. Jackson, Ed., *Algae, Man, and the Environment.* Syracuse University Press, Syracuse, N.Y. pp. 429–439.

Sakaguchi, K., Kotani, S., Suginaka, H., Hirachi, Y., Kashiba, S., and Ozawa, Y. 1966. *Agr. Biol. Chem.*, **30**, 1097–1101.

Seidler, R. J., and Starr, M. P. 1969. *J. Bacteriol.*, **97**, 912–923.

Sequeira, L. 1962. *Phytopathology*, **52**, 976–982.

Shilo, M., and Bruff, B. 1965. *J. Gen. Microbiol.*, **40**, 317–328.

Shilo, M., and Shilo, M. 1962. *J. Gen. Microbiol.*, **29**, 645–658.

Sonneborn, T. M. 1959. *Advan. Virus Res.*, **6**, 229–356.

Sparrow, F. K., and Ellison, B. 1949. *Mycologia*, **41**, 28–35.

Stabler, R. M., and Chen, T.-T. 1936. *Biol. Bull.*, **70**, 56–71.

Stewart, J. R., and Brown, R. M. 1969. *Science*, **164**, 1523–1524.

Stolp, H., and Starr, M. P. 1963. *Antonie van Leeuwenhoek J. Microbiol. Serol.*, **29**, 217–248.

Tikhonenko, A. S., and Zavarzina, N. B. 1966. *Mikrobiologiya*, **35**, 850–852.

Tribe, H. T. 1957. *Trans. Brit. Mycol. Soc.*, **40**, 489–499.

Van Wagtendonk, W. J., Clark, J. A. D., and Godoy, G. A. 1963. *Proc. Natl. Acad. Sci. U.S.,* **50,** 835–838.

Warren, J. R., and Winstead, N. N. 1965. *Phytopathology,* **55,** 244–245.

Watson, W. 1948. *Trans. Brit. Mycol. Soc.,* **31,** 305–339.

Willoughby, L. G. 1956. *Trans. Brit. Mycol. Soc.,* **39,** 125–141.

15

Predation

In predation one organism, the predator, feeds on a second and commonly causes death of the unicellular prey or destruction of part or all of the multicellular prey. As a rule, cells of the species devoured, if unicellular, are both smaller in size and more numerous than the exploiting individual. The population of the organism consumed is deleteriously affected, but the predator derives benefit in the form of carbonaceous nutrients and growth factors. The prey in essence frequently converts an area unable to support the predator into a favorable habitat, and it does this by manufacturing consumable products, namely, its own cellular constituents, from nutrients unavailable for growth of the feeding organism. The classical predator exhibits phagotrophic feeding, getting its nutrients by ingesting particulate matter, in this instance as living organisms, although some predators lyse the prey individuals and assimilate only soluble constituents.

KINDS OF PREDATORS

The predatory habit characterizes groups of wholly dissimilar microorganisms. Many protozoa spend their entire lives as phagotrophs, largely using viable organisms for sustenance. At stages in their life cycles, myxobacteria, Myxomycetes, and fungi classified among the Acrasiales rely entirely on the availability of suitable prey. Unique fungi are peculiarly fitted to trap and devour nematodes, and though attacking and consuming a larger individual, these fungi, too, are usually designated predacious. A variety of chlorophyll-containing organisms such as dinoflagellates and other marine and freshwater algae also live at the expense of small-celled neighbors

369

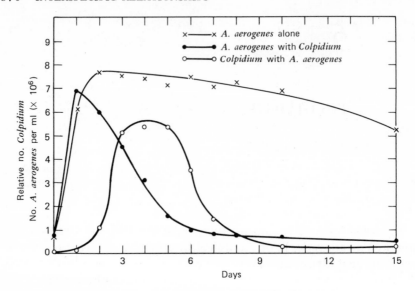

Figure 15.1   Effect of *Colpidium* on the multiplication of *Aerobacter aerogenes* (Butterfield and Purdy, 1931). By permission of the American Chemical Society.

(Pringsheim, 1959). Certain chrysophyceans, *Ochromonas* being a well-known representative, are brown-pigmented freshwater and marine flagellates which, though containing chlorophyll, likewise may ingest bacteria, small algae, or yeasts. *Peranema,* a colorless flagellate classified in Euglenophyta, swallows small organisms whole, but can also attack larger individuals by cutting them open and sucking out their contents (Chen, 1950). Phagotrophy also typifies the phagocytes of metazoa that function in protecting the animal against microbial invaders. The present discussion will deal with intermicrobial predation; the grazing on microorganisms by aquatic and terrestrial metazoa, which is of enormous importance, will be considered later. A typical relationship between a predatory protozoan, *Colpidium,* and its bacterial food is depicted in Fig. 15.1.

Though numerous protozoa are capable of reproducing at the expense of inorganic or organic nutrients in solution, a large percentage of these animals are restricted to and dependent on a predatory mode of nutrition, ingesting and digesting living or dead organisms, or both. They sustain themselves by hunting, and many are remarkably voracious and quite active as they seek out edible cells. These microscopic animals do not grow unless their particular prey is present. Availability of the food organisms is frequently not enough, however, since the protozoa often need a reasonably large number of edible individuals in order to maintain vegetative development; hence the qualitative and quantitative composition of the community, especially of consumable microorganisms, is of paramount importance to them. Predation is typical of Mastigophora, Ciliata, and Sarcodina residing in aquatic and terrestrial ecosystems as well as within the animal's alimentary tract.

Myxobacteria dissolve intact bacterial cells by means of extracellular enzymes, and the products solubilized enter and satisfy the nutrient demands of myxobacteria like *Myxococcus* and *Chondrococcus*. *Dictyostelium* and other cellular slime molds, by contrast, feed directly on bacterial cells. The myxamebal stages of these fungi need suitable prey for their development, and the bacteria they successfully capture are ingested, destroyed, and ultimately digested. In addition, the plasmodia of certain Myxomycetes consume a multitude of bacteria and, in vitro at least, occasional yeasts and filamentous fungi. Aquatic dinoflagellates like *Oxyrrhis* and *Gymnodinium* attack a wide range of algae, including dinoflagellates, in laboratory media and also in their natural habitats.

Fungi preying on soil nematodes are readily demonstrable, and their effect is extremely dramatic. Some groups of fungi possess structural features suiting them to this way of life, for they produce traps admirably suited for capture of the nematodes. Some traps rely on an adhesive substance for their operation, a few ensnare worms that enter and become so tightly wedged in the ring-shaped structure that they are unable to withdraw, while the most interesting traps are those that spring suddenly upon a nematode unfortunate enough to have inserted its head within the ring. The fungal hyphae then penetrate the animal and assimilate its contents. Another group of fungi, the Zoopagales, form no traps but instead excrete a sticky fluid that adheres to the worm; though it may struggle violently for a while, the nematode ultimately dies, and the mycelium enters and ramifies through its body, assimilating its contents (Duddington, 1957). Nematode-capturing fungi behave in many ways like those feeding upon protozoa, particularly the amebae, and indeed they are taxonomically related. The latter have been considered in the previous discussion on parasitism, but they behave to a significant degree as predators and might just as well be considered as such.

A biological relationship closely akin to the amebal predator–bacterial prey association is localized in the blood and tissues of animals, wherein invading bacteria are constantly being engulfed by the hordes of scavenging, ameba-like phagocytes. Phagocytes provide an extremely effective defense against bacteria as well as yeasts, protozoa, and fungi. Indeed, the resistance or susceptibility of a microbial invader to phagocytosis frequently determines the outcome of an infection in man and animals, and virulence is often correlated with the inability of phagocytes to prey upon and destroy the invader. By contrast with many free-living predators, however, the ability of such scavenger cells to eliminate their quarry is not simply governed by whether engulfment is successful; ingested bacteria, like *Mycobacterium tuberculosis,* may survive and even thrive within the phagocytes, ultimately doing serious injury to the infected animal (Hirsch, 1965).

ENVIRONMENTS INHABITED BY PREDATORS

Considerable attention has been given to the distribution, abundance, and function of predators in selected ecosystems, notably in sewage, polluted waters, untainted fresh water, estuaries, the open sea, soil, and within the animal's gastrointestinal tract. The presence of predators in additional regions or ecosystems has been recorded, but the information is far too sketchy to permit a definition of their ecological role.

Bodies of water receiving organic pollutants teem with bacteria, organisms that may be introduced together with the pollutants or indigenous species that participate in the decomposition of the carbonaceous matter. Provided that O_2 is in ample supply and the pollutants are not highly toxic, the water soon undergoes self-purification, and both the polluting organics and the alien populations disappear. It is generally believed that protozoa contribute materially to the bacterial decline as the water becomes purified, and many studies have recorded the rise in protozoan abundance in waters receiving sewage and organic remains. A variety of ciliates and flagellates predominate in this bacterial feeding, and genera like *Colpidium, Oikomonas,* and *Paramecium,* among others, are notably voracious in this activity, although rarely is a single species overwhelmingly dominant in recently polluted areas.

Ciliates and flagellates have likewise been assigned a major role in the elimination of bacteria participating in the destruction of sewage by the activated sludge process. In the initial phase of the activated sludge system the population densities of all organisms are low, but the bacteria promptly flourish on the array of organic nutrients in

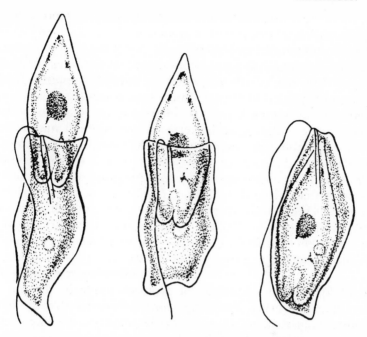

Figure 15.2 *Peranema* engulfing *Euglena* (Curtis, 1968).

the incoming sewage. As the numbers of bacteria increase, so do the flagellates ingesting them, but the latter are soon replaced in the succession of predators by bacteria-consuming ciliates. The ciliates feed principally on cells in the liquid phase of the sewage treatment unit and presumably reduce their abundance, contributing thereby to purification of the water (McKinney and Gram, 1956). Predatory protozoa, especially large ciliates and colorless flagellates, also occur in lakes, estuaries, the open sea, rivers, and streams containing sufficient bacterial or small protozoan prey (Fig. 15.2).

Soils likewise are well endowed with a multitude of microorganisms unquestionably active in consuming their small-celled neighbors. Bacteria-feeding protozoa are quite common, with densities reaching tens of thousands per gram in many soils. Flagellates and amebae are the more widely encountered predators, while comparatively few ciliates are observed. Myxobacterales like *Myxococcus*, *Chondrococcus*, and *Archangium* are frequently observed, but Myxomycetes are also widely distributed, though possibly they are not as numerous. *Dictyostelium* and related cellular slime molds are common, too. These predacious groups exhibit a distinct preference for the true bacteria, and they thus undoubtedly have an influence on terrestrial communities of bacteria. At one time myxobacteria and fungi like *Dictyostelium* were considered to be dung inhabitants, owing to their obvious development on animal droppings lying on the ground for a period of time, but it now appears instead that they are soil residents taking advantage, where possible, of the opportunity to exploit the millions of bacteria in the animal wastes (Singh, 1947a). The habitat *par excellence* for the nematode-trapping fungi is below ground because it is here that their animal prey abounds.

The dominant and possibly the sole predators of the rumen are protozoa. Ciliates like *Entodinium, Diplodinium,* and *Isotricha* are among the chief protozoa of the adult ruminant, although small flagellates are not rare. The major groups of ciliates are capable of, or are restricted to, feeding on bacteria, of which there never is a short supply, and it is considered axiomatic that the indigenous bacteria are perpetually under attack by the microscopic animals (Hungate, 1966). Immense numbers of protozoa reside in the intestinal tract of man and other vertebrates, too, and presumably the bulk of the nutritional demand of these predators is satisfied by the consumption of intestinal bacteria.

SELECTION OF PREY

Discrimination in the choice of prey is the rule among most predators, but the degree of selectivity varies enormously. For any given feeder, the food sources generally are of three kinds: (a) prey populations consumed rapidly and with little difficulty; (b) species that are consumed slowly or are rejected if a choice is possible; and (c) microorganisms totally immune from attack by that particular predator population. What is repugnant to one predator may be a delectable morsel for another, however, so that escape from the first marauder does not necessarily signify resistance against the second. Thus strains or species of bacteria inedible by one protozoan population are freely consumed by another. Nevertheless, selective feeding by the totality of predators is a usual occurrence, and as a result individuals of less readily eaten groups of microorganisms remain while susceptible categories either are reduced in density or fail to proliferate as rapidly as they would in predator-free habitats.

A wide range of microorganisms are devoured by occasional predators. In one study, for example, 83% of the bacterial species tested

Table 15.1

Food Choices of Predators with Broad and Narrow Prey Specificities

Predator	Prey	Reference
	Nonfastidious Predators	
Dictyostelium discoideum	*Aerobacter, Bacillus, Flavobacterium, Micrococcus, Pseudomonas*	Raper (1937)
Dimorpha	Flagellates, unicellular algae	Sandon (1932)
Mayorella bigemma	Ciliates, diatoms, flagellates, nematodes, rhizopods, rotifers	Sandon (1932)
Noctiluca	Diatoms, dinoflagellates, metazoa	Sandon (1932)
Oxyrrhis marina	Chrysophyta, Chlorophyta, Cryptophyta, Rhodophyta, Bacillariophyta	Droop (1966)
Uronychia transfuga	Algae, bacteria, ciliates	Webb (1956)
	Fastidious Ciliates	
Actinobolina radians	*Halteria*	Luck et al. (1931)
Didinium nasutum	*Paramecium*	Luck et al. (1931)
Woodruffia metabolica	*Paramecium*	Salt (1967)
Nassula citrea	*Oscillatoria* filaments	Webb (1956)

were ingested and supported growth of the flagellate *Oikomonas termo* (Hardin, 1944), and more than half of those surveyed were digested by *Dictyostelium* and by the myxobacterium, *Myxococcus fulvus* (Anscombe and Singh, 1948). The omnivorous predator may reach for its sustenance beyond a single broad taxonomic category and find all adjacent individuals nearly equally acceptable as prey. Table 15.1 shows that *Dictyostelium,* the dinoflagellates *Oxyrrhis* and *Noctiluca,* and the protozoa *Dimorpha, Mayorella,* and *Uronychia* are potentially capable of satisfying their nutrient demands at the expense of an assortment of organisms. Still the frequency with which a given predator ingests individuals of one group will be governed not only by their availability in a habitat but also by their relative abundance, the edible prey species that are numerous or are in the vicinity of the feeder making up the major part of the diet.

Dietary fastidiousness is rare, but it is known (Table 15.1). A select few ciliates and flagellates are thus reported to be extremely exacting and feed on a solitary genus. Such fastidious organisms are denoted *stenophagous,* in contrast with the *euryphagous* or polyphagous mode of nutrition of the omnivorous forms. The majority of predators probably fall between the extreme of stenophagy and the highly omnivorous pattern of nutrition. Furthermore, stenophagy commonly is not absolute among the fastidious heterotrophs, though one preference is paramount; that is, there is markedly preferential but not exclusive ingestion of members of one population. A clear distinction must also be made between the organism that is innately stenophagous—one restricted to a very limited range of prey under all circumstances—and the predator that has a potential for consuming a number of different prey types but is compelled to live on a single species because no other food source is immediately available. Exclusive dependence on a single species or genus denotes a tenuous existence, for the predator is thereby confined completely to those microenvironments inhabited by its solitary prey.

Most bacterial genera indigenous to water and soil are, so far as is known, susceptible to phagotrophic protozoa of one or another genus, although any single protozoan population is generally capable of feeding on only part of the bacterial community in its immediate environs. Available to individual populations of these minute animals are not only an impressive collection of gram-positive and gram-negative rods and cocci but, in addition, algae, yeasts, fungal spores, viruses, nematodes, and rotifers, as well as dissimilar protozoa. The growth rate of the protozoan, however, depends on the particular kind of individual consumed. Myxobacteria and myxomycete plasmodia may obtain their nutriment by attacking bacteria, filamentous fungi, and yeasts, while Acrasiales sustain themselves at the expense of a reasonable cross-section of gram-negative and gram-positive rods and cocci. Yet a few microorganisms or stages in their life cycles apparently are not prone to predation, and they do not succumb readily, if at all. For example, though the vegetative cells of *Bacillus* spp. are by and large edible to soil amebae, their spores are not. Some *Mycobacterium* strains are perfectly acceptable, but quite a number are rejected by terrestrial protozoa. The reproduction of soil amebae, moreover, is not supported by a high proportion of actinomycetes, filamentous fungi, and algae (Heal and Felton, 1965).

Dinoflagellates eat dissimilar dinoflagellates, and protozoa consume members of different protozoan genera, but cannibalism—the eating of individuals of the same species—sometimes occurs, too. Protozoa such as *Amoeba, Pleurotricha,* and *Stentor* are known to resort to this mode of nutrition on occasion. The factors responsible for or contributing to microbial cannibalism are largely unknown, but some instances are associated with a depletion of the food supply.

The availability of suitable prey regulates the distribution of most predators, and the density of such phagotrophs is directly correlated with the abundance of digestible cells. Some phagotrophs unquestionably are able to combine development on inanimate matter, either particulate or soluble, with the predatory habit, and they may take animate and inanimate food simultaneously or at different times. These opportunists prey on their coresidents of the ecosystem when conditions are favorable, but they do not rely on a phagotrophic life. Obligate predators, on the other hand, are restricted entirely to feeding on nearby microorganisms. Prey is apparently essential for the characteristic development of particular myxobacteria, and many terrestrial and aquatic protozoa fail to develop in the absence of digestible cells. Similarly, various pathogenic protozoa will not multiply in their natural environments unless edible bacteria are accessible, a fact attested to by the failure of *Entamoeba histolytica* to survive when introduced into germfree guinea pigs, by contrast with its behavior when inoculated into conventional guinea pigs (Phillips and Wolfe, 1959). In like fashion, though their spores may germinate and myxamebae may be formed, myxomycetes do not complete their life cycles unless cells of a food species are in the vicinity (Alexopoulos, 1963).

The nutritional requirements of a few phagotrophs have been established, and occasionally the needs are surprisingly simple. Far more often the list of required growth factors is quite long. Proliferation of these fastidious populations proceeds only when they are supplied with a mixture containing B vitamins, amino acids, simple organic acids, long-chain fatty acids, nucleotides, or sterols. Nevertheless, a vast number of protozoa, and undoubtedly representatives of additional categories of phagotrophs, will not replicate without prey, either viable or dead, and indeed most isolates investigated to date still cannot be grown in synthetic media or even in axenic culture. In time, many of these microorganisms probably will be propagated axenically, but they still must be considered ecologically obligate predators, either because those having a simple nutrition are poor competitors with heterotrophs or because those more numerous organisms needing complex mixtures of growth factors cannot have their wants satisfied except by consuming cells containing all the requisite compounds in a concentrated form.

Despite the ease and frequency of observing discrimination in feeding, little information has been gathered on the morphological or biochemical bases for prey preferences. Depending on the predator, feeding may entail searching out the prey (although often the contact is fortuitous), capturing the individual, ingestion, and digestion, and the choice and acceptance or rejection may be determined at any one of these stages. The failure of the prey to be engulfed and to support growth may be ascribed, therefore, to its avoidance of the predator, its resistance to ingestion, or its nutritional insufficiency. More specifically, cells too large to permit engulfment, a filamentous habit, peculiarities of shape that deter ingestion, surface structures or capsular materials that ward off the feeder, inhibitory exudates, toxic cell surface components, and innate resistance to digestive enzymes may account for a species eluding predation.

Size contributes appreciably to food selection. This is shown by the preferential consumption of small diatoms by the ciliate *Diophrys appendiculatus* and of small bacilli and yeasts by *Bodo*, and the lesser suitability of large cells (Sandon, 1932; Webb, 1956). Yet size alone cannot explain the discrimination of numerous predators, and indeed *Didinium* will ingest whole paramecia larger than itself, the didinia then appearing as thin protoplasmic films surrounding the victimized ciliate. Shape may condition the acceptability of a potential prey, and the long filaments of algae, fungi, and actinomycetes may present an insurmountable difficulty because of their strandlike morphology. A striking effect, apparently attributable to shape, is seen with *Entodinium vorax*, which can devour the smaller cells of the spiny protozoan *Entodinium caudatum* in the goat rumen, although it strongly prefers the spineless or the small-spined *E. caudatum* to cells with large spines (Fig. 15.3).

Chemical composition, rather than size or shape, may be responsible for prey selection. Constituents of the walls of bacterial endospores, which may be ingested but remain unharmed, and of certain fungal spores may contribute to their resistance to digestion. Extracellular gum, slime, or capsular polysaccharides produced by bacteria have been proposed as the reason for the resistance of strains of *Rhizobium*, *Azotobacter*, and *Aerobacter* to attack by protozoa and *Dictyostelium* myxamebae, the resistance presumably resulting from shielding of the edible bacteria by the coating around the cells or colonies. Some populations may not provide an adequate diet because they lack components essential for the fastidious predator, but neither qualitative nor quantitative analyses of microbial cells provide a sufficient basis for assessing nutrient sufficiencies or deficiencies of potential prey.

Mention has already been made of the fact that capsules or surface-

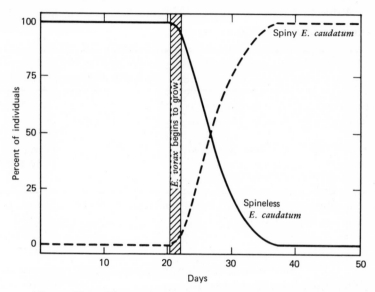

Figure 15.3 The preferential consumption of spineless *Entodinium caudatum* by *Entodinium vorax* in a mixture containing spined and nonspined prey (Strelkoff and Poljanskij, 1937).

located structures of particular bacteria make them resistant to phago-cytosis and contribute to their successful establishment within the animal body. Thus encapsulated strains of *Bacillus anthracis, Diplo-coccus pneumoniae, Pasteurella pestis, Staphylococcus,* and *Strepto-coccus* are phagocytized far more slowly than nonencapsulated strains. Whether protection by capsules and surface components is wide-spread in habitats other than that of the animal body and whether this shielding contributes materially to food discrimination by preda-tors await to be determined.

In notable instances, however, evidence exists that specific products do explain the inability of protozoa to consume and destroy bacteria. The products concerned are toxins, and the cell excreting an inhibi-tor or containing a harmful protoplasmic component is protected by virtue of the injurious metabolite. In this way an extracellular com-pound elaborated by *Pseudomonas fluorescens* (Knorr, 1960), an in-hibitor bound within the cells of *Mycobacterium* (Dudziak, 1962), and apparently the pigments of *Serratia marcescens* and *Chromobac-terium violaceum* (Groscop and Brent, 1964) retard or ward off proto-zoa attempting to make a meal of the bacteria. The formation of

Table 15.2

Predator Consumption Rates

Predator	Prey	No. of Prey Consumed	Reference
Amoeba proteus	Tetrahymena pyriformis	28–47/hr	Salt (1968)
Didinium nasutum	Paramecium aurelia	3/cell division[a]	Butzel and Bolten (1968)
Leucophrys patus	Glaucoma pyriformis	50/cell division	Brown (1940)
Paramecium caudatum	Bacillus subtilis	18,000/cell division	Ludwig (1928)

[a] The number of cells consumed in the time required for an individual predator to give rise to two daughters.

deleterious compounds or constituents does not explain the inedibility of countless bacteria which, though not consumed, are nontoxic. The elaboration of an antiprotozoan inhibitor does not ensure immunity from predation, moreover, inasmuch as bacteria harmful to one protozoan species are not only nontoxic but may well be excellent food sources for another.

PREDATOR-PREY RELATIONSHIP

The rate of prey consumption is governed by the predator in question, the identity of the food species, and environmental conditions. As illustrated by the data of Table 15.2, protozoa can have voracious appetites for bacteria in vitro and presumably in nature as well, and each protozoan may consume in excess of 10,000 cells before it divides. Because of the enormous quantity of bacteria that must be engulfed to make possible a single protozoan cell division, communities in ecosystems containing many bacteria-feeding protozoa—as in sewage, water, soil, rumen, or the intestine—probably are greatly affected by these predacious organisms. One rough calculation indicates, for instance, that the amebae in 1 square meter of soil consume 500 g of microorganisms a year (Heal, 1967). The rate of consumption of large-celled prey, as expected, is much less than for the bacteria, but nevertheless they are devoured at a pace sufficiently rapid to suggest that their enemies indeed have an appreciable, if not devastating, impact in vivo.

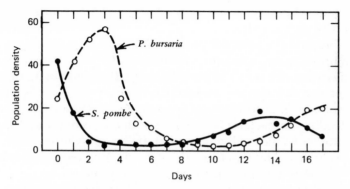

Figure 15.4 Fluctuations in population density of *Paramecium bursaria* feeding on *Schizosaccharomyces pombe* in vitro (Gause, 1934). By permission of The Williams & Wilkins Co., Baltimore, Md.

Since the abundance of ecologically obligate predators is regulated by the food they obtain in the form of prey cells, the more numerous are the individuals that constitute their nutrient supply, the more frequent generally are the feeders. However, the rate of development of the predator population depends on the kinds of organisms it consumes, some cells supporting luxuriant and others poor growth of the feeder. Furthermore, not only is the identity of the prey of concern but also its nutritional status, a fact well illustrated by observations that starved paramecia eaten by *Didinium nasutum* decrease the fission rates of the ciliate and cause it to lose its ability to make cysts, whereas well-nourished prey cells promote high rates of fission and cyst formation (Butzel and Bolten, 1968).

Environmental factors greatly altering the physiology of either of the two interacting populations will likely have an effect on the predation rate, in terms of the number of cells consumed per unit of time or per predator individual, or sometimes both. This is clearly exemplified by temperature changes, which often modify, for reasons as yet uncertain, the number of prey consumed by each predator. The tolerance range of the predator is at the same time tied intimately to the tolerance of the food species to extremes of temperature, osmotic pressure, inhibitors, pH, and sunlight.

It has been suggested that periodic fluctuations occasionally take place in the relative density of predator and prey. This is illustrated in Fig. 15.4. The abundance of predators rises when food is plentiful, but when they become sufficiently numerous the density of their prey falls, since replication of the prey cannot keep up with the pace of destruction. However, as the viable food supply becomes sparse, the probability of an individual predator encountering an edible cell diminishes, and the predator population therefore declines, too. This in turn provides an opportunity for the prey to proliferate once again. As a consequence a series of periodic oscillations involving the abundance of the two species is observed. Such cycles have been noted with protozoa in culture, with myxamebae of *Dictyostelium discoideum* feeding on bacteria in laboratory media (Bungay and Bungay, 1968), and with amebae living on bacteria in soil (Cutler and Crump, 1920). Whether fluctuations of this sort are frequent or common in vivo remains to be ascertained. Should there indeed be cycles of alternating dominance between prey and predator in nature, they would be more likely to involve an obligate predator restricted to a single kind of prey rather than an omnivorous organism, which, when one food population became scarce in the microenvironment, could merely alter its diet in favor of still prominent species. Nonetheless, sequential oscillations between consumer and consumed are not readily discernible or, indeed, may not exist in most habitats supporting predation. The relative frequency of the two species remains substantially the same, as has been found for *Streptococcus bovis*, a prey of rumen ciliates; the numbers of *S. bovis* are reasonably fixed, indicating that the predation rate and the bacterial growth rate are approximately the same (Jarvis, 1968).

Prey species are frequently not eliminated by their enemies, for, were they to be eradicated, so would be the obligate predators, particularly the stenophagous types. Predation does not usually go unchecked, and the predator-prey relationship does not necessarily lead to self-annihilation of the interactants. Possibly, it rarely does so. Extreme oscillations or examples of mutual extinction are readily demonstrable in laboratory media, but the environment here is characteristically homogeneous, refuges for the prey are absent, and the predator itself has no enemies threatening its continued existence, situations far different from those usually encountered in nature.

A balance between predator and prey, a near steady-state condition, is probably more often the rule. The basis for the steady state and the reason why predators do not generally overwhelm their prey are not firmly established, but unquestionably an overly aggressive consumer that abolishes the resource on which it relies wholly would disappear in the course of evolution. The prudent predator allows for the survival of enough of its viable food supply to maintain the nutrient reserve; natural selection probably would eliminate rapidly and effectively all voracious obligate predators that do away with their food store while simultaneously preserving their more prudent counterparts (Slobodkin, 1968).

Several tentative hypotheses have been advanced, in addition, to explain the inability of most predators to eradicate their prey species. (a) It is probable that a smaller proportion of individuals is captured at low than at high densities of the prey population, a density-dependent mortality that makes it ever more difficult for complete eradication to occur as the food source becomes scarcer and harder to find. (b) The ability of the consumer to find its food is markedly reduced by the availability of refuges for the prey and microenvironments inaccessible to large-celled predators. The minute pores in soil or water sediments and the floc in the activated sludge type of sewage treatment, though surrounded by protozoa and other microbial feeders, can protect a small-celled species from extinction. Particulate matter offers a hiding place and prevents total exploitation. The predator too large to penetrate continues to live off those cells that leave the refuge. (c) The predator itself is not devoid of enemies that, by governing its abundance, allow for survival of the quarry. Microorganisms may parasitize or prey upon the hunter or produce toxins affecting its well-being. Furthermore, the predators are eaten by metazoa inhabiting the soil, oceans, or fresh water. (d) Although the vegetative stage of a prey population is totally destroyed, a situation not unlikely in microenvironments, the species still may not be eliminated from the habitat, because the depopulated area receives new colonists from adjacent microhabitats or the prey has a resting stage resistant to digestion, as with bacterial endospores, protozoan cysts, and some fungal spores.

Nevertheless, occasional prey species are probably totally eliminated in selected ecosystems, a destruction that could lead to the subsequent demise of the consumer. Streams receiving organic materials or soils treated with carbonaceous remains are colonized by an enormous collection of bacteria, and the decline and eradication of one or more populations in these short-lived communities, which are quite common in nature, might well result from predation. However, there is no evidence in support of such an activity.

ECOLOGICAL SIGNIFICANCE OF PREDATION

Predation represents a unique way of life in which one individual uses another to concentrate the nutrients present in a dilute solution, to synthesize essential metabolites, and to package these various substances in a convenient, edible form. As expected, this kind of interaction has a marked impact on the perennial struggle for existence. Natural selection operates among predators inhabiting the same microcosm, and their nutritional peculiarities, ability to find food, and growth rates contribute to a successful outcome in the interspecific selection. When several predator species make simultaneous use of a common prey population present in numbers too small to satisfy all contenders, it is probable that competition will ensue and that the species multiplying more rapidly under the prevailing conditions will assume dominance (Evans, 1958). Conversely the identity of the prey will have a marked influence on the final outcome of the selection.

The various prey organisms, too, are clearly subject to natural selection by the very process of differential feeding, the agent of selection being the predator. Biochemical or morphological traits responsible for resistance to attack, or its avoidance, underlie the success of an organism not only relative to those attempting to capture and ingest it but also to those related microorganisms with which it must compete.

Predation may affect a community in three nonmutually exclusive ways: quantitatively, by affecting the numbers of organisms; qualitatively, by altering the biological composition of the ecosystem; and by changing the metabolic activity of the community. The importance of predation may be appreciable, or it may be of little consequence, depending on the vulnerability of the prey and on the relative abundance of both prey and predators. Knowledge of the impact of one group on the second is based almost entirely on laboratory investigations, and even these have been largely conducted with one category of predators, the protozoa, so that comments on ecological significance must be taken solely as first approximations.

It is generally assumed that predators limit the size of populations they consume. Such a regulatory effect on bacteria has been attributed to the protozoan inhabitants of soil, sewage, and polluted bodies of water. These microscopic animals do indeed have voracious appetites, they must digest enormous numbers of cells to maintain themselves, and their abundance in these environments is great enough to influence the bacterial density appreciably. Model experiments involving the addition to sterile soil of strains of *Dimastigamoeba* and *Cercomonas* have lent credence to the contention that protozoa in fact keep in check the indigenous bacteria (Fig. 15.5). Similar studies in which *Dictyostelium* was added to sterile soil together with an edible bacterium showed that this fungus also actively consumes bac-

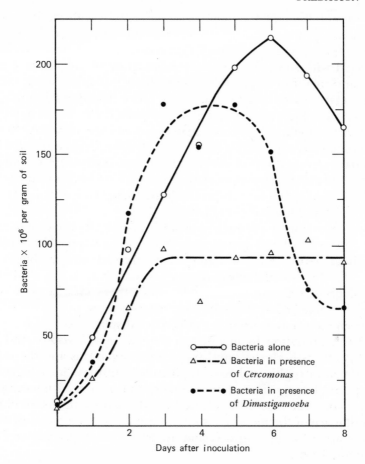

Figure 15.5 Suppression of bacteria introduced into sterile soil together with *Dimastigamoeba gruberi* or *Cercomonas crassicauda* (Cutler, 1923).

teria (Singh, 1947*b*). Ciliates in the rumen also have been reported to decrease the bacterial concentration, and the introduction of ciliates into a protozoan-free rumen leads to a diminution in bacterial density (Kurihara et al., 1968). Additional kinds of prey and organisms resident in different environments may be expected to react in an analogous fashion.

Qualitative modifications induced by predation may be of great consequence. A prey that is inedible or escapes being eaten is favored, while those edible forms unable to find a refuge suffer in the selection process. The uneaten or slowly devoured strains presumably become more numerous, and the population of edible cells declines to a minor place in the community. Views of this sort are attractive, but data to support them are sparse. To be sure, experiments have been conducted with sterile soil into which mixtures of amebae and both edible and inedible bacteria were introduced; as anticipated, the inedible organisms dominated their less lucky neighbors as the amebae sated their hunger, but extrapolation from such results to natural conditions is foolhardy. More meaningful are the observation that prey of *Polyplastron multivesiculatum* decline in frequency when this ciliate assumes dominance in the rumen (Eadie, 1967) and the finding that the composition of the bacterial community in the rumen is markedly different in protozoa-free sheep and in animals supporting ciliates (Kurihara et al., 1968).

Monopolization of key environmental resources by a few organisms may be minimized and species diversity of the ecosystem maximized through predation. Species under attack probably are unable to exploit a locality totally when they are themselves being exploited, and organisms that are not effective competitors may be preserved, if they are not prone to predation, whereas they otherwise would have been eliminated through competitive displacement.

Selective predation by metazoa in the sea, inland waters, and soil probably affects both the numbers of microorganisms and the makeup of the autotrophic or heterotrophic communities. Some algae are quite prone to grazing aquatic animals, and terrestrial microorganisms are constantly being swallowed by all manner of invertebrates. The influence of the metazoa, though rarely placed on a quantitative or qualitative footing, on occasion appears to be especially striking (Raymont, 1966).

Evidence also exists that predatory protozoa accelerate biochemical processes, including the decomposition of organic matter, catalyzed by residents of polluted waters or marine habitats (Javornicky and Prokesova, 1963; Johannes, 1965). This presumably results, in some instances at least, from the protozoa keeping the bacteria from reaching self-limiting numbers and maintaining them in a prolonged state of high metabolic activity.

References

REVIEWS

Duddington, C. L. 1957. The predacious fungi and their place in microbial ecology. *In* R. E. O. Williams and C. C. Spicer, Eds., *Microbial Ecology.* Cambridge University Press, London. pp. 218–237.

Hirsch, J. G. 1965. Phagocytosis. *Ann. Rev. Microbiol.,* **19,** 339–350.

Pringsheim, E. G. 1959. Phagotrophie. *In* W. Ruhland, Ed., *Encyclopedia of Plant Physiology,* Vol. 11. Springer-Verlag, Berlin. pp. 179–197.

Slobodkin, L. B. 1968. How to be a predator. *Am. Zoologist,* **8,** 43–51.

OTHER LITERATURE CITED

Alexopoulos, C. J. 1963. *Bot. Rev.,* **29,** 1–78.

Anscombe, F. J., and Singh, B. N. 1948. *Nature,* **161,** 140–141.

Brown, M. G. 1940. *Physiol. Zool.,* **13,** 277–282.

Bungay, H. R., and Bungay, M. L. 1968. *Advan. Appl. Microbiol.,* **10,** 269–290.

Butterfield, C. T., and Purdy, W. C. 1931. *Ind. Eng. Chem.,* **23,** 213–218.

Butzel, H. M., and Bolten, A. B. 1968. *J. Protozool.,* **15,** 256–258.

Chen, Y. T. 1950. *Quart. J. Microscop. Sci.,* **91,** 279–308.

Curtis, H. 1968. *The Marvelous Animals.* Natural History Press, Garden City, N.Y.

Cutler, D. W. 1923. *Ann. Appl. Biol.,* **10,** 137–141.

Cutler, D. W., and Crump, L. M. 1920. *Ann. Appl. Biol.,* **7,** 11–24.

Droop, M. R. 1966. *In* H. Barnes, Ed., *Some Contemporary Studies in Marine Science.* Allen and Unwin, London. pp. 269–282.

Dudziak, B. 1962. *Acta Microbiol. Polon.,* **11,** 223–244.

Eadie, J. M. 1967. *J. Gen. Microbiol.,* **49,** 175–194.

Evans, F. R. 1958. *Trans. Am. Microscop. Soc.,* **77,** 390–395.

Gause, G. F. 1934. *The Struggle for Existence*. Williams & Wilkins, Baltimore.

Groscop, J. A., and Brent, M. M. 1964. *Can. J. Microbiol.*, **10**, 579–584.

Hardin, G. 1944. *Ecology*, **25**, 192–201.

Heal, O. W. 1967. *In* O. Graff and J. E. Satchell, Eds., *Progress in Soil Biology*. North Holland Publishing Co., Amsterdam. pp. 120–125.

Heal, O. W., and Felton, M. J. 1965. *In Progress in Protozoology*. Excerpta Medica, Amsterdam. p. 121.

Hungate, R. E. 1966. *The Rumen and Its Microbes*. Academic Press, New York.

Jarvis, B. D. W. 1968. *Appl. Microbiol.*, **16**, 714–723.

Javornicky, P., and Prokesova, V. 1963. *Intern. Rev. Ges. Hydrobiol. Hydrogr.*, **48**, 335–350.

Johannes, R. 1965. *Limnol. Oceanog.*, **10**, 434–442.

Knorr, M. 1960. *Schweiz. Z. Hydrol.*, **22**, 493–502.

Kurihara, Y., Eadie, J. M., Hobson, P. N., and Mann, S. O. 1968. *J. Gen. Microbiol.*, **51**, 267–288.

Luck, J. M., Sheets, G., and Thomas, J. O. 1931. *Quart. Rev. Biol.*, **6**, 46–58.

Ludwig, W. 1928. *Arch. Protistenk.*, **62**, 12–40.

McKinney, R. E., and Gram, A. 1956. *Sewage Ind. Wastes*, **28**, 1219–1231.

Phillips, B. P., and Wolfe, P. A. 1959. *Ann. N.Y. Acad. Sci.*, **78**, 308–314.

Raper, K. B. 1937. *J. Agr. Res.*, **55**, 289–316.

Raymont, J. E. G. 1966. *Advan. Ecol. Res.*, **3**, 117–205.

Salt, G. W. 1967. *Ecol. Monogr.*, **37**, 113–144.

Salt, G. W. 1968. *J. Protozool.*, **15**, 275–280.

Sandon, H. 1932. *The Food of Protozoa*. Egyptian University Faculty of Science, Cairo. Publication No. 1.

Singh, B. N. 1947a. *J. Gen. Microbiol.*, **1**, 11–21.

Singh, B. N. 1947b. *J. Gen. Microbiol.*, **1**, 361–367.

Strelkoff, A., and Poljanskij, J. 1937. *Zool. Zh.*, **16**, 77–87.

Webb, M. G. 1956. *J. Animal Ecol.*, **25**, 148–175.

Part 3

EFFECT OF MICROORGANISMS ON THEIR SURROUNDINGS

16

Microorganisms and Biogeochemistry: I

Microorganisms are deemed to be of importance either because they directly affect man's well-being or because they are fascinating objects for biological inquiry. Commonly overlooked is the vast impact these microscopic forms have on the chemistry of the earth and on geochemical transformations that are, indirectly to be sure, critical or absolutely essential for the continued existence of higher animals and plants. A variety of geochemical processes are modified and, occasionally, entirely governed by the biochemical processes catalyzed by living organisms, and these activities and their influence on the environment constitute the subject matter of *biogeochemistry*. The transformations of major concern in biogeochemistry occur in soil, aquatic habitats, or both. Terrestrial regions where microorganisms play a crucial role include forests, cropped land, grasslands, and even the deserts and tundra, while the aquatic areas include both the water phase and the bottom sediments of fresh water and the sea.

Microbiological reactions of geochemical prominence take place within the *biosphere,* that portion of the outer surface of the earth that supports life. The biosphere is taken to extend upward into the lower part of the atmosphere and downward to depths of some 2000 meters below the soil surface and to about 11,000 meters in the seas. Though only a small percentage of the mass of the biosphere is composed of living microorganisms, they are still indispensable agents of geochemical change. Each cell and every species, in one way or another, modify the chemistry of their own small segments of the biosphere, but generally it is only when community metabolism, the sum total of all the processes conducted by the indigenous populations, is considered that a true picture of the magnitude of the geochemical alteration emerges. Occasionally, however, a particular genus or a collection of physiologically related genera stands out as the chief transforming agent because of intrinsic metabolic capacities that it, but not adjacent populations in the community, possesses.

Biogeochemical cycles are concerned with the circulation of an element from the inorganic or inanimate organic form into protoplasmic combinations and then back into the abiotic portion of the habitat, and numerous microbial conversions are segments of these cycles. With specific elements, two or more oxidation states participate in the cyclic sequences, and in certain instances the substance undergoing a biologically induced change may not be essential for or even penetrate the cell affecting the alteration. Cyclic processes, as a rule, bring about no long-term upsets in the chemistry of the earth's surface, but noncyclic transformations do indeed cause lasting or essentially permanent modifications in the distribution of an element or in the compounds wherein the element is found.

Elements participating in biogeochemical cycles exist in inorganic and organic pools, the sizes of which range from vast to minute. The compounds in the pool can be readily accessible and completely available, but pools containing components that are poorly available to viable organisms are not rare, and the rate at which an element in a poorly available pool is utilized or transformed is invariably slow. Nevertheless some elements, though readily available, are never cycled rapidly, regardless of the size of the populations capable of effecting the conversion. The pool size of many elements in natural ecosystems is relatively constant during the course of time, not because compounds containing the element are not utilized microbiologically but rather because steady-state conditions exist, the depletion being just balanced by the influx. For example, the organic carbon content of virgin land is reasonably constant because the efflux of carbon as CO_2 is just balanced by the return to the soil of organic carbon in the form of plant litter and dead root tissue. Similarly, a steady-state condition characterizes the nitrogen, phosphorus, and silicon concentration of the oceans, the loss of these elements by deposition in marine sediments, or by other means, being compensated for by the influx from terrestrial sources, rainfall, or algal N_2 fixation.

BIOGEOCHEMICAL PROCESSES

Microbial activities of geochemical significance may be subdivided into several broad categories. These groupings are not mutually exclusive, a single kind of conversion sometimes being included in more than one category.

(a) *Mineralization,* the conversion of an organic form of an element to the inorganic state. Mineralization is essentially a means whereby microorganisms decrease the biochemical complexity of the ecosystem. Inasmuch as some of the elements thus acted on are essential for life and are assimilated by numerous species only when in the inorganic state, mineralization is sometimes designated *nutrient regeneration,* a term introduced in Chapter 5.

(b) *Immobilization,* the conversion of an inorganic nutrient element to an organic complex as a result of the assimilation of the element by microbial cells and its incorporation into protoplasmic combination. The population gains a substance necessary for its multiplication, but at the same time it increases the biochemical complexity of the ecosystem. Mineralization and immobilization are, in a sense, opposing processes, and the two together form a biogeochemical cycle. Every element that is incorporated into microbial cells may be said to be immobilized.

(c) *Oxidation.* A few microbiologically induced oxidations are linked with the cell's energy metabolism, the reaction sequence providing the energy needed for growth; for example, the oxidation of organic compounds by heterotrophs and of inorganic nitrogen, sulfur, iron, and hydrogen by chemoautotrophic bacteria. Other oxidations are not directly linked with the organism's energy metabolism, and the reactions provide little or none of the energy necessary for reproduction.

(d) *Reduction.* Reductions, too, may be directly linked with the energy metabolism of the cell, the microorganism using the substance as the terminal electron acceptor for growth, and the acceptor in turn being reduced. Reductions not directly coupled with the individual's energy metabolism commonly are brought about by modifications in the environment attributable to microbial proliferation, as by the consumption of O_2, the fall in E_h associated with the accumulation of reduced products, or the formation of acids.

(e) *Volatilization* or *fixation,* transformations changing the quantity of an element in the ecosystem. Fixation is a term employed to denote the conversion of a gaseous form of an element to a nongaseous compound. Mineralization may result in a net loss of carbon as CO_2 or CH_4 is evolved, and the nitrogen supply is invariably depleted as denitrifiers reduce nitrate to N_2 or N_2O. H_2S and NH_3 are lost in special circumstances, depleting the reserve of sulfur and nitrogen, but conversely the volatilization of H_2 and O_2 does not materially alter the quantity of hydrogen and oxygen in the environment. Carbon is fixed and returned to the ecosystem by algal, and usually to a much lesser extent by bacterial, photosynthesis, and nitrogen accretion may be pronounced where particular blue-green algae, bacteria, or symbiotic associations flourish.

(f) Formation of geological deposits. Diverse geological formations arise, to a greater or lesser extent, from microbial oxidations or reductions, abiotic reactions involving microbial products, or the accumulation of cellular remains. The genesis of sulfur and sulfide deposits, petroleum, and coal all appear to be in one way or another dependent on microorganisms.

(g) Production of organic chelating or complexing agents that solubilize relatively insoluble inorganic substances or maintain various compounds in soluble form.

(h) Serving as foci for adsorption of inorganic substances. Accumulations on microbial surfaces can be quite pronounced in the case of ferric and manganic oxides or hydroxides. Not yet resolved is whether the accumulation is passive, resulting merely from the presence of a suitable surface, or whether physiological processes are implicated.

(i) *Isotope fractionation,* the selective and more rapid use of one in a mixture of naturally occurring isotopes. Because of the fractionation, the microbial product containing the element in question has a different isotope ratio from that of the initial substrate.

Though most of the processes cited above are enzymatically catalyzed, some clearly are not, involving instead a nonbiological reaction with an enzymatically synthesized product as one of the reactants. The genesis of organic acids by heterotrophs, of nitric or sulfuric acid by chemoautotrophs, or of carbonic acid from the metabolically released CO_2 may bring about dramatic changes: the solubilization of phosphorus present in $Ca_3(PO_4)_2$, the conversion of insoluble ferric oxide and manganese dioxide to the soluble ferrous and manganous state, and the weathering of rocks, to mention a few. Conversely, a rise in the pH upon the release of ammonia from amino acids tends to favor spontaneous precipitation of iron and manganese. Solubilization of poorly soluble substances may be promoted indirectly in additional ways, as by the metabolic removal of O_2 or the excretion of products with chelating properties, whereas precipitation of soluble ions results from the liberation of O_2 in photosynthesis, the making of sulfide from sulfate by *Desulfovibrio* and biochemically similar anaerobes, or the heterotrophic evolution of CO_2 in solutions containing cations that yield relatively insoluble carbonates.

In Table 16.1 is presented a partial list of elements contained in

Table 16.1

Elements in Compounds Metabolized by Microorganisms

Transformation	Element
Mineralization	C, N, P, S, K, Si, Fe, etc.
Immobilization	C, N, P, S, K, Si, Fe, etc.
Oxidation	C, N[a], P, S[a], H[a], Fe[a], Mn, As, Se
Reduction	C, N, P, S, Fe, Mn, Cl
Solubilization	P, S, Fe, K, Ca, Si, Mn, Mg, Al, Cu, Zn, Co
Precipitation	S, Fe, Ca, Mg, Mn
Fixation	C, N, H, O
Volatilization	C, N, S, H, O, Se

[a] Oxidation performed by both chemoautotrophs and heterotrophs.

compounds that are transformed microbiologically. The various reactions tabulated have been demonstrated in samples from natural ecosystems, or their existence is inferred on the basis of sound evidence. In addition, studies with axenic or enrichment cultures or enzyme preparations derived from microbial cells have revealed that ions or compounds containing arsenic, bismuth, cadmium, cobalt, copper, molybdenum, nickel, silver, tellurium, uranium, vanadium, and zinc can be precipitated, reduced, or oxidized (Silverman and Ehrlich, 1964), but the ecological or geochemical significance, if any, of these reactions remains to be ascertained. Some elements, arsenic and mercury being notable instances, are converted microbiologically to methyl derivatives, and such methylation reactions are of particular concern because of the products' hazard to man.

In mineralization, carbonaceous molecules are degraded, and the elements they contain are released in an inorganic state. In this way CO_2, ammonium, nitrate, orthophosphate, sulfate, and soluble silicates are liberated from plant, animal, or microbial remains under aerobic conditions. Anaerobically, nitrate and often sulfate are not formed, but appreciable quantities of CH_4, H_2, ammonium, and sulfide as well as CO_2 and orthophosphate may appear. So far as is known, every element incorporated enzymatically into cellular complexes is subject to mineralization. In the upper zone of the soil and in the topmost layers of aquatic habitats where plant roots and algae develop, nutrients are withdrawn and enter into protoplasmic combination, and for these elements to return to forms assimilable by terrestrial plants and algae, the green organisms as well as the animals that feed upon them must be decomposed. Microorganisms also tie up an array of nutrients, and the elements thereby removed from ready circulation must be replenished. Thus the substrates for mineralization include plant remains, the cells of algae and heterotrophic microorganisms, dead animals, and soluble compounds excreted or solid wastes defecated by higher animals. Industry and cities likewise generate a variety of natural and synthetic organic compounds, and these, whether processed through sewage treatment operations or entering bodies of water as pollutants, are acted upon microbiologically. The inorganic ions ultimately released are then again available for uptake by photosynthetic organisms in lakes, the oceans, or on land.

Mineralization may take place essentially where the organic complexes are first made accessible to the microflora, as in soil receiving dead tissues or excretion products of plants and animals. Much of the nutrient regeneration in oceans and lakes also occurs in the surface regions inhabited by the phytoplankton and animals grazing on it. Alternatively, mineralization may be effected at some distance from the place where the organic matter was formed or released; this is well illustrated in marine ecosystems, wherein a significant proportion of the nutrient regeneration takes place in deep water layers, near the bottom, or in the sediments. The locus of mineralization is below the layers where the organic matter was synthesized by algae or modified by animal grazing, because of the sinking of dead cellular material and the vertical migration of living organisms. The downward movement of organic materials depletes the ocean surface of algal nutrients and, upon decay, enriches the underlying, heterotrophic regions. Vertical mixing of deep, rich waters is necessary for the nutrient elements mineralized by the heterotrophs, largely bacteria, residing near or at the bottom to become once again available to sustain algal life in the upper regions, where light penetration is adequate for photoautotrophy.

Nutrient regeneration in situ is frequently essential for continued rapid photosynthesis in soil and water. Though soils often contain a plentiful reserve of nitrogen, phosphorus, and sulfur, most or a considerable percentage of the three elements is usually in organic combination and hence not assimilable to a significant extent by angiosperms and gymnosperms, and therefore the rate at which they are mineralized, especially the elements whose available forms are present in limited amounts, may regulate plant development. Not uncommonly it is nitrogen mineralization that is the growth-limiting process in soil. Similarly, when algal proliferation proceeds apace in lakes or oceans, little inorganic nitrogen or phosphate exists in the surface water, and vertical transport from nutrient-rich deep layers may be so slow that regeneration processes in the upper stratum of the water become of paramount importance. Phosphorus mineralization is often, but not always, the rate-limiting microbial conversion in these circumstances.

Considerable attention has been given to the participants in mineralization. Usually the transformations entail a sequence of steps rather than a single discrete process, and the responsible organisms are categorized on the basis of the particular reactions they carry out. Nevertheless, a great assortment of microorganisms is involved in each of the separate stages, and these are ubiquitous and generally numerous. Bacteria of many genera seem to be dominant in water, but they are major participants in mineralization in nonaquatic environments, too. Fungi and actinomycetes are quite active in terrestrial ecosystems, and protozoa may contribute to mineralization, especially by preying upon bacteria and thereby promoting further turnover of the nutrient elements. Terrestrial and marine metazoa must not be overlooked, moreover, since the food they digest is eventually converted to inorganic or simpler organic compounds.

Immobilization is a term used to describe part of what happens in the ecosystem as microorganisms assimilate their needed inorganic nutrients. The creation of new cells or filaments requires the uptake not just of carbon but of nitrogen, phosphorus, potassium, sulfur, and additional elements, usually as some appropriate inorganic anion or cation. As long as growth proceeds, whether heterotrophic or autotrophic, immobilization occurs. This assimilation brings about the biosynthesis of the complex from the simple, essentially the reverse of mineralization, and simultaneously modifies the environment by concentrating within the intracellular milieu elements that may have been in dilute solution in the nonliving surroundings, the concentration factor ranging from several to more than 1000-fold. Whether or not the immobilization is of consequence to the chemistry of the environment depends on the demand for the element and the amount available. The supply of many nutrients, though small in actual values, far exceeds the need, so that immobilization has no effect on nearby organisms; conversely, the reserve of certain essential elements is at times deficient. For photosynthetic populations, chiefly algae in water and rooted plants on land, the elements that generally stand out are phosphorus and nitrogen; which of the two reaches short supply first, or whether it is a third element, depends on local

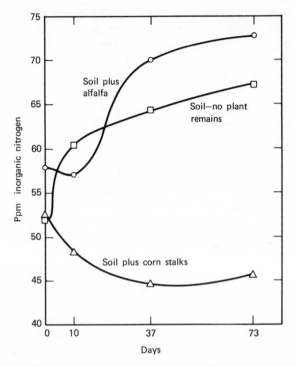

Figure 16.1 The influence of plant remains on the content in soil of nitrogen directly available to plants (Bartholomew, 1965).

circumstances. In soils receiving ample carbonaceous materials, the explosive development of heterotrophs causes a marked immobilization of inorganic nutrients and thereby induces a microflora-plant competition, with the frequent appearance of deficiency symptoms in the plants. Undoubtedly a similar intermicrobial competition is prominent in waters receiving organic substances as heterotrophs assimilate and immobilize nitrogen, phosphorus, and other nutrient elements.

Whenever there is mineralization, immobilization proceeds simultaneously, for the proliferation associated with the decomposition entails the microbial uptake of essential elements. The former process results in the generation of available nutrients; the latter, in a diminution of the reserve (Fig. 16.1). The two sequences must be considered together, therefore, particularly since the net accumulation or disappearance of each nutrient reflects a balance between these opposing transformations. Where there is mineralization of tissue remains or organic pollutants, the net accumulation of the inorganic ion—be it nitrogen, phosphorus, or another constituent element—is governed by the concentration of the element in the heterogeneous substrate in excess of the microbial demand; only the amount that is not needed will be liberated, essentially as a metabolic waste. Simple compounds or complex natural materials are commonly rich in carbon but contain insufficient levels of one or more essential elements for the mineralizing heterotrophs; in these circumstances there is no increase in the inorganic form of that element in the habitat, but rather a net immobilization as the active populations turn to the surroundings to make up for the inadequate concentration in the substance they are degrading.

The extent of assimilation of nutrient elements from any environment is governed by the elementary composition of the protoplasmic material of the indigenous community, that is, of the biomass, and the ratio of the elements disappearing because of microbial immobilization generally mirrors the ratio of the same elements in the biomass. Thus, from the amount of carbon assimilated by heterotrophs and the C:N, C:P, or C:S ratio of their cells, the magnitude of nitrogen, phosphorus, and sulfur immobilization can be estimated. Similarly, for algae, the ratio of the amounts of each element removed from water should reflect the cellular ratios; that this is indeed so is evident from data showing that in sea water microbiologically effected changes in carbon, nitrogen, and phosphorus occur in an atomic ratio of 105:15:1, figures essentially identical with the atomic ratio of 106:16:1 for the plankton protoplasmic components (Redfield, 1958). Inasmuch as 2 oxygen atoms are evolved, as O_2, in the photosynthetic assimilation of each carbon atom, an atomic ratio of 212 O:106 C:16 N:1 P would be expected during algal photosynthesis. The O_2 would be volatilized, whereas the other elements would be immobilized. A similar ratio should be observed in mineralization of the phytoplankton, but in this process the carbon, nitrogen, and phosphorus would be regenerated and the O_2 consumed; in aerobic mineralization, however, 4 additional oxygen atoms are consumed for every nitrogen because the ammonium nitrogen released in decomposition would be oxidized to nitrate,

$$NH_4^+ + 2\,O_2 \rightarrow NO_3^- + 2\,H^+ + H_2O,$$

so that the ratio should be 276 O:106 C:16 N:1 P. In this regard diatoms assume great interest, as they contain considerable silicon. The Si:P atomic ratio varies from 16:1 to 50:1, according to the diatom species. In view of these figures, it is noteworthy that the microbiologically induced changes in certain oceanic waters inhabited by sizable diatom communities have been reported, on the basis of atomic ratios, as 272 O:15 Si: 16 N:1 P (Redfield et al., 1963).

Analogous but probably not identical atomic ratios unquestionably characterize mineralization and immobilization in nonaquatic ecosystems, the ratios being determined by the chemical composition of the respective communities. Thus the biomass of a community dominated by fungi will have a different C:N ratio from that where bacteria are prominent, because fungal hyphae are commonly poorer in nitrogen. Furthermore, the composition of autotrophic and heterotrophic cells and filaments is markedly affected by the nutrient status of the habitat; at least, such are the indications from investigations in vitro. Mineralization in soil also typically shows fixed proportions among the various elements released in inorganic form, and the ratios, so far as they are known, are reminiscent of those in the sea (Thompson et al., 1954).

A surprisingly large number of elements existing in more than a single oxidation state are subject to microbial oxidation or reduction, either direct or indirect. The oxidation or reduction may be an essential feature of the metabolism of the autotroph or heterotroph responsible for the transformation, or it may be entirely incidental. Many of the changes, moreover, are of profound ecological, geochemical, agricultural, or economic significance and hence have deservedly received considerable study; others may be important in nature, but so little information is available that meaningful generalizations or conclusions are impossible.

Microorganisms oxidize inorganic ions or compounds by one of three general mechanisms. First, chemoautotrophic bacteria may catalyze the reaction, using the energy released in the oxidation in order to permit their own multiplication. In this way *Nitrosomonas, Nitrobacter, Thiobacillus, Ferrobacillus, Hydrogenomonas,* and metabolically related bacteria grow at the expense of ammonium, nitrite, sulfur, H_2S, metallic sulfides, ferrous iron, or H_2. Because these transformations are the sole means employed by the organisms for obtaining energy when growing autotrophically, relatively large amounts of product are formed per cell, and when the energy yield in the oxidation is notably small, as in ferrous or nitrite oxidation, the amount of product made by each cell is truly immense. Second, selected fungi and bacteria can bring about a heterotrophic but nonetheless enzymatically catalyzed oxidation of certain elements; for example, the formation of sulfate in soil from organic sulfur often appears to be a heterotrophic process, and nitrate and sulfate synthesis is readily demonstrable in axenic cultures of suitable fungi.

The responsible populations gain little if any energy from the oxidations, and they must have an organic compound to sustain their development. Third, some oxidations are effected indirectly, the process involving a nonenzymatic reaction between a microbiological product and an inorganic ion. The O_2 evolved in the light by algae, for instance, causes an oxidation of soluble ferrous and manganous ions to the poorly soluble ferric and manganic state, or the alkali produced by heterotrophs shifts the ferrous-ferric and manganous-manganic equilibria to favor the oxidized forms of the elements. Organic metabolites excreted by heterotrophs also may be responsible for such conversions, as with the hydroxy acids that effect the oxidation of manganous ions. Several oxidative transformations are quite evident and important in nature: the autotrophic release of copious quantities of sulfuric acid when sulfide ores are exposed to the air, the formation of the nitrate that is avidly assimilated by terrestrial plants, or the precipitation of ferric and manganic salts that occasionally impede water flow through pipes. Some elements are potentially oxidizable in nature, as iodide is converted to I_2 in samples of sea water or arsenite to arsenate in soil, and some, like molybdenum and selenium, can be oxidized in culture, but it is not clear whether these are merely intriguing curiosities or whether the conversions are ecologically consequential.

Populations capable of reducing inorganic or organic substances are everywhere. Three mechanisms apparently account for the reductions they bring about. (a) The reduction may be directly linked with the cell's energy metabolism, the oxidant serving as the terminal electron acceptor. Aerobes thus reduce O_2 to water, species of *Clostridium* and yeasts convert organic acids and aldehydes to corresponding reduced compounds, *Desulfovibrio* makes sulfide from sulfate, denitrifying bacteria form N_2 and N_2O from nitrate and nitrite, and unique anaerobes convert CO_2 to methane. (b) The reduction may result from a fall in E_h as O_2 is consumed, as reducing compounds are excreted, or both. Such nonenzymatic reductions take place in lake and ocean bottoms, waterlogged soil, deoxygenated streams, and heaps of decomposing organic materials that become anaerobic because of microbial activity. In this manner trivalent iron and tetravalent manganese are converted to the divalent cations. These reductions characteristically require anaerobiosis and an energy source, usually organic but sometimes H_2. (c) The reduction may be attributable to acid formation. Ferrous or manganous ions are produced spontaneously from the ferric and manganic state with falling pH.

Elements like phosphorus and silicon are stable in their oxidized state in anaerobic environments, that is, as phosphate and silicate, while nitrogen and iron, to mention two, are readily reduced where-ever there is appreciable microbial activity and O_2 is used up or never enters. Conversely, some elements may exist in anaerobic cir-cumstances in both an oxidized and a reduced condition—CO_2 and CH_4, sulfate and sulfide, and organic acids and the corresponding alcohols; whether or not the reduced compound is generated in these situations, and to what extent, depends on the presence of the re-sponsible organisms, the availability of suitable energy sources, and the existence of proper environmental conditions. The enormous number of reductions that microorganisms may be able to catalyze is suggested by the finding that an enzyme preparation derived from *Micrococcus lactilyticus* catalyzes the reduction of arsenate, bismuth-ate, vanadate, selenite, tellurite, tellurate, molybdate, and cupric and auric ions, as well as elemental selenium, lead dioxide, osmium dioxide, and hexavalent uranium (Table 16.2). However, informa-tion showing that such alterations occur in nature is largely nonex-istent.

Modifications in pH arising from autotrophic or heterotrophic growth can be effective agents of geochemical change. In water, soil, or ore deposits, chemoautotrophic bacteria of the *Thiobacillus-Ferrobacillus* group oxidize iron and other sulfides, and the nitrifying genera oxidize ammonium to yield considerable quantities of sulfuric and nitric acids. Scores of heterotrophic and photosynthetic organisms produce organic acids or, from metabolically derived CO_2, carbonic acid. In addition, the consumption of CO_2 during photosynthesis and its release in the respiration of large masses of aquatic algae appreci-ably shift the equilibrium of the CO_2-bicarbonate-carbonate buffer system of water, so that the pH may rise dramatically on a sunny day and then fall abruptly at night when photosynthesis ceases. The acids generated by these various means have been noted to solubilize phosphate, silicate, potassium, aluminum, mag-nesium, calcium, iron, and manganese originally in insoluble mate-rials, to decompose clay minerals, to weather rocks, and to be major factors in the leaching through surface soil and the subsequent deposition in subsoil of the calcium or magnesium solubilized bio-logically. In soil, the acid-induced solubilization of inorganic nu-trients increases their availability to plants; on the other hand, excessive iron and aluminum may be brought concurrently into solution, with phytotoxicity the outcome. Acid mine drainage bear-ing the sulfuric acid from sulfide ores attacked by chemoautotrophs often has a disastrous impact on plant, animal, and microbial life

Table 16.2

Reactions Catalyzed by an Enzyme Preparation Made from
Cells of *Micrococcus lactilyticus* [a]

Compound or Ion	Reaction
Selenium	$Se + H_2 \longrightarrow HSe^- + H^+$
Molybdate	$Mo_6O_{21}{}^{6-} + 3\,H_2 \longrightarrow 6\,MoO_{2.5} + 6\,OH^-$
Uranyl hydroxide	$UO_2(OH)_2 + H_2 \longrightarrow U(OH)_4$
Sulfur	$S + H_2 \longrightarrow H_2S$
Bismuth hydroxide	$Bi(OH)_3 + 3/2\,H_2 \longrightarrow Bi + 3\,H_2O$
Ferric hydroxide	$Fe(OH)_3 + 1/2\,H_2 \longrightarrow Fe(OH)_2 + H_2O$
Tetrathionate	$S_4O_6{}^{2-} + H_2 \longrightarrow 2\,S_2O_3{}^{2-} + 2\,H^+$
Tellurite	$HTeO_3{}^- + 2\,H_2 \longrightarrow Te + 2\,H_2O + OH^-$
Arsenate	$H_2AsO_4{}^- + H_2 \longrightarrow HAsO_2 + OH^- + H_2O$
Vanadate	$H_2VO_4{}^- + 1/2H_2 \longrightarrow VO(OH)_2 + OH^-$
Selenite	$HSeO_3{}^- + 2\,H_2 \longrightarrow Se + 2\,H_2O + OH^-$
Osmium dioxide	$OsO_2 + 2\,H_2 \longrightarrow Os + 2\,H_2O$
Nitrite	$NO_2{}^- + 3\,H_2 \longrightarrow NH_4{}^+ + 2\,OH^-$
Cupric hydroxide	$Cu(OH)_2 + 1/2\,H_2 \longrightarrow CuOH + H_2O$
Osmium tetroxide	$OsO_4 + H_2 \longrightarrow OsO_4{}^{2-} + 2\,H^+$
Manganese dioxide	$MnO_2 + H_2 \longrightarrow Mn^{2+} + 2\,OH^-$
Nitrate	$NO_3{}^- + H_2 \longrightarrow NO_2{}^- + H_2O$
Tellurate	$HTeO_4{}^- + 3\,H_2 \longrightarrow Te + 3\,H_2O + OH^-$
Dithionite	$S_2O_4{}^{2-} + H_2 \longrightarrow S_2O_3{}^{2-} + H_2O$
Lead dioxide	$PbO_2 + H_2 \longrightarrow Pb(OH)_2$
Oxygen	$O_2 + 2\,H_2 \longrightarrow 2\,H_2O$
Bismuthate	$BiO_3{}^- + 5/2\,H_2 \longrightarrow Bi + 2\,H_2O + OH^-$
Hydroxylamine	$NH_3OH^+ + H_2 \longrightarrow NH_4{}^+ + H_2O$

[a] From Woolfolk and Whitely (1962).

in streams receiving the acid waters (Alexander, 1961). The diurnal
oscillation of pH engendered by algal mats as they assimilate CO_2
in the daylight and liberate it at night, a fluctuation in pH extending sometimes from 7.4 to 9.2, may cause the solution or precipitation
of carbonate and quartz particles (Oppenheimer and Master, 1965).
Increases in alkalinity arising through the cleavage of ammonia from
organic nitrogenous compounds and possibly in the reduction of
nitrate to N_2 might be expected to have the opposite effect of acidifi-
cation.

Metabolites excreted during active proliferation combine with
inorganic substances of low solubility to form soluble, stable com-

plexes. Microorganisms thus may solubilize or act to prevent precipitation of metallic ions. Evidence is available in soil and water for the presence of complexes, the organic moieties of which probably have a biological origin, that maintain metal ions in the aqueous phase. Iron, calcium, magnesium, zinc, manganese, and cobalt may exist in such organic combinations. Among the chelating and complexing agents excreted in axenic cultures are 2-ketogluconic acid, polypeptides, amino acids, and additional carboxylic acids, and these might well be the classes of molecules capturing and holding the insoluble substances in nature. Organic chelating and complexing agents have been assigned a role not only in bringing poorly soluble ions into solution but also in the vertical migration of metallic elements and in the weathering of rocks. At some stage, however, the organo-metal complexes are attacked by populations able to use the organic moieties of the molecules, and the metal ions released thereby precipitate out of solution.

A major consequence of the transformations mentioned above is the transport of elements from one region to another. The migration may be short or, alternatively, the distance may be great. Microbial reduction of insoluble ferric oxide or hydroxide or of MnO_2 yields the soluble divalent cation, and this can be translocated in soil or in marine sediments to a point a distance away, where O_2 or sulfide promotes a reprecipitation. The oxidation of insoluble metallic sulfides by *Thiobacillus* gives rise to sulfate and solubilizes the cation, and the sulfate and soluble cation are at times carried with water to far-off locations, where deposition takes place. The formation of organo-metal complexes also is a frequent prelude to the migration of selected elements, as when iron combines with products of organic matter decomposition to give soluble compounds that are carried downward through soil, only to be precipitated in a region supporting a microflora functioning in degradation of the organic portion of the migrating molecules.

Novel microorganisms inhabiting swamps, mineral springs, iron-rich waters, and pipes conveying potable water accumulate ferric or manganic oxides or hydroxides. Several of the ferric iron accumulators have achieved a degree of notoriety, for they are responsible for the fouling of water pipes, a result of the enormous quantities of ferric salts deposited on or embedded in the cellular material. The organisms involved include the bacterium *Gallionella*, sheathed bacteria classified among the Chlamydobacteriales, flagellated protozoa, and a number of algae. In some instances deposition may result from oxidation of divalent iron or manganese in solution. In other cases, however, the encrustation probably is an outcome of the passive adsorption of cationic iron or manganese that is already oxidized, possibly on a microbial surface with a peculiar binding capacity for ferric or manganic salts. Accumulations of oxides on algal surfaces are not rare, and these may owe their origin to the O_2 generated by the photoautotrophs (Silverman and Ehrlich, 1964).

A peculiar capacity of a broad spectrum of microorganisms is the differential rate of utilization of isotopes of the same element. Experimentation in vitro has revealed that living organisms discriminate between the isotopes of hydrogen, carbon, oxygen, and sulfur, the lighter isotope being metabolized more readily than the heavier one. Fractionation of the oxygen isotopes ^{16}O and ^{18}O proceeds, for example, during algal photosynthesis, and the O_2 released is richer in ^{16}O than the water from which it was formed. Excellent evidence for fractionation in nature, as well as in the laboratory, has been obtained in the case of sulfur, which has two abundant naturally occurring isotopes, ^{32}S and ^{34}S. In a few bacterial transformations involving the oxidation or reduction of inorganic sulfur compounds in vitro and in the course of formation of the sulfur deposits off the coast of Louisiana and Texas, deposits that seem to have a microbial origin, sulfur is acted upon enzymatically to yield products relatively richer in ^{32}S and poorer in ^{34}S than the initial substrates (Jones and Starkey, 1957). Fractionation is not universal, however, and numerous populations are unable to discriminate among the isotopes, at least not to an extent detectable by current techniques.

BIOGEOCHEMISTRY OF CARBON

The geochemistry of carbon has a special attraction by reason of the key role of this element in protoplasmic structure and its essentiality in the energy metabolism of heterotrophs. Its biogeochemistry is made still more interesting because of the vast array of organic molecules that are implicated and the cyclic sequence embracing both these compounds and inorganic carbon, a cycle that describes the movement of carbon from the inorganic to the organic state and back to the inorganic again. A very diagrammatic sketch of the carbon cycle is presented in Fig. 16.2.

The driving force for the operation of this cycle, and the chief energy source for life in general, is sunlight, and photosynthesis provides not only terrestrial plants and aquatic algae but ultimately animals and heterotrophic microorganisms with the energy and

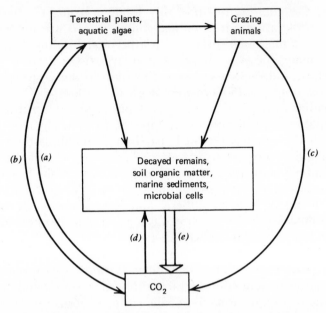

Figure 16.2 The carbon cycle. (*a*) Photosynthesis; (*b*) respiration of algae and higher plants; (*c*) animal respiration; (*d*) CO_2 fixation by autotrophic bacteria; (*e*) microbial mineralization.

nearly all the carbon needed for biosynthetic purposes. The amount of CO_2 fixed by chemoautotrophic and photosynthetic bacteria in nature is unknown, but except in limited areas the quantity is undoubtedly quite small, especially in global terms. The photosynthetic process, in effect, traps the carbon in a large volume of air, concentrates it many, many fold, reduces it to the oxidation state typifying protoplasmic material, and leads to the biosynthesis of a highly complex product that has no equivalent in the abiotic world. A portion of the CO_2 thus assimilated is withdrawn from the active phase of the carbon cycle by virtue of its deposition in organic form in marine sediments, and part is preserved and is biologically unavailable as coal, oil, or peat; the carbon so withdrawn either is of the sort that is not readily degradable or it is removed from the biological mainstream and preserved before significant decomposition takes place. A part of the carbon assimilated by terrestrial plants is also entrapped, but in an entirely distinctive chemical complex, a heterogeneous organic material known as humus. Humus is generated in soil from incompletely decayed plant material and the associated microbial cells, and after its synthesis it generally tends to persist for long periods. However, once the concentration characteristic of that land area is attained, the rates of formation and destruction of humus are equal, so that no additional net carbon loss to the humus reserve is encountered in the cycle.

The bulk of the CO_2 carbon fixed biologically has a short residence in the tissues of rooted plants and algae, for it is quickly returned to the atmosphere as CO_2. This cycling back and forth between the inorganic and the organic state is surprisingly rapid, and a sizable percentage of the atmospheric carbon is each year captured by green plants and likewise much of the organic carbon in living matter is each year returned to the atmosphere. Animals presumably mineralize little of the carbon that plants have fixed, so that the annual contribution of heterotrophic microorganisms to mineralizing the carbon acquired from the atmosphere by land and aquatic plants must be appreciable. By thus decomposing the organic matter created by plants or consumed by animals, the heterotrophs play an essential role in nature, replenishing the relatively small supply of CO_2 available for photosynthesis. Both mineralization and respiration are the forces opposing carbon fixation, functioning to regenerate an essential nutrient that otherwise would be wholly tied up.

To appreciate somewhat the magnitude of the transformations, a carbon balance is necessary. The data presented in Table 16.3 for biological processes and constituents are surely not completely accurate and debate continues about the precise values, but the figures can be considered as probably within one order of magnitude or less of the correct values. Even granting their lack of absolute reliability, the values in Table 16.3 indicate that CO_2 is indeed in short supply in the atmosphere relative to the rate of its consumption, about 5% of the total being used annually. The actual reserve is not this small, however, because an appreciable amount of CO_2 is dissolved in the oceans. Almost all of the carbon in the sea is present as CO_2 or as bicarbonate and carbonate, with organic matter accounting for a mere fraction of the total; though the precise CO_2 level of the oceans remains in dispute, it is generally conceded that the seas have 30–70 times more than the air and that, by comparison, only a minute reserve is retained in inland waters. Notwithstanding both the gaseous and dissolved CO_2, clearly a significant portion of the global supply is assimilated biologically each year. In view of the limited reserve, therefore, CO_2 must be turned over quite rapidly, and this turnover is essential for plant, and hence animal, life. The combustion of coal and oil and fires in the forest and prairie con-

Table 16.3

Geochemical Distribution of Carbon [a]

Distribution or Process	Carbon (kg)
Photosynthesis rate	
Net CO_2 fixed/yr, land plants	1.9×10^{13}
Net CO_2 fixed/yr, marine plants	1.6×10^{13}
Living organisms	2.8×10^{14}
CO_2 in atmosphere	6.4×10^{14}
CH_4 in atmosphere	4×10^{12}
Total carbon in ocean	4×10^{16}
Earth's sedimentary crust[b]	1.5×10^{19}
Coal and petroleum[b]	6×10^{18}

[a] Based on Borchert (1951), Redfield (1958), and Stee-man Nielsen (1952).

[b] Biochemically largely inert.

tribute little to CO_2 regeneration. It is specifically in regard to this regeneration that heterotrophs stand out most dramatically, re-creating CO_2 and maintaining the reserve. Undoubtedly a steady state now exists between CO_2 use and production, between photosynthesis and mineralization, although in the early history of the earth photosynthesis must have exceeded mineralization, with the quantity in excess of the balance accumulating as coal, petroleum, peat, and related deposits.

Photosynthesis in the sea accounts for a major fraction of the CO_2 assimilated per annum (Table 16.3), and some authorities believe that marine fixation exceeds that on land. Most of the activity in the sea is attributable to algae; hence it is microbial. The data presented also show that the CO_2 carbon utilized biologically each year is a sizable percentage of that present in living organisms on the earth's surface. It is also apparent that a vast amount of carbon is in deposits that are largely inert biochemically or, at best, inaccessible to viable organisms, and thus only a minute segment of the earth's carbon is actively undergoing recycling at any one time.

The rate of energy storage by a community as a consequence of photosynthesis or chemoautotrophic activity is known as *primary productivity*. The energy accumulated as organic molecules created from CO_2 by the autotrophs is ultimately available to nearby animal species and heterotrophic microorganisms. Gross primary production refers to all the carbon fixed, including that used in respiration of the autotrophs and returned to the atmosphere as CO_2, but of greater interest is frequently the net primary production; that is, the amount of energy, or carbon, stored in excess of that consumed in respiration of the autotrophs. Net primary productivity is thus the energy input phase of the carbon cycle and of the biosphere in general. Algae in aquatic habitats and rooted plants on land dominate in primary production. Heterotrophic microorganisms do fix small amounts of CO_2 during their growth, but it is unlikely that this fixation ever accounts for much carbon gain in natural ecosystems, except possibly in unusual circumstances.

Marine algae are, as the data of Table 16.3 indicate, major contributors to primary production in the biosphere. Oceans cover about 71% of the globe, and the algae they contain are undeniably the chief agents of CO_2 assimilation in the sea. It is the microscopic algae of the phytoplankton—diatoms, dinoflagellates, and, to a smaller extent, the less widely distributed blue-green and green algae—that dominate in vast expanses of the sea, whereas the large red and brown algae common near the shore contribute less to global primary production, though their regional significance is immense. Freshwater algae, too, have an enormous impact on primary production in lakes, but their global significance is minimal because little of the earth's surface is covered by lakes. On land and in many shallow waters, by contrast, higher plants govern photosynthetic activity largely or entirely. Terrestrial algae play an extremely minor role in production in the biosphere, but because they generate organic matter from inorganic substances, they may indeed have a striking, albeit local, influence in colonizing denuded, barren, or eroded land areas.

The rate of primary production varies enormously. A common method of expressing it is in terms of the quantity of organic carbon synthesized in a given area in a unit of time. Using this kind of expression, the rate may be as low as 24 mg of carbon per square meter per day in a lake in Antarctica sealed under clear ice and supporting a sparse phytoplankton, while the figure may be as high as 7200 mg of carbon per square meter per day in a productive shallow lake with periodic algal blooms (Goldman, 1968). In open regions of tropical oceans, where the activity may be low because of the impoverished nutrient status of surface waters, production may be as little as 50–150 mg of carbon per square meter per day, while in sections of a cooler, more fertile sea the figure may be in the vicinity of 500 mg/day. Values greater than 1000 mg are not unknown in shallow sea water, whereas kelps flourishing along the California coast show staggering production rates, as high as 33,000 mg of carbon per square meter per day (Ryther, 1963). When placed on a seasonal or annual basis, these photosynthetic rates compare quite favorably with agricultural crops grown on fertile land under the best of conditions; for marine algae the annual quantity of carbon fixed may range from 18 to 350 g of carbon per square meter, although lower figures are known in the Arctic.

Ignoring temporal fluctuations, there is no long-term increase or decrease in the mass of the algal crop in many oceanic localities, because an amount of carbon equivalent to that newly fixed is used by nearby organisms. Part of the organic matter synthesized is decomposed by heterotrophic bacteria and fungi, either when the carbonaceous substances are excreted or upon death of the primary producer. A significant portion of the recently fixed carbon in the algal crop is consumed by herbivorous animals, which in turn are eaten by carnivores, and so on up the marine food chain. A very small percentage of the carbon is returned to the land with the fish catch. The bulk of the organic carbon generated in the sea, as in lakes, is ultimately returned to the CO_2 pool, either through respiration of the herbivores and carnivores or via mineralization processes catalyzed by heterotrophic microorganisms.

Primary producers inhabiting the sea, inland waters, and terrestrial habitats provide most of the organic matter serving as substrates for the heterotrophic populations concerned in carbon mineralization. Nearly all the plant material produced on the land, except for that devoured by herbivores or removed for human food, falls to the ground or is already below the surface as root material; either on or within the soil, this multitude of carbonaceous substrates is attacked and destroyed by a heterogeneous collection of fungi, bacteria, and actinomycetes, as well as by subterranean animals. In water, too, microbial mineralization results in the breakdown of much of the cellular material of the primary producers, either near the surface, in subsurface and deep waters, or at the bottom; though herbivorous metazoa ranging from small floating animals to vigorously swimming vertebrates participate in consuming much of the newly fixed carbon, their tissues will in time be likewise subject to microbial decay. In addition, microorganisms mineralize the humus of soil and aquatic sediments, excreta of land and aquatic animals, the organic matter introduced into municipal sewage systems, and—too often in streams, rivers, and estuaries—untreated domestic, agricultural, and industrial wastes.

A fantastic number of distinct chemical entities are thus encountered by microbial communities, and most are mineralized readily. Terrestrial and aquatic plants provide enormous amounts of cellulose, hemicelluloses, and dissimilar polysaccharides as substrates. Polyaromatics enter the mineralization phases of the carbon cycle as lignin, unrelated plant products, and humus. Lesser amounts of hydrocarbons, lipids, proteins, and nucleic acids of cellular origin come into one or another ecosystem, there to be utilized as sources of carbon, energy, or both and, in time, to be returned to the CO_2 pool, the continued existence of which rests so precariously on the functions of the mineralizing heterotrophs. Care should be taken not to assume mistakenly that mineralization is wholly microbiological, for an appreciable amount of this activity, in addition to that attributable to rooted plant and phytoplankton respiration, is effected by animals residing in marine, freshwater, and subterranean environments.

The versatility of natural communities in carbon mineralization cannot fail to impress. Because few if any of the large assortment of biologically produced substances accumulate in aerated environments, and they are destroyed and their carbon regenerated, one must assume—and it may be taken as a general principle—that for every organic compound synthesized biologically, an organism exists in soil, aquatic muds, the open water, sewage, or some less carefully explored habitat with the capacity to effect a partial or complete degradation. A single species may simply initiate the decomposition, but the resulting products will in turn be metabolized by a different species in aerated circumstances. The best evidence for this fully acknowledged microbial versatility is the fact that, with the notable exception of humus and possibly a few additional polyaromatics and heteropolymers, biologically formed organic compounds do not accumulate in surroundings where O_2 is present and conditions are conducive to active life. Either the locality harbors species able to act on the natural product or a readily transported invader is capable of gaining a foothold, growing on the substrate not utilized by the indigenous populations.

Mineralization caused by any single population is never complete, even if O_2 is plentiful and conditions for the organism's replication are ideal. The oxidation of organic carbon has a special metabolic function for the heterotroph: it provides energy for growth. Part of this energy, moreover, is used to make new cells, cells that obtain the carbon they require from the substrate being mineralized or, more properly, from products of substrate degradation. Therefore, though much of the substrate carbon may end up as CO_2 or low-molecular-weight organic compounds, a significant portion is incorporated into new protoplasm. The extent of this assimilation varies with the particular organism and environmental circumstances, but figures of 5–40% of the substrate carbon ending up in heterotrophic cells are not uncommon. The carbon thus transformed into fresh complexes is not, in turn, immune from attack, and it, too, will be mineralized by succeeding heterotrophic microorganisms or consumed by aquatic or terrestrial metazoa.

The microbial impact on the geochemistry and transformations of carbon is quite different when O_2 is absent or its diffusion rate into the habitat is insufficient to meet the demand of the resident community, as is true in marine and lake sediments, swamps, marshes, flooded soils, and decaying plant remains, or in the anaerobic digestion of animal or human wastes. The rate of carbon mineralization is slower and the amount of microbial cells formed per unit of substrate metabolized is less than when O_2 is plentiful. Simple organic compounds like fatty acids and amines accumulate, sometimes in traces but, where the quantity of readily fermentable carbon is large, occasionally in abundance. Needless to say, the composition of the community is different, a fact attested to by a wealth of research on major taxonomic categories of heterotrophs.

Organic matter decomposition under anaerobiosis or limiting aeration is directly responsible for a variety of changes of considerable geochemical or local importance. Some of these are clearly illustrated in Fig. 16.3 for a soil that has been flooded. Here the diffusion rate does not meet the O_2 need arising from the metabolism of readily utilizable organic substrates, and the little O_2 initially present is consumed and the E_h falls. Nitrate disappears soon thereafter, and ammonium formed in nitrogen mineralization accumulates, for it cannot be oxidized by the nitrifying bacteria, which are obligate aerobes. Soluble manganous and ferrous ions appear in solution following reduction of the insoluble ferric iron and manganic oxides. After the drop in E_h, *Desulfovibrio* and possibly physiologically related sulfate reducers as well as methane-forming bacteria proliferate, and the sulfate and CO_2 they use as electron acceptors are converted to H_2S and CH_4. H_2 is formed and, though some is lost to the atmosphere, much is probably utilized by the methane-generating bacteria in the reduction of CO_2. Methyl groups of a few simple fatty acids excreted by cells destroying complex substrates also end up as CH_4. The pattern of changes seems to follow a reasonably fixed sequence: O_2 goes first, then nitrate, followed by appreciable manganous accumulation; at this stage, when the E_h is low, ferrous iron appears

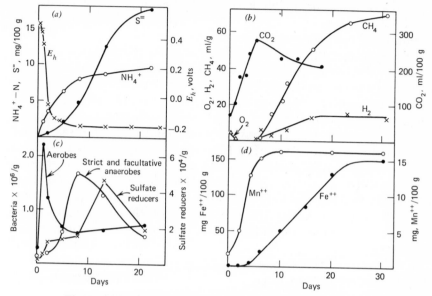

Figure 16.3 Changes occurring in a soil that becomes flooded. (d) represents a separate experiment (Takai and Kamura, 1966).

in large amounts; finally, and generally only after much ferric iron has been reduced, CH_4, H_2, and considerable H_2S are evolved (Takai and Kamura, 1966). The sequence of processes is that expected on the basis of what is currently known about the physiology of the responsible bacteria. Entirely analogous transformations are characteristic of marine and lake bottoms as O_2 is depleted. The various conversions are all of microbiological origin, since none occurs in samples of environments that are taken into the laboratory and sterilized. Anaerobic decomposition of plant remains in soil, and presumably in other habitats, also leads to the solubilization of insoluble oxides and complexes of trace elements to a greater or lesser extent; thus zinc, cobalt, molybdenum, chromium, and copper appear in solution during the anaerobic decay of carbonaceous matter in waterlogged land (Ng and Bloomfield, 1961, 1962). The flooding of soils rich in fermentable substrates also may lead to a rise in the concentration of aluminum, calcium, magnesium, and phosphorus in the water.

In regard to their potential for carbon mineralization, natural communities are unquestionably versatile, indeed impressively so, but they are far from infallible. Frequently a normally biodegradable compound is surprisingly persistent by reason of local conditions unfavorable to its destruction. On the other hand, numerous compounds are extremely resistant to microbial mineralization under all circumstances, although they slowly disappear, and a select group possibly are wholly nonbiodegradable; these two kinds of refractory chemicals have been termed *recalcitrant* molecules. Normally biodegradable materials that owe their longevity exclusively to the place where they are deposited include submerged tree stumps, logs buried in peat, peat itself, and humus, each of these having endured, albeit in a condition greatly modified from its original state, for periods in excess of one and sometimes more than 20 millenia. Furthermore, carbon dating of components of terrestrial humus and of the organic fraction of lake or sea bottom muds reveals ages considerably in excess of 1000 years. The constituents of petroleum, too, are quite obviously resistant to degradation in situ; otherwise the petroleum deposits would long since have been mineralized. These natural materials are all biodegradable, and at reasonable rates, in circumstances different from those where they are found, and their seemingly anomalous longevity can probably be attributed to one or more of the following reasons: deposition of the potential substrate in an inaccessible microenvironment or in a region not suitable for microbial growth owing to the presence of toxic or antimicrobial agents; adsorption and inactivation of extracellular enzymes required to initiate the degradation, the enzyme being retained or inactivated by abiotic components of the locality; or the absence from the site of water, nitrogen, a suitable electron acceptor, or some other factor essential for growth (Alexander, 1965a).

Considerable amounts of organic carbon are refractory and either are essentially outside the biochemical carbon cycle or participate in it excruciatingly slowly. Humus, a well-known example of refractory organic matter, is found in terrestrial mineral soils in concentrations ranging from a trace to 5% or more, a quantity that is appreciable on a biosphere scale. The carbon in marine sediments, peats, and petroleum hydrocarbons is to a large extent untapped by living things, since it is not readily susceptible to mineralization in situ. A large portion of the organic carbon in ancient sedimentary rocks, areas inhospitable and inaccessible to microbial life, is present as kerogen, an insoluble complex, high-molecular-weight polymer. Experimental trials have disclosed that the longevity of humus, peat, and petroleum is not an intrinsic property of the materials but rather is a result of circumstances at the particular place where the substances are found, conditions that preserve the potential substrate and render it less available for mineralization.

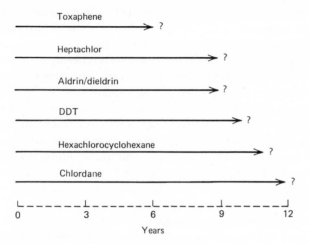

Figure 16.4 Persistence of chlorinated hydrocarbon insecticides in soil. The question mark indicates that the compound persists longer than the time shown (Alexander, 1965*b*).

The existence of truly recalcitrant molecules, compounds refractory to enzymatic degradation and ultimate mineralization in all environments, is of more than academic interest because some have been, or still are, pollutants of regional or global concern. Although it was stated above that all biological products are potentially mineralizable, occasional ones persist for long periods; for example, the surface structures of fungal sclerotia and chlamydospores, protozoan cysts, and bacterial endospores are responsible in part for the long life of these resistant bodies in habitats swarming with potential parasites. With time, however, the resting structures are destroyed. Particular attention has been given to synthetic chemicals, which are finding ever-increasing uses in industry, agriculture, and the home. They are not enzymatically produced, and it is not unreasonable to assume that some are not prone to enzymatic mineralization, at least not at rates sufficiently rapid to minimize or prevent large-scale pollution of water and soil. Recalcitrant molecules of this sort include the branched alkylbenzene sulfonate detergents, which at one time caused foaming problems in watercourses where they exhibited remarkable persistence, an array of substituted aromatic compounds introduced as industrial wastes into rivers, and chlorinated hydrocarbon insectides like DDT, aldrin, chlordane, and hexachlorocyclohexane (Fig. 16.4). Years after the date of application, high concentrations of chlorinated hydrocarbon insecticides are still found in soils supporting heterogeneous and biochemically active communities, attesting to the resistance of these molecules. The durability of selected insecticides is of especial concern on account of their toxicity to fish, birds, and possibly other wildlife and also because of their widespread distribution, DDT having reached sites quite remote from the original place of application (Alexander, 1965a, 1965b).

Recalcitrant, synthetic compounds are frequently subject to *cometabolism*, however. Cometabolism refers to a microbiologically induced change in a molecule that modifies the compound somewhat, but not to an extent sufficient for the responsible populations to utilize the substrate as a source of energy or of any nutrient element it contains. Hence, despite the alteration in the molecule, it is not mineralized, and products of the partial transformation accumulate in the environment. Cometabolism has been investigated solely in bacterial and fungal cultures in vitro, particularly with halogen-substituted aromatic compounds, but the phenomenon probably accounts for the microbiologically effected changes in DDT, dieldrin, and related chlorinated hydrocarbon insecticides so widely dispersed in the biosphere. That the changes these molecules undergo in soil are of microbial origin is suggested by findings that one or another reaction proceeds in natural conditions but not in a sterilized soil sample. Nevertheless, from the sample no organism can be isolated that is capable of using the pesticide as a source of energy or of an element contained in the molecule, observations suggesting the functioning of cometabolism. The explanations for cometabolism are as yet obscure, but a reasonable hypothesis is that the exotic substrate is acted on by an enzyme normally having a different but related function, and the product of this enzyme-catalyzed reaction is not utilized for growth either because it cannot be further metabolized, at least not to yield a compound that is assimilated, or because it is toxic and inhibits further proliferation.

References

REVIEWS

Alexander, M. 1961. *Introduction to Soil Microbiology*. Wiley, New York.

Alexander, M. 1965*a*. Biodegradation: Problems of molecular recalcitrance and microbial fallibility. *Advan. Appl. Microbiol., 7*, 35–80.

Delwiche, C. C. 1965. The cycling of carbon and nitrogen in the biosphere. *In* C. M. Gilmour and O. N. Allen, Eds., *Microbiology and Soil Fertility*. Oregon State University Press, Corvallis, Ore. pp. 29–58.

Goldman, C. R. 1968. Aquatic primary production. *Am. Zoologist, 8*, 31–42.

Raymont, J. E. G. 1966. The production of marine plankton. *Advan. Ecol. Res., 3*, 117–205.

Redfield, A. C., Ketchum, B. H., and Richards, F. A. 1963. The influence of organisms on the composition of sea-water. *In* M. N. Hill, Ed., *The Sea*, Vol. 2. Wiley, New York. pp. 26–77.

Ryther, J. H. 1963. Geographic variation in productivity. *In* M. N. Hill, Ed., *The Sea*, Vol. 2. Wiley, New York. pp. 347–380.

Silverman, M. P., and Ehrlich, H. L. 1964. Microbial formation and degradation of minerals. *Advan. Appl. Microbiol., 6*, 153–206.

OTHER LITERATURE CITED

Alexander, M. 1965*b*. *Proc. Soil Sci. Soc. Am., 29*, 1–7.

Bartholomew, W. V. 1965. *In* W. V. Bartholomew and F. E. Clark, Eds., *Soil Nitrogen*. American Society of Agronomy, Madison, Wisc. pp. 285–306.

Borchert, H. 1951. *Geochim. Cosmochim. Acta, 2*, 62–75.

Jones, G. E., and Starkey, R. L. 1957. *Appl. Microbiol., 5*, 111–118.

Ng, S. K., and Bloomfield, C. 1961. *Geochim. Cosmochim. Acta, 24*, 206–225.

Ng, S. K., and Bloomfield, C. 1962. *Plant Soil, 16*, 108–135.

417

Oppenheimer, C. H., and Master, M. 1965. Z. *Allgem. Mikrobiol.,* 5, 48–51.

Redfield, A. C. 1958. *Am. Scientist,* 46, 205–221.

Steeman Nielsen, E. 1952. *Nature,* 169, 956–957.

Takai, Y., and Kamura, T. 1966. *Folia Microbiol.,* 11, 304–313.

Thompson, L. M., Black, C. A., and Zoellner, J. A. 1954. *Soil Sci.,* 77, 185–196.

Woolfolk, C. A., and Whitely, H. R. 1962. *J. Bacteriol.,* 84, 647–658.

17

Microorganisms and Biogeochemistry: II

BIOGEOCHEMISTRY OF NITROGEN

The biogeochemistry of nitrogen stands out among the elements that are acted upon microbiologically for two reasons: it is an essential element and, of even greater importance, its transformations in the biosphere are regulated almost completely by terrestrial and aquatic microorganisms. As with carbon, the supply of nitrogen readily assimilable by higher plants, and therefore animals indirectly, is under tight microbiological control. In regard to man, moreover, a deficiency of available nitrogen in soil often limits food production and is thus a major contributor to malnutrition and starvation, despite the thousands of tons of N_2 above each hectare of land.

On the basis of current knowledge, nitrogen passes through a nearly perfect cycle. Modest losses to marine sediments and slight gains of ammonia from volcanic action do occur, but the bulk of the nitrogen in the biosphere participates in a regular cycle, one involving innumerable organic nitrogen compounds, ammonium, nitrate, atmospheric N_2, and, though their steady-state concentrations are never high, nitrite and N_2O. A schematic diagram of the cycle in water and on land is depicted in Fig. 17.1. The outline of the cycle is essentially identical in oceanic, freshwater, and terrestrial habitats, the differences lying not in the kinds of biological transformations but rather in the entirely dissimilar micro- and macroorganisms participating, the translocation of organic nitrogen in water as living matter settles to the ocean or lake bottoms, and the translocation of inorganic nitrogen as nitrate is leached out of soil and enters adjacent bodies of water.

419

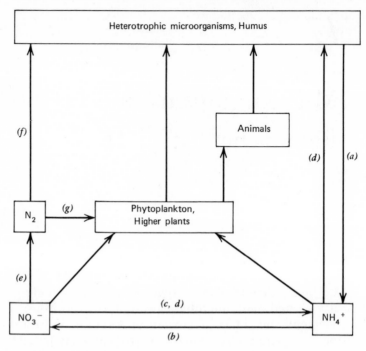

Figure 17.1 The nitrogen cycle. The chief microbiological processes are mineralization (*a*), nitrification (*b*), nitrate reduction (*c*), immobilization (*d*), denitrification (*e*), nonsymbiotic N_2 fixation (*f*), and symbiotic N_2 fixation (*g*).

Higher plants and algae of the phytoplankton assimilate nitrogen predominantly as ammonium or nitrate, and in turn they synthesize diverse organic nitrogenous compounds, including amino acids, proteins, and nucleic acids. Part of this plant nitrogen is consumed by animals. The dead tissues of the rooted plants, algae, and animals, plant excretions, and animal wastes constitute the substrates for nitrogen mineralization, the end product of which is ammonium. A fraction of the ammonium is used by terrestrial plants, but a goodly portion is assimilated or immobilized by aquatic algae and by terrestrial and aquatic heterotrophs, giving rise to complexes that allow these organisms and their predators to grow. A fraction of the ammonium is oxidized by the ubiquitous chemoautotrophs responsible for *nitrification,* to yield nitrate as the final product. This anion is readily assimilated by photosynthetic life on land or water, but because it is not readily retained by soil colloids, nitrate is washed downward through soil and into the ground water. By contrast with ammonium, nitrate functions as an electron acceptor in the metabolism of widespread and abundant bacteria that reduce the anion to N_2, a process known as *denitrification.*

Denitrification results in a net loss of the element from the biosphere, but there is likewise a significant depletion of the readily available reserve as nitrogenous substances settle out and enter permanent sea and lake bottom sediments. Large-scale translocation of the element takes place in a number of ways: nitrate is leached out of soil and finds its way ultimately to the sea or lakes, erosion carries nitrogen off the land, agricultural crops are harvested and removed from farms, sewage originating on land is disposed of in bodies of water, and streams or ocean currents transport the element from place to place. Yet these sorts of translocation do not deplete the biosphere reserve, whereas denitrification and losses to the sediments do. The magnitude of the actual annual depletion, either by microbiological denitrification or by deposition in the sediments, is not accurately known, but the figure for the earth as a whole is assumed to be enormous. To maintain a steady-state level of nitrogen in the biosphere, losses must equal gains, and the leakage from the biosphere is apparently made up by the unique organisms that bring about *nitrogen fixation,* species that alone or in symbiotic association assimilate N_2, commonly a biologically inert gas, for the purposes of synthesizing their own cell constituents. The nitrogen thus fixed, when mineralized or excreted, is available to the less versatile surrounding populations.

Numerous attempts have been made to establish a global nitrogen balance, and controversy rages still on the exact quantities in the various pools and the rates of individual processes. Figures for the biological transformations usually differ by no more than an order of magnitude, however, and the data for the pool sizes may generally be regarded as quite precise. Typical values are given in Table 17.1. The amount of nitrogen in rocks greatly exceeds that in the atmosphere and biosphere combined, but this is biochemically inert and is effectively outside the cycle. Since the rate of N_2 fixation is of the order of 10^{11} kg/year and the earth's atmosphere contains 3.9×10^{18} kg, clearly a far smaller percentage of the atmospheric pool of the element is turned over than that of carbon. Nevertheless, despite the 10 or more million years necessary for consumption of all the nitrogen in the air, the return effected by the denitrifying bacteria is essential in terms of geological time scales, because life has existed on this planet for considerably longer periods. Indeed, if the nitrogen content of the biosphere is at a steady state, as generally believed,

Table 17.1

Geochemistry of Nitrogen

Distribution or Process	Nitrogen (kg)	Reference
Primary rocks	10^{20}	Stevenson (1960)
Atmospheric N_2	3.9×10^{18}	Mason (1966)
Atmospheric N_2O	4.0×10^{12}	Mason (1966)
N_2 in oceans	2.2×10^{16}	Emery et al. (1955)
Oceans (other than N_2)	9.2×10^{14}	Emery et al. (1955)
Biological N_2 fixation, annual		
Terrestrial	$7–36 \times 10^{10}$	Hutchinson (1944)
Marine	$2–9 \times 10^{10}$	Hutchinson (1944)
Biological N_2 fixation, annual[a]	10^{11}	Donald (1960)
Marine denitrification, annual	7×10^{10}	Emery et al. (1955)

[a] Another estimate of the rate.

then inflow must equal outflow. Though nitrogen enters the biosphere as ammonia from volcanoes and as nitrogen oxides created by lightning discharges and falling to earth with precipitation, and though some is lost to the sediments, values for these gains are presumably small by comparison with denitrification and N_2 fixation. Hence, global denitrification and fixation must, if steady-state conditions apply, be about equal.

Reasonable estimates of the nitrogen mineralized per year can be made in several ways for small or large ecosystems or for the biosphere as a whole. Since the biogeochemistry, and biochemistry, of carbon is intimately linked with that of nitrogen as well as with that of every element that is a constituent of protoplasmic material, a common procedure for estimating the mineralization rate utilizes the figures for carbon turnover. In a soil, for example, if the amount of CO_2 fixed by plants per year and the mean percentage of carbon and nitrogen in the vegetation are known, the annual plant uptake of nitrogen is obtained by multiplying the value for annual carbon fixed by the N:C ratio of the vegetation. At the steady state there is neither a net gain nor loss, so that the quantity of nitrogen assimilated by the vegetation each year is roughly the same as the nitrogen mineralized microbiologically. The result is not quite correct, for nitrogen is lost from soil by leaching and denitrification and some returns by biological N_2 fixation and in precipitation, but leaching losses and gains from precipitation, although not biological fixation or volatilization, are easily measured. Nevertheless, the figures give a general approximation of annual mineralization rates. Similar estimates can be made for the mineralization of phosphorus, potassium, and sulfur.

Mineralization leads to the disappearance of organic nitrogen and the accumulation of ammonium and, where O_2 is present, nitrate. This transformation is the basis for nutrient regeneration, inasmuch as higher plants and phytoplankton communities probably get little of their nitrogen from complex or simple organic compounds in nature. In the sea the concentration of inorganic nitrogen (other than N_2), which is present largely as nitrate, is about 60 times that assimilated annually by the phytoplankton (Emery et al., 1955), so that nutrient regeneration in and below the zone of photosynthesis is critical, not so much to allow for development of the current phytoplankton mass as to assure continued primary production in the oceans. In surface soil, by contrast, nearly all the nitrogen is bound in complex molecules and hence is unavailable for plant use; consequently growth of the currently standing terrestrial flora is dependent on the degradation of organic nitrogen in the rooting zone, a process usually liberating no more than 1–4% of the humus nitrogen each year in the temperate zone (Alexander, 1961).

The nitrogenous substrates for mineralization are multitudinous. Proteins are hydrolyzed by extracellular enzymes, ultimately to yield amino acids, and the latter then are deaminated to set free ammonium. Nucleic acids are cleaved by well-characterized enzymes, and the nitrogen in their purine and pyrimidine bases is also liberated as ammonium. Animal excreta contain numerous simple constituents that are likewise readily deaminated. By contrast, the nitrogen in humus and diverse natural products is regenerated very slowly, and neither the responsible enzymes nor the biochemical intermediates have been characterized. Decomposition processes of these kinds may entail a single enzyme produced by one species or a sequence of separate reactions catalyzed by a score of dissimilar organisms, but as a rule the organic intermediates do not accumulate in nature when O_2 is plentiful, indicating that their formation is rate-limiting in mineralization.

Mineralization almost invariably is brought about by growing populations, and therefore inorganic nitrogen is assimilated concomitantly with its production. New proteins, nucleic acids, and minor nitrogenous constituents of microbial cells are synthesized just as preexisting equivalents are destroyed. The extent of this immobilization can be calculated if the amount of substrate carbon and the mean

C:N ratio of cells in the populations active on that substrate are known. Should the substrate or natural product have less nitrogen than needed by the microbiota, a net disappearance of ammonium or nitrate from the surroundings will be observed; this can have disastrous effects on higher plants, because microorganisms then compete with the roots for a nutrient element possibly already in short supply. If the environment does not have enough readily available nitrogen to satisfy the microbial need, the rate of substrate decomposition will be slow and further decay will only proceed as the originally immobilized nitrogen is regenerated through decomposition of cells participating in the initial metabolism of the carbonaceous material. However, should the substrate have more nitrogen than is required by the populations causing its mineralization, as is true with protein-rich substances, this excess will appear as ammonium.

The ammonium thus set free is an excellent nitrogen source for most algae and rooted plants. Alternatively, the cation may provide, by its oxidation, energy for *Nitrosomonas* and physiologically related bacteria that are common in soil, ocean bottoms, marine muds, and freshwater habitats, and these chemoautotrophs oxidize the ammonium to nitrite. Nitrite is the sole energy source for *Nitrobacter* and related autotrophs, bacteria apparently perpetually accompanying the ammonium oxidizers, and hence the final product of the two-step, chemoautotrophic nitrification sequence is nitrate. Nitrification is almost universal where ammonium is present, provided that O_2 is available, the E_h and pH are high, and toxins affecting these very sensitive organisms are absent. Nitrate is formed in cool waters, even in lakes under an ice cover, and in soils that are not frozen, and the nitrate thus generated during the cold part of the year is available for a spring outburst of algae in water or for higher plant use on land. Surprisingly, the population density of the nitrifying autotrophs is quite low in the upper layers of the ocean at distances remote from land, but nitrification does occur there, albeit probably slowly, and also in lake water. The oxidation is particularly rapid in soil and aerobic sewage treatment operations. Recent research has established that fungi and other heterotrophs are able to oxidize the nitrogen in organic and inorganic compounds in vitro, but still the weight of evidence, unfortunately far from unequivocal, suggests that the dominant populations in soil, water, and sewage are autotrophic.

A unique group of bacteria, all seemingly capable of aerobic growth, use nitrate as an electron acceptor and do so exclusively when the microenvironment is devoid of O_2. Occasionally it appears that they utilize the nitrate when O_2 is still present, but the gross analytic

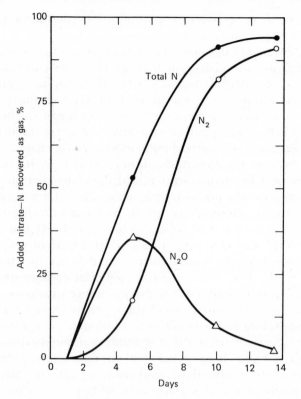

Figure 17.2   Liberation of N_2 and N_2O from soil treated with nitrate and plant remains (Wijler and Delwiche, 1954).

techniques employed undoubtedly fail to show the actual anaerobiosis prevailing at the microsite inhabited by the organisms. Denitrification requires, in addition to anaerobiosis, an energy source to sustain the bacteria, which must expend energy to reduce nitrate. The product of denitrification is initially nitrite, and this in turn is reduced to the N_2 that escapes into the atmosphere. For many and possibly all denitrifiers, N_2O is a precursor of N_2, and traces of the former gas may escape before it is reduced further to N_2, the latter being both the major and the final product of the reaction sequence (Fig. 17.2). N_2O has been found in minute quantities in surface and deep ocean water and also as a metabolite generated in soil during the process of denitrification, and the hypothesis has been advanced that the trace of this gas in the atmosphere (Table 17.1) originates from bacterial denitrification (Adel, 1951). The steady-state level above the earth is presumably maintained by the decomposition of N_2O in the upper atmosphere.

Only a few genera of bacteria can denitrify, but representatives of these genera are prominent in nature. They need not denitrify to proliferate, but when circumstances are right for this activity—the essential conditions being the absence of O_2 and the presence of nitrate and an energy source, nearly always an organic substrate—the population is frequently sufficiently large to convert the nitrate readily to N_2. An apparent contradiction exists in this phase of the nitrogen cycle, because the environment must contain O_2 for nitrification and be free of it for denitrification, but the contradiction is not real inasmuch as adjacent microenvironments often differ completely in their O_2 status or, alternatively, nitrate may leach downward through soil from an oxygenated to an anaerobic zone. Furthermore, the reductive process may take place in waters at a time of year when the O_2 supply has been dissipated entirely, and hence denitrification is not suppressed. Many microorganisms generate copious quantities of ammonium from nitrate, but the disappearance of nitrate that is not attributable to assimilatory use of the anion is, in certain lakes and in soil at least, largely the result of its reduction to N_2 rather than to ammonium. Assessments of the magnitude of denitrification in nature are difficult to make, because the chief product of the reduction, N_2, exists everywhere in abundance and because, until the advent of gas chromatography, the analyses were difficult to perform. Still its potential impact can be appreciated from a report that 11% of the nitrogen annually entering a moderately eutrophic lake is lost as N_2 (Brezonik and Lee, 1968). Marine denitrification (Table 17.1) and that in soil, particularly in agricultural land receiving nitrogenous fertilizers, is also believed to be quite significant.

Denitrification represents a leak in the cycle of nitrogen in a particular region, a leak in the sense that the element is no longer present for use by most residents of the community. In many ecosystems the loss presumably is just made up for by biological N_2 fixation. A number of genera of blue-green algae and free-living bacteria make use of N_2 in vitro, but the significance of most of these organisms to nitrogen gains in nature is unknown. The mere fact that large populations of N_2 fixers exist in a habitat is not adequate evidence for appreciable fixation, as growth may proceed at the expense of nongaseous nitrogen compounds or the activity may be too low to make a meaningful contribution. Nevertheless some species, without doubt, vigorously metabolize N_2 in vivo, acquiring the element not only for themselves but also—on account of nitrogen excretion, the grazing by aquatic animals on blue-green algae, and the release of the newly acquired nitrogen upon cellular lysis and mineralization—for adjacent populations. Considerable attention has not unexpectedly been given to nonsymbiotic fixation in soil, since this element often limits crop and food production, but the data are still sparse. Measurements are made difficult by the relatively small gains of nitrogen when compared with the immense store already in the soil, albeit in largely unavailable complexes, and the expense and problems of analysis when the stable isotope, ^{15}N, is employed in these studies. New, sensitive, and inexpensive techniques have recently been devised, however. Current data indicate that fixation by free-living heterotrophs does occur in soils of the temperate zone, but the gains are usually not appreciable at any individual site and the identities of the active heterotrophs can only be guessed; still, even where the rate is less than, say, 10 pounds per acre per year, this may be quite important in a forest or grassland ecosystem where the nitrogen loss or removal is modest. In intensively cropped land such low rates are well-nigh inconsequential on account of the high demand for and loss of nitrogen. Summation of the low rates for the entire earth's surface, on the other hand, gives a large value, and though precise figures cannot be cited, it is reasonable to assume that nonsymbiotic fixation by heterotrophic bacteria is geochemically consequential.

Photosynthetic N_2 fixers have an advantage not possessed by their heterotrophic counterparts: where light is available, they have a plentiful source of energy and hence can proliferate in localities not suitable for heterotrophs. Their lack of dependence on organic carbon and fixed nitrogen allows the N_2-assimilating algae—a physiological group restricted to individual genera of the blue-greens—occasionally to become prolific, develop dense blooms, and fix much N_2. Fixation, sometimes at notably high rates, has been recorded at sites in the open ocean inhabited by *Trichodesmium,* offshore oceanic waters containing *Calothrix,* lakes and ponds supporting *Anabaena,* and hot springs with growths of *Mastigocladus,* as well as on sand dunes, rocks, and soil surfaces colonized by blue-greens living alone or in a lichen symbiosis. Blue-greens likewise are frequently important in flooded paddy fields, where they may flourish and fix much N_2. Indeed, it is commonly believed that the reason rice has been cultivated successfully for centuries without additions of nitrogen-containing fertilizer is specifically a result of the blue-greens developing in the water of the paddy field.

Energy is not limiting for symbiotic N_2 fixation either, and the microbial partner, whether *Rhizobium* or the microsymbiont of non-

Table 17.2

Quantities of Nitrogen Fixed by Free-Living Microorganisms and Symbiotic Associations [a]

Group	Species or Habitat	N_2 Fixed per Acre per Year (lb)
Nodulated legumes	Alfalfa: *Medicago sativa*	113–297
	Red clover: *Trifolium pratense*	75–171
	Soybean: *Glycine max*	57–105
	Black locust trees: *Robinia pseudoacacia*	600[b]
Nodulated nonlegumes	*Alnus glutinosa*	200
	Casuarina equisetifolia	52
	Hippophae rhamnoides	62
	Myrica gale	8
Blue-green algae	Arid soil in Australia	3
	Paddy field in India	30
	Aulosira in paddy field	48
	Cylindrospermum in paddy field	80
Free-living heterotrophs	Soil under wheat	14
	Soil under grass	22
	Soil and litter under pine	32
	Rain forest in Nigeria	65

[a] Based on Alexander (1961), Ike and Stone (1958), Moore (1966), and Stewart (1966).

[b] Nitrogen fixed after 16–20 years, not annual rate.

legumes, is protected in its plant-created microhabitat from microbial competitors. Under these circumstances the heterotroph benefits from the photosynthetic capabilities of its associate, and the fixation in terrestrial communities containing effectively nodulated plants may be marked and of real consequence on a local as well as a geochemical scale. Representative data on magnitude of the symbiotic gains in nature, as well as figures for the nonsymbiotic process, are given in Table 17.2. Some of the values given are at the top of the range of likely fixation in nature, and much lower figures are to be expected where the density or activity of the fixer or the symbiotic association is low. Owing to this biological transformation, soil under effectively nodulated legumes and nonlegumes is often richer in nitrogen than soil under non-nodulated plants, and the foliage of grasses and trees living together with nodulated plants frequently has a higher percentage of nitrogen than leaves and needles of plants not so fortunately situated. Furthermore, plants on land previously supporting a nodulated symbiont do better than individuals of the same species grown in adjoining fields with no such prior population. The nitrogen acquired from the air, moreover, not only may affect the underlying soil but be transported to adjacent bodies of water; for example, *Alnus* species inhabit the margins of lakes and rivers, and the nitrogen gained by a well-nodulated stand of *Alnus tenuifolia* trees has been found to be transferred in substantial quantities to the lake around which the trees stand (Goldman, 1961).

BIOGEOCHEMISTRY OF OXYGEN, PHOSPHORUS, AND SULFUR

Microorganisms are essential participants in the cycling of oxygen in the oceans, lakes, atmosphere, and on land. A substantial portion of the oxygen at the earth's surface is bound in rocks and in carbonate deposits that are largely or wholly biologically unavailable. However, the vast reserves of oxygen in water and in the atmosphere as O_2, as well as the smaller amounts in nitrate, phosphate, sulfate, and organic combinations, are involved in microbial metabolism. Geochemists are of the opinion that aquatic algae and terrestrial plants are the source of atmospheric O_2, the gas being liberated in splitting of the water molecule during photosynthesis; oxygen biogeochemistry is thus linked directly with the CO_2 fixation portion of the carbon cycle. The consumption of O_2 by aerobes and its resulting loss from the atmosphere likewise represent a tie between the carbon and oxygen cycles, inasmuch as O_2 utilization is largely a direct consequence of the aerobic mineralization of carbon. With one notable exception, chemoautotrophs forming sulfate and nitrate have an obligatory O_2 requirement, and since the oceans and land areas are rich in these two anions, both of which are apparently largely of biological origin, the microbial oxidation of reduced sulfur and nitrogen compounds similarly alters the geochemistry of oxygen. The reduction of nitrate, sulfate, and metallic oxides, the formation of microbial cell constituents containing oxygen, the subsequent degradation of these protoplasmic complexes, and the mineralization of components of plant and animal tissues represent additional ways in which microorganisms regulate the geochemistry of this element.

On a local level algae contribute to the oxygen cycle in aquatic habitats by releasing O_2 during photosynthesis, the result of which is a rise in the dissolved O_2 concentration and a loss of oxygen to the atmosphere. The oxygenation in the daytime is pronounced in the

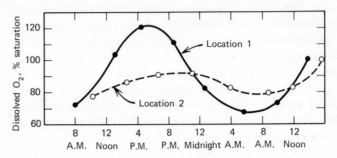

Figure 17.3 Dissolved O_2 in creek water at various times of day (Bartsch, 1960).

sea, fresh water, flooded soils, waste stabilization ponds, and polluted streams. In these very same habitats the concentration of dissolved O_2 falls at night as algal respiration and the metabolism of heterotrophic populations consume the O_2 already present in solution and that diffusing in from the overlying air (Fig. 17.3). Aerobes are invariably deoxygenators, and their proliferation causes a depletion in the O_2 supply whenever the rate of reoxygenation from the atmosphere fails to keep pace with the biological demand.

The atmosphere contains 12×10^{17} kg of O_2, and the oceans about 0.14×10^{17} kg (Redfield, 1958). From the generalized equation,

$$CO_2 + H_2O \longrightarrow (CH_2O) + O_2$$

it is to be expected that 1 molecule of O_2 would be liberated for each carbon atom fixed. Information presented previously indicates that the annual rate of CO_2 fixation is in the vicinity of 3.5×10^{13} kg of carbon per annum, and hence O_2 evolution would proceed at $32/12$ times this rate; that is, about 10^{14} kg of O_2 would be produced each year. A high proportion of this O_2 biogenesis is no doubt attributable to marine algae. Moreover, if O_2 exists in the atmosphere in a steady-state concentration, some 10^{14} kg of O_2 would be consumed each year, and this probably can be ascribed in large measure to microbial respiration. The value of 10^{14} is 0.01% of the O_2 level of the atmosphere, and thus there must be a reasonably rapid turnover of the gas. Calculations of this sort should always be taken as gross approximations of the actual values, but they do give an intimation of the microbial impact on geochemical processes.

Phosphorus, an element essential for all living things, undergoes a cyclic series of transformations in terrestrial and aquatic environments, a set of conversions with microorganisms as critical participants. The element is found in reasonable levels in rocks, and it is abundant in soil, too, as aluminum, iron, and calcium phosphates. Though present in water often in mere traces, it is concentrated many fold by algae and then is transferred in organic combination to animals grazing on the phytoplankton. Tissues of these aquatic plants and animals, the corresponding land forms, water-soluble inorganic phosphate, and the insoluble aluminum, iron, and calcium phosphates of land and sediments are the chief substrates of concern in the three major kinds of reactions by which microbial communities regulate the phosphorus cycle: (a) heterotrophic mineralization, mainly by bacteria and fungi, of organic phosphorus compounds and the regeneration of orthophosphate; (b) immobilization of inorganic phosphorus by the proliferation of photoautotrophs and heterotrophs and the resultant decrease in the available phosphorus supply; and (c) solubilization of insoluble inorganic phosphates. Solubilization may be effected through the biogenesis of organic, nitric, or sulfuric acids—which, for example, render soluble the phosphorus in Ca_3-$(PO_4)_2$—or by the production of H_2S, which may dissolve the phosphorus in ferric phosphates of bottom sediments and soil. The rate of solubilization may limit primary production by algae or rooted plants where the level of phosphorus readily available to photosynthetic organisms is low and, in turn, affect fish production or the density of terrestrial animals feeding on green plants. By contrast with carbon, nitrogen, and sulfur, phosphorus exists in the biosphere largely or entirely in a single oxidation state, as inorganic or organic phosphates, and although microorganisms can both oxidize and reduce inorganic phosphorus in vitro, such reactions have little, and possibly no, importance in nature.

The phosphorus cycle is complicated by the retention of phosphate by inanimate components of soil and bottom sediments, the movement of the anion from the land ultimately to the sea, and the downward movement of the element through bodies of water as dead organisms and particulate organic materials settle out. The amount of phosphorus entering the seas, mainly as a result of soil erosion, is nearly the same as that deposited on the ocean floor, as indicated by the following figures (Emery et al., 1955):

Annual phosphorus transfer from land to sea	1.4×10^{10} kg
Annual phosphorus loss to sediments	1.3×10^{10} kg
Annual phosphorus consumption by phytoplankton	1.3×10^{12} kg
Phosphorus reserve in ocean	1.2×10^{14} kg

The transport of phosphorus to the sea is principally unidirectional, as only a small amount is returned to the land, primarily by the deposition of guano by marine birds. The figures above also reveal the acuteness of the need for phosphorus regeneration, since each year the marine phytoplankton consumes 1% of the phosphorus in the oceans.

Heterotrophs responsible for phosphorus regeneration are ubiquitous and numerous. Bacteria in soil and water and terrestrial fungi dominate in this type of transformation, but soil actinomycetes, too, can split phosphate from organic molecules. The mineralization of most phosphorus-containing constituents of microbial and plant cells is rapid, except if they are adsorbed onto inanimate materials or complexed in a manner that increases resistance of the molecule to enzymatic breakdown; this type of protection of otherwise highly susceptible substrates is particularly prominent in soil. Autolysis of dying phytoplankton cells and plant tissue components contributes to the mineralization.

As an element essential for cell synthesis, phosphorus is assimilated by the same populations responsible for its mineralization, so that inorganic phosphate release is perpetually accompanied by immobilization. Other populations, autotrophic and heterotrophic, simultaneously make use of the phosphate that is regenerated, and the immobilization is rapid whenever conditions are favorable for microbial proliferation. Indeed, the uptake may be so fast and so complete that, in water, no free inorganic phosphate can be detected or, on land, symptoms of severe phosphorus deficiency appear in plants whose roots are in competition with the subterranean community.

The biogeochemistry of sulfur is unique because, in addition to the transformations expected of any element that enters into cell structure, the conversions that sulfur undergoes have a dramatic effect on the behavior and reactions of other elements and, by way of the biologically generated H_2S and H_2SO_4, on natural communities. Four types of reaction sequences are prominent: mineralization of the organic sulfur of cellular components or of synthetic origin; assimilation of inorganic sulfur compounds and their incorporation into microbial bodies; oxidation of inorganic and of amino acid sulfur; and reduction of sulfate and elemental sulfur to sulfide. These processes are outlined in the schematic cycle of Fig. 17.4. The sulfur in animal and plant tissues and in humus is mainly but not exclusively in the reduced state, as in the amino acids cysteine and methionine, and upon the degradation of these carbonaceous substrates the proteins are hydrolyzed to amino acids, which, together with other

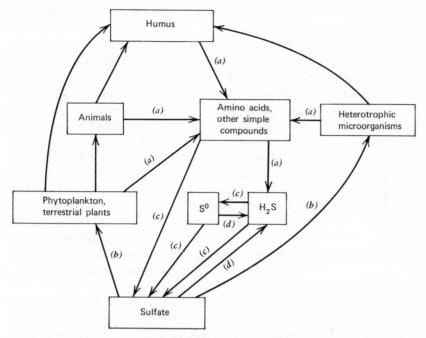

Figure 17.4 A simplified sulfur cycle. (*a*) Mineralization; (*b*) immobilization; (*c*) oxidation; (*d*) reduction.

simple sulfur-containing compounds, may be converted to H_2S. When O_2 is present, the H_2S is readily oxidized microbiologically to sulfate. However, selected fungi in axenic culture and unidentified components of the soil community oxidize the sulfur of amino acids to sulfate by a pathway not involving H_2S (Freney, 1960; Stahl et al., 1949); hence H_2S is not an obligate intermediate in the genesis of sulfate from organic materials. Microorganisms are the principal agents of sulfur mineralization and are apparently solely responsible for regenerating sulfate in nature. Autotrophs and heterotrophs in turn obtain much of their sulfur from sulfate, although some populations can satisfy their requirement for this element by using H_2S or amino acids. Sulfur is thereby removed from immediate circulation.

The oxidation of sulfides in lakes, ponds, reservoirs, the deeper part of the ocean, and soil is effected by one of three categories of physiologically dissimilar bacteria. *Thiobacillus,* the most widely distributed group, flourishes in environments having O_2 and either sulfide or elemental sulfur, these chemoautotrophs converting their energy source to sulfuric acid. Photosynthetic green and purple sulfur bacteria, principally residents of shallow waters, are able to oxidize H_2S to elemental sulfur or sulfate, but they have a local rather than a widespread geochemical influence. The third group, represented by *Beggiatoa* and related bacteria, lives in aquatic surroundings and is likewise only of local significance in generating elemental sulfur or sulfate from H_2S. *Beggiatoa* and sulfur-oxidizing photosynthetic bacteria do, however, occasionally become quite abundant in lake waters having a continuous source of H_2S.

A striking effect of *Thiobacillus* or the closely related *Ferrobacillus* is evident when sulfide-containing ores are exposed to the air. The bacteria rapidly colonize the ore surfaces and contribute to oxidation of the sulfide, a reaction proceeding by abiotic mechanisms as well. In the case of sulfide minerals associated with coal deposits, the reaction yields enormous quantities of sulfuric acid, and the resulting acid mine drainage pollutes adjoining streams and rivers, completely disrupting the aquatic communities and destroying much of the flora along the banks. The magnitude of this potential for pollution is evident from an old estimate that in a single year 3 million tons of sulfuric acid are discharged into the Ohio River and its tributaries (Hodge, 1937). Sulfur-containing lakes in volcanic craters may similarly develop extremely high sulfate concentrations and pH values of less than 2.0 (Yoshimura, 1933).

The oxidation of iron sulfide ores, like pyrite and marcasite, has been thoroughly explored. This transformation leads to the appearance of sulfate and quite low pH values. A *Thiobacillus*-effected oxidation of such sulfides in land reclaimed from sea bottoms by the Netherlands, for example, has resulted in rapid increases in acidity and the accumulation of colored masses of ferric sulfate. The thiobacilli also act on sulfide minerals containing antimony, arsenic, copper, lead, molybdenum, nickel, and zinc (Silverman and Ehrlich, 1964). Some of the reactions proceed slowly under sterile conditions, but the rate is enhanced appreciably by the thiobacilli or related chemoautotrophs. The products often remain in solution, and solubilization of metallic sulfides may thus set the stage for a migration of the elements contained in the mineral and their deposition some distance away. The microorganisms may also cause, indirectly or directly, oxidation of the element associated with the sulfur, as with the production of hexavalent and pentavalent molybdenum from the molybdenite (MoS_2) metabolized by autotrophic bacteria (Bryner and Anderson, 1957).

Notorious for its ubiquity as well as for its easily discernible activity is *Desulfovibrio*, a genus of anaerobic bacteria using sulfate as their terminal electron acceptor and thereby producing copious quantities of H_2S. The population density of *Desulfovibrio* can be extremely high in lake and sea bottoms, polluted estuaries and streams, marshes, black muds, tidal basins, swamps, flooded soils, and decomposing sewage, and the large cell numbers provide unquestionable evidence for active sulfate reduction because sulfate is the sole electron acceptor for growth of these anaerobes. As a rule, whenever sulfate, organic matter, and anaerobiosis coexist in a nonacid locality, there *Desulfovibrio* and sulfide biogenesis will be discovered. Anaerobic bacteria physiologically akin to *Desulfovibrio* have been isolated and characterized, but current information suggests that none is as widespread or numerous. *Desulfovibrio* is especially interesting by reason of its involvement in the formation of metallic sulfides and elemental sulfur, the corrosion of iron and steel pipes, and the genesis of a potent toxin, H_2S. Without enzymatic intervention, the H_2S formed by *Desulfovibrio* will react with heavy metals in solution to yield insoluble metal sulfides. Thus much of the H_2S that appears in O_2-free muds, soil, and water containing ferrous iron combines immediately with the iron, and H_2S may not even accumulate in appreciable quantities until all the iron has been precipitated as iron sulfide. The final product is often pyrite, FeS_2 (Barghoorn and Nichols, 1961). Sulfides of nickel, cobalt, copper, silver, calcium, lead, molybdenum, and arsenic are prevalent in the earth's crust, many having been commercially mined, and it is plausible to suggest that some of the deposits originated when soluble ions of the element reacted with and were precipitated by H_2S. In fact, evidence in vitro exists to show that bacterial reduction of sulfate in an artificial marine ecosystem is a prelude to the formation of copper, silver, lead, and zinc sulfides (Baas Becking and Moore, 1961). In addition, the reducing conditions engendered by the activity of *Desulfovibrio* in sulfate-rich waters have been implicated in reduction of the soluble uranyl ion and precipitation of UO_2 in natural uranium deposits (Jensen, 1958).

A portion of the sulfide made by bacteria in anaerobic ecosystems may come not from sulfate but from organic compounds. This source of H_2S is probably the major one in sewage, in decaying masses of plant or animal remains, and wherever much proteinaceous matter and little sulfate are present. In the sea, black muds, and estuarine and lake sediments, sulfate and not organic sulfur appears to be the chief precursor of H_2S; the microflora metabolizing the carbonaceous substrates in these environments creates reducing conditions, and when the organic nutrients are degraded to simple molecules the latter serve as energy sources for sulfate reducers.

The formation of a number of deposits of elemental sulfur has been ascribed to *Desulfovibrio* and other genera. The deposits investigated include several in brackish lakes of Cyrenaica, in the salt domes along the shores of the Gulf of Mexico, and in lakes of volcanic origin. The deposits are occasionally quite thick, and those along the coast of the Gulf of Mexico are located at considerable depths below sea level. The experimental findings indicate that the first step is bacterial reduction of sulfate in the water to sulfide. The sulfide is subsequently oxidized to elemental sulfur by photosynthetic bacteria such as *Chromatium* and *Chlorobium,* by *Beggiatoa* or related organisms, by species of *Thiobacillus,* or by nonbiological means, the particular populations and the mechanism of the oxidation depending on the site (Butlin and Postgate, 1954; Jones and Starkey, 1957; Silverman and Ehrlich, 1964).

TRANSFORMATIONS OF OTHER ELEMENTS

The geochemistry of various elements in addition to those cited above is altered by microbial communities. In some instances the mechanisms underlying the changes are similar to those already presented and in other instances they are quite different, but unfortunately too often the way in which microorganisms affect the geochemistry of an element or the significance in vivo of a process observed in the laboratory is still not adequately defined.

One biochemical transformation that probably is of great importance is the destruction of silicate minerals. Numerous species can grow in media containing a potassium aluminosilicate as the lone potassium source, not only satisfying their own requirements for assimilatory purposes but simultaneously solubilizing an appreciable amount of the element. An analogous destruction of these minerals seems to occur in soil, and possibly in aquatic habitats too, providing to higher plants the previously unavailable potassium. The solubilization commonly is beneficial, since the level of readily available potassium often limits plant growth. Dissolution of the potassium results from the release into the environment of organic, nitric, sulfuric, and carbonic acids, the last of which is derived from CO_2 generated in heterotrophic metabolism.

Silicon itself is rarely a nutrient, with the notable exception of selected algae. Diatoms characteristically are surrounded by a sili-

ceous wall, but silica-rich structures are also found in marine silicofla-
gellates and in scattered genera of Chrysophyta and Xanthophyta.
The siliceous walls of the diatoms are made up of hydrated amor-
phous silica with traces of elements in addition to silicon and oxygen.
The diatoms have an absolute requirement for silicon, so that their
replication necessarily leads to immobilization of this element in the
marine and freshwater areas where they multiply. If the siliceous surface structures are precipitated, massive accumulations of diatoma-
ceous earth or oozes may arise, and such deposits are widespread in
ocean or lake bottoms. The mechanism of silicon resolubilization
differs totally from the mineralization of carbon, nitrogen, or phosphorus, in that it is apparently, or at least presumed to be, entirely
or largely a nonenzymatic hydrolysis (Golterman, 1960). Notwithstanding the deposition of considerable amounts of the element in
marine sediments, the dissolved silicon level of the oceans far exceeds
this drain on the supply, and more continuously enters in the form
of mineral grains washed from the land and carried by rivers to the
sea.

The regeneration of many elements relies on microbial activities.
In addition to those cited in the preceding sections, for example, iron
is mineralized directly through the metabolism of populations deriving carbon and energy from the carbonaceous moieties of water-
soluble organic iron salts, the iron being liberated as a waste in the
process of substrate oxidation. Additional elements are undoubtedly
mineralized in an identical way. On the other hand, a few essential
nutrients are regenerated as an indirect consequence of the decomposition of tissues. Thus potassium exists in plant cells mainly in
the inorganic form, and its liberation results not from mineralization
of potassium-containing organic substrates but rather from the destruction and disintegration of tissues bearing the element. Whether
by direct attack or indirect release, however, the microflora is essential for the cycling of a score of elements as plant litter, animal
remains, or microbial cells undergo decay. These changes have been
most intensively studied in regard to decomposing plant litter. In a
grassland or forest at a steady state with reference to nutrient cycling in the plant-soil system, the rate of uptake of elements by the
plant community will probably be nearly equal to that deposited on
and regenerated in the underlying soil, with losses by leaching,
cropping, or forestry practices detracting from the equivalence.
Among the elements incorporated into plant tissues and subsequently
regenerated are those listed in Table 17.3, which shows the annual
return of several elements to the soil. The extent of the microbial

Table 17.3

Return of Elements to Soil with Litter Fall [a]

Type of Plant Community	Rate of Return (kg/hectare/yr)				
	Si	Ca	K	Mg	Fe
Arctic tundra	2	4	4	3	0.9
Beech forest	61	101	35	13	7
Meadow steppe	170–243	25–130	56–116	14–31	4–9
Semishrub desert	2	14–22	3–7	2	0.5
Tropical rainforest	770	181–307	53–84	58–72	63

[a] From Rodin and Bazilevich (1967).

contribution, direct or indirect, to the regeneration and the relative importance of abiotic release is still completely obscure.

Microorganisms alter the geochemistry of iron and manganese by quite similar means. Though both are subject to mineralization and immobilization, these processes are as a rule inconsequential in nature, by contrast with the far more pronounced oxidation and reduction reactions. Microbial biosynthesis of organic or inorganic acids, consumption of O_2, and the creation of reducing conditions shift the ferrous-ferric and manganous-tetravalent manganic equilibria to favor the reduced cations, which are the more soluble; alkali formation and O_2 evolution in algal photosynthesis favor the oxidized cations, ferric and tetravalent manganese (MnO_2), which are far less soluble and thus, if present in reasonable concentrations, are lost from solution. Reduction is common whenever the O_2 demand is high, as when organic substrates are actively undergoing attack, and the rate of O_2 entry into the locality is insufficient to satisfy the need (Alexander, 1961). The soluble ferrous and manganous ions are quite mobile and may be leached or otherwise transported by moving water, only to be precipitated and deposited by biological oxidation or, probably more usually, spontaneously when an O_2-rich site is encountered. Reductions and oxidations of these sorts are typical in soil, lakes, rivers, estuaries, and the open sea.

Specific sheathed and stalked bacteria, algae, protozoan flagellates, and true bacteria amass ferric oxides and MnO_2 around themselves. Only one or two of the groups associated with this type of ferric iron formation (the *Ferrobacillus-Thiobacillus* group and possibly *Gallionella*) and none of those accumulating MnO_2 are chemoautotrophic. Whether the iron or manganese precipitation by heterotrophs multiplying at near neutral pH reflects a biologically mediated change or results solely from passive accumulation remains problematic, posing a dilemma whose solution is made difficult because ferrous iron in particular is readily autooxidizable at near neutral pH values. The actions of these bacteria, algae, and protozoa are easily observed in waters where they thrive, deposit insoluble oxides, and occasionally cause appreciable discoloration (Pringsheim, 1949). In addition, a role has been postulated for certain manganese-oxidizing bacteria, such as *Metallogenium*, in the genesis of oxidized ores in manganese deposits (Sokolova-Dubinina and Deryugina, 1966).

Calcium, too, is markedly affected by microbial metabolism. The autotrophic oxidation of ammonium to nitric acid in soil, the formation of organic acids by heterotrophs, and the conversion of sulfide to sulfuric acid by *Thiobacillus* multiplying in drained marine soils, as in the Netherlands land reclamation program, result in solubilization of insoluble calcium salts, and the calcium in turn is carried downward with the flow of percolating water. Should the formation of 2-ketogluconic acid and other chelating agents be a reality in nature, as it is in the laboratory, insoluble calcium will be dissolved in microenvironments where these metabolites are excreted. Furthermore, the lime crusts coating rocks in arid regions adjacent to the Mediterranean Sea and dissimilar calcareous accumulations in saline land have been assigned a biological origin, and algal and possibly heterotrophic residents of the sea and lakes apparently cause precipitation of soluble calcium salts to give rise to marls, limestone, and other calcareous deposits.

A hydrogen cycle exists in nature, but the in vitro information undeniably far exceeds the in vivo knowledge. Bacteria produce considerable H_2 during their anaerobic metabolism in sewage, marshes, waterlogged soil, the gastrointestinal system of countless animals, and presumably in ocean and lake bottoms. Much of this H_2 never escapes, because it is immediately utilized in the reduction of CO_2 by methane-forming bacteria or in the bacterial reduction of sulfate, nitrate, and organic compounds. A part of the H_2 will find its way to oxygenated microenvironments or overlying strata, where it serves as an energy source for *Hydrogenomonas* or is oxidized by heterotrophic bacteria. Algae assimilate vast amounts of the element from water during the photosynthetic reduction of CO_2, and the algal mass of the seas, as well as heterotrophic cells, ties up in its cells an immense reserve of organically bound hydrogen, which is subsequently mineralized and returned to the water when the protoplasmic material is degraded enzymatically. A significant portion of the hydrogen assimilated, however, is removed from rapid circulation when it is deposited as organic hydrogen in the form of humus, petroleum, and peat.

Evidence is at hand for the cycling of selenium, too, but details of the individual transformations are still vague. Occasional plants growing in semiarid soils rich in selenium are notorious accumulators of the element, and the organic selenium compounds they make are very likely mineralized and the selenium released in inorganic form when the tissues are subject to microbial decay. Moreover, fungi and bacteria in culture and unidentified components of the soil community reduce selenate and selenite and liberate volatile selenium compounds (Abu-Erreish et al., 1968; Shrift, 1964).

PEDOGENESIS AND PETROLEUM FORMATION

Microorganisms representing a heterogeneous collection of classes, families, and genera participate in the *weathering* of rocks, that is, in their disintegration and destruction. Algae, lichens, actinomycetes, fungi, and heterotrophic and chemoautotrophic bacteria contribute to weathering, the dominant taxonomic category varying from place to place. Lichens often play an especially prominent role in the disintegration and, where circumstances permit, so too do nitrifying and sulfur-oxidizing autotrophs through the acids they generate. This destruction can be induced to proceed rapidly in the laboratory by incubating ground stone in liquid media fortified with a simple carbon source and a soil inoculum; the fact that the disintegration can thus be made rapid prompts the view that breakdown of rocks is slow in nature by reason of the lack of organic carbon, the paucity of water, and the small surface area exposed, as well as the temperature extremes common to the surfaces of rocks in many regions.

Because rocks contain little or no organic carbon to support heterotrophs, algae and lichens usually flourish first, using light and CO_2 for energy and carbon, and they initiate both the colonization and the corrosion. The organic substances synthesized by the photoautotrophs permit establishment of heterotrophs, which further the disintegration. The developing community obtains its inorganic nutrients from the rocks, although N_2-fixing algae or alga-fungus associations may acquire nitrogen from the atmosphere. Colonization of this sort is evident on exposed stones and boulders on mountain sides, in the forest or open field, and along the shoreline. The compounds bringing about biological weathering are ill defined, but it is likely that organic acids, chelating agents, and biologically produced inorganic acids are implicated. Weathering, partly microbiological, partly the result of abiotic physical and chemical changes, is the first step in *pedogenesis,* the formation of soils. In pedogenesis, autotrophs and heterotrophs not only destroy the physical structure of the stony materials but also create de novo a multitude of organic substances typical of terrestrial soils. As it promotes weathering, whether through the excretion of acids, chelating agents, or other metabolites, the microbial community destroys silicate minerals in the rocks and thereby aids in the release of silicon, aluminum, iron, potassium, sodium, magnesium, or manganese bound within the minerals.

Terrestrial communities also serve as indispensable agents in the formation of podzolic soils, a process known as *podzolization.* Podzolic soils are generally acidic, they contain a reasonably high content of iron, aluminum, and humus at lower depths, and they are located in forests of the humid temperate zones. The organic substrates for podzolization are the leaves falling to the ground beneath trees, there to decay. In this decomposition, compounds with chelating properties like carboxylic acids and simple or polyphenolic substances are formed by the litter and underlying communities. These compounds apparently react with iron-containing and additional components of the upper layers of the parent soil material to yield soluble organo-metal complexes that are leached downward. At a deeper site the organic moieties of these organo-metal complexes are degraded, and the cations that migrated vertically are set free, for example, as the iron oxide. These changes in time give rise to the podzolic soil type (Coulson et al., 1960; Stobbe and Wright, 1959).

By the polysaccharides produced by bacteria, the cementing agents elaborated by innumerable heterotrophs, and the filaments of fungi and actinomycetes, microorganisms are largely responsible for the establishment and maintenance of soil aggregates (Fig. 17.5). An aggregate is a collection of sand, silt, and clay particles bound together into stable granules. A soil possessing good aggregation is usually well aerated, drains rapidly following rainfall, and permits unhindered development of roots and root hairs. The polysaccharides and capsular materials synthesized by heterotrophs bind the small particles directly, whereas their filaments permeate, surround, and mechanically hold together the minute granules. Members of the underground metazoan fauna, notably earthworms, as well as abiotic forces also contribute to aggregate formation (Harris et al., 1966).

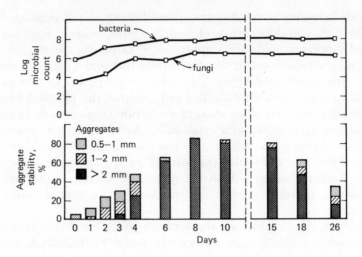

Figure 17.5 Aggregation and microbial growth in sucrose-amended soil. Aggregate stability is here expressed as the percentage of soil material bound into 0.5–1.0, 1.0–2.0, and >2.0 mm of water-stable aggregates (Harris et al., 1964). By permission of the publisher.

Algae at the soil surface, where light is available to allow for extensive proliferation, occasionally affect physical properties of the uppermost soil layers. Such activities are sporadic and highly localized because light and adequate moisture do not universally coexist at the very surface, but where conditions are favorable the algal crusts that develop do indeed have a profound importance by binding small particles together, stabilizing soil structure, and ostensibly reducing erosion losses. Even in semiarid or desert regions, a bloom of algae may emerge following the very infrequent rains, a photosynthetic crust adding organic matter to the surface, making substrates for fungi and bacteria, and, if the bloom contains blue-greens, possibly fixing N_2.

Vast land areas are covered by peats, the formation of which necessitates microbial action. A critical requirement for this geochemical transformation is anaerobiosis, and though aerobes may be abundant initially, the final stages of peat formation are carried out solely by anaerobic bacteria. In the primary phase in the genesis of peats, plants in the water or in the water-saturated soil die and are subject to decay, primarily at first by O_2-requiring bacteria and fungi. However, the deposition of dead vegetation is so great that the rate of O_2 diffusion cannot keep pace with demands of the microbial community, and the consequent shift to anaerobiosis results in an anaerobic biota and a marked diminution in the rate of decomposition. Ultimately the destruction of dead plant matter becomes slower than the influx of new vegetation, so that partially decayed, loosely consolidated plant materials and microbial products accumulate. For reasons as yet unclear, bacterial activity slows down and, in time, largely ceases. The peat thus reaches a stage of self-preservation, one that is maintained as long as reoxygenation is prevented. When a peat is drained, on the other hand, aerobic conditions are reinitiated, and mineralization progresses forthwith. The anaerobic conversions associated with peat formation lead to a decrease in the relative abundance of cellulose and other plant-derived polysaccharides and a rise in the percentage of humic substances.

Through their role in peat formation, microorganisms also are implicated in the genesis of coal. The magnitude of this transformation can be appreciated from the estimate made several years ago of the coal resources situated at workable depths, a figure of about 10^{15} kg. Peats once sealed beneath an impervious sediment may be exposed to high temperatures and pressures, and should these abiotic factors act for a prolonged period, a sequence of events occurs that results in the appearance of bituminous and finally anthracite coal (van Krevelen, 1963).

The synthesis of humus, one of the major organic complexes of geochemical importance, is attributable to microbial action. The precise quantity of this product on the earth's surface has not been accurately determined, but some estimates suggest that more organic carbon is found in humus than in all living organisms of the biosphere. This highly complex, heterogeneous material, or at least substances designated humus, is generated in many environments, but the bulk of information on its biosynthesis and chemistry is derived from investigations of well-aerated, terrestrial soils. Humus is essentially the organic fraction of soil, that portion which has been modified by the subterranean community to an extent that its plant origin is no longer recognizable. By the time the physical identity of the initial source material has been lost, however, the conversion has gone so far that few if any of the vegetation-derived substrates remain. The actual polymeric material in soil must therefore be considered as composed of building blocks, a portion of which was in the plant debris but has been liberated through microbial hydrolysis, and a part of which is completely microbial. The biochemical alteration from a plant chemistry to one dictated by the soil microflora is reflected in the rapid mineralization of readily degradable components of the plant litter, such as low-molecular-weight compounds and starch, the slower turnover of plant cellulose and hemicelluloses, and the accumulation of polyaromatics and other refractory classes of compounds. The resistant substances may originate in three ways: biosynthetic reactions performed by the bacteria, fungi, and actinomycetes causing decay of the plant remains; products of heterotrophs feeding on cellular components of the pioneers in this colonization; and plant lignins. Data concerning the changes during humus formation in marine and fresh waters are quite sparse, but the few published studies suggest that transformations corresponding to those on land take place at the bottom of bodies of water; in this instance the primary sources of organic carbon are the algae rather than terrestrial vegetation.

Controversy has raged for years regarding how petroleum was formed. It is generally believed that petroleum hydrocarbons have a biological origin, but the identity of the organisms and their subsequent fate are uncertain. Both microorganisms and higher plants contain hydrocarbons that, although different from those in petroleum, might be their precursors. On the other hand, all living things are rich in fatty acids, and these could give rise to aliphatic hydrocarbons by a simple enzymatic or nonbiological decarboxylation or by a reduction of the carboxyl groups. According to one of the more plausible hypotheses concerning the first stages of petroleum formation, the starting point was marine plankton that had settled to the ocean floor and reached the anaerobic sediments. This organic debris was then metabolized, but only partially so, by anaerobic bacteria, and the decomposition resulted in a diminution in the relative abundance of nitrogen, oxygen, and phosphorus to yield a material more akin to petroleum; deamination, decarboxylation, and dephosphorylation, for example, would effect a loss of these three elements. Concomitantly, the aliphatic hydrocarbons, fatty acids, and waxes either generated microbiologically or present initially would tend to be preserved owing to their intrinsic resistance to anaerobic decomposition. In the final phases leading to petroleum, however, the various compounds quite likely were modified to a lesser or greater extent by abiotic mechanisms (Davis, 1967).

References

REVIEWS

Alexander, M. 1961. *Introduction to Soil Microbiology*. Wiley, New York.

Davis, J. B. 1967. *Petroleum Microbiology*. Elsevier, Amsterdam.

Johannes, R. E. 1968. Nutrient regeneration in lakes and oceans. *In* M. R. Droop and E. J. F. Wood, Eds., *Advances in Microbiology of the Sea,* Vol. 1. Academic Press, New York, pp. 203–213.

Mason, B. 1966. *Principles of Geochemistry*. Wiley, New York.

Silverman, M. P., and Ehrlich, H. L. 1964. Microbial formation and degradation of minerals. *Advan. Appl. Microbiol.,* 6, 153–206.

Wood, E. J. F. 1967. *Microbiology of Oceans and Estuaries*. American Elsevier, New York.

OTHER LITERATURE CITED

Abu-Erreish, G. M., Whitehead, E. I., and Olson, O. E. 1968. *Soil Sci.,* 106, 415–420.

Adel, A. 1951. *Science,* 113, 624–625.

Baas Becking, L. G. M., and Moore, D. 1961. *Econ. Geol.,* 56, 259–272.

Barghoorn, E. S., and Nichols, R. L. 1961. *Science,* 134, 190.

Bartsch, A. F. 1960. *In* C. A. Tryon and R. T. Hartman, Eds., *The Ecology of Algae.* Pymatuning Laboratory of Field Biology, University of Pittsburgh, Pittsburgh, Pa. pp. 56–71.

Brezonik, P. L., and Lee, G. F. 1968. *Environ. Sci. Technol.,* 2, 120–125.

Bryner, L. C., and Anderson, R. 1957. *Ind. Eng. Chem.,* 49, 1721–1724.

Butlin, K. R., and Postgate, J. R. 1954. *In* J. L. Cloudsley-Thompson, Ed., *Biology of Deserts.* Institute of Biology, London. pp. 112–122.

Coulson, C. B., Davies, R. I., and Lewis, D. A. 1960. *J. Soil Sci.,* 11, 30–44.

Donald, C. M. 1960. *J. Australian Inst. Agr. Sci.*, **26**, 319–338.

Emery, K. O., Orr, W. L., and Rittenberg, S. C. 1955. *In Essays in the Natural Sciences in Honor of Captain Allan Hancock.* University of Southern California Press, Los Angeles. pp. 299–309.

Freney, J. R. 1960. *Australian J. Biol. Sci.,* **13**, 387–392.

Goldman, C. R. 1961. *Ecology,* **42**, 282–288.

Golterman, H. L. 1960. *Acta Bot. Neerl.,* **9**, 1–58.

Harris, R. F., Chesters, G., and Allen, O. N. 1966. *Advan. Agron.,* **18**, 107–169.

Harris, R. F., Chesters, G., Allen, O. N., and Attoe, O. J. 1964. *Proc. Soil Sci. Soc. Am.,* **28**, 529–532.

Hodge, W. W. 1937. *Ind. Eng. Chem.,* **29**, 1048–1055.

Hutchinson, G. E. 1944. *Am. Scientist,* **32**, 178–195.

Ike, A. F., and Stone, E. L. 1958. *Proc. Soil Sci. Soc. Am.,* **22**, 346–349.

Jensen, M. L. 1958. *Econ. Geol.,* **53**, 598–616.

Jones, G. E., and Starkey, R. L. 1957. *Appl. Microbiol.,* **5**, 111–118.

Moore, A. W. 1966. *Soils Fertilizers,* **29**, 113–128.

Pringsheim, E. G. 1949. *Biol. Rev.,* **24**, 200–245.

Redfield, A. C. 1958. *Am. Scientist,* **46**, 205–221.

Rodin, L. E., and Bazilevich, N. I. 1967. *Production and Mineral Cycling in Terrestrial Vegetation.* Oliver and Boyd, Edinburgh.

Shrift, A. 1964. *Nature,* **201**, 1304–1305.

Sokolova-Dubinina, G. A., and Deryugina, Z. P. 1966. *Mikrobiologiya,* **35**, 344–349.

Stahl, W. H., McQue, B., Mandels, G. R., and Siu, R. G. H. 1949. *Arch. Biochem.,* **20**, 422–432.

Stevenson, F. J. 1960. *Geochim. Cosmochim. Acta,* **19**, 261–271.

Stewart, W. D. P. 1966. *Nitrogen Fixation in Plants.* Athlone Press, University of London, London.

Stobbe, P. C., and Wright, J. R. 1959. *Proc. Soil Sci. Soc. Am.,* **23**, 161–164.

van Krevelen, D. W. 1963. *In* I. A. Breger, Ed., *Organic Geochemistry.* Pergamon Press, New York. pp. 183–247.

Wijler, J., and Delwiche, C. C. 1954. *Plant Soil,* **5**, 155–169.

Yoshimura, S. 1933. *Arch. Hydrobiol.,* **26**, 197–202.

18

Effect of Microorganisms on Animals and Plants

Higher animals and plants serve as environments upon and within which microbial populations and communities develop, and the activities, feeding habits, and tissue constituents of metazoa and rooted plants provide an assortment of nutrients that sustain microscopic living things. In turn, microorganisms may perform essential or beneficial functions, act in a detrimental way, or cause marked morphological, biochemical, or pathological responses in the individual with which they are associated. This interplay between macro- and microorganism frequently is remarkably intimate, and the behavior of the large and small partners tends to be highly interdependent because of their close physical proximity. On occasion an animal or plant fares poorly in nature if it lacks its normal microbial inhabitants, and some species are entirely unable to live or reproduce unless accompanied and colonized by distinct and specific microsymbionts.

Microorganisms residing in, on, or in close proximity to higher organisms have a favorable, a detrimental, or no influence on their growth, development, and reproduction. The microbial reactions may occur in the rhizosphere, on leaf surfaces, or in soil at a small distance from the plant, or they may be effected in the gastrointestinal system, on the skin, or at a point away from the animal. In addition, food consumed by metazoa may be modified, long before it is ingested, in a manner that materially alters the food value of the substance. Various microbiological activities, moreover, take place directly within the organs, tissues, and cells of macroorganisms, and special attention has been given to those that lead to physiological disturbances and communicable disease. Furthermore, distinct biochemical and physiological properties usually attributed to plants and animals in natural habitats are in fact not characteristics of these individuals but are the result of the microscopic communities with which they are closely linked.

A precise categorization of the ways in which microorganisms influence higher animals and plants is not yet feasible. Too little information is available, and far too few mechanisms have been adequately delineated. As research in the appropriate subject matter areas progresses, however, it becomes increasingly evident that the processes and mechanisms by which microbial populations affect other species are few in number, so that what now appears to be a confusing array of influences should ultimately fit into well-defined reaction sequences, environmental alterations, and biochemical mechanisms. For the purpose of clarity, our still limited knowledge will be discussed in the present chapter in terms of (a) general morphological and physiological changes; (b) microorganisms as food; (c) metabolites influencing animals and plants; (d) effects on mineral nutrition and metabolism; (e) changes resulting from metabolism of O_2 and CO_2; (f) protection against and predisposition to disease; and (g) effects associated with pathogenicity.

GENERAL MORPHOLOGICAL AND PHYSIOLOGICAL CHANGES

Diverse genera of bacteria, fungi, algae, and protozoa are necessary for the normal growth and development of innumerable higher organisms, evolutionary modifications during the passage of time having made the latter dependent on processes or metabolites of the former. Despite this dependency, the ubiquity and ready dispersal of the microbial partner make less tenuous the continued existence of the non-self-reliant metazoan or rooted plant. Conversely, though microorganisms are often helpful, some unquestionably compromise the general well-being of their large companions.

Possibly the best approach, but not the sole one, for revealing a morphological or physiological response induced by a particular population involves a comparison of germfree and normal, sometimes called conventional, animals and plants. The preparation of germfree animals or plants is accomplished by a variety of techniques: surface sterilization of eggs or seeds with disinfectants, surgical removal of the young or of tissues that will propagate to give mature individuals, heating to destroy the resident microorganisms, or the use of massive quantities of antibiotics to bring about sterilization. The resultant germfree animal often develops more slowly than the conventional animal, as shown in studies of rats, cockroaches, and beetles. With

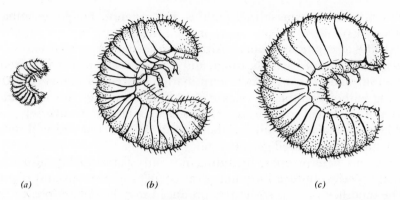

(a)   *(b)*   *(c)*

Figure 18.1   Effect of yeast symbionts on the bread bug *Sitodrepa panicea*. The larvae are all 10 weeks old. (*a*) Symbiont-free larva kept on a normal diet; (*b*) symbiont-free larva on highly enriched diet; (*c*) symbiont-containing larva on a normal diet (Buchner, 1965).

some insects bearing microsymbionts, the germfree individual is almost incapable of growth, despite the provision of a diet fully adequate for the normal animal; when the symbiont-free insect is infected, however, its rate of growth improves appreciably (Fig. 18.1). Corals harboring zoochlorellae as symbionts develop faster, too, when compared with those deprived of their algae. In like fashion, plants emerging from surface-sterilized seeds and cultured aseptically in a sterile environment containing abundant nutrients sometimes grow more rapidly, are taller, and weigh more when inoculated with selected bacteria or fungi than when maintained axenically. Trees characteristically bearing mycorrhizae are less vigorous in the field and look less healthy if their roots are devoid of fungal symbionts, while germfree orchid seedlings may not develop at all; successful inoculation overcomes these difficulties.

Conversely, a microbial population may have a dramatic deleterious influence. Growth of animals or plants, or of a single tissue or organ, is thus depressed when the individuals are exposed to parasites and even particular nonparasitic species. Either discrete inhibitory metabolites or generalized infections may be responsible for reduction in the rate or extent of development. Often the harmful organism is itself incapable of invading and colonizing the tissues it damages; thus germfree chickens and angiosperms, though their tissues may not be penetrated, are injured through actions of normal intestinal or root-inhabiting bacteria.

Seed germination may be under direct or indirect microbiological control. Seeds of specialized orchids, for instance, fail to germinate unless penetrated by the appropriate fungal symbiont or supplied with the associate's products. Although the significance in nature is unsure, strains of bacteria, fungi, and actinomycetes can synthesize in vitro substances that promote germination of seeds of plants typically living without symbionts (Koaze et al., 1957); the active compounds are not restricted to the classical auxins or vitamins. Conversely, soluble microbial metabolites elaborated in culture and soil will delay or totally prevent seed germination.

Various heterotrophs, including not only parasites but also free-living species, induce morphological changes in animals and plants. The modified or new structures produced may be deleterious to the organism, but they need not be. The kind and extent of the morphogenetic effect are governed by the interacting macro- and microorganisms and by prevailing environmental conditions, but the precise biochemical bases for the resulting upset of metabolism, differentiation, and regulation of physiological processes are not at all understood. Well known are the local inflammations initiated in vertebrates following bacterial penetration, a response that may be of profound importance in infection but one also elaborated by chemicals wholly unrelated to microbial metabolism. Research on germfree animals has disclosed the surprising fact that indigenous communities control a few of the morphological features of their usual vertebrate partners; thus certain species of germfree animals have a grossly enlarged cecum, a smaller spleen, a poorly developed lymphatic system, and a small intestine of different length and weight from that of their colonized counterparts (Luckey, 1965).

Assorted morphological responses occur, too, in plants colonized by free-living or parasitic organisms, or in plants exposed to discrete metabolites of these two microbial groups. Free-living bacteria may reduce or stimulate root elongation, inhibit or incite the formation of root hairs, favor the development of lateral roots, or deform root hairs. Occasional parasites excrete substances, some possibly identical with those formed in uninfected plant tissues, that cause minor or appreciable distortions in the invaded regions, and new structures often are thereby created that display either a degree of differentiation or exist as irregularly shaped masses with no structural organization.

Among the more thoroughly explored morphological alterations are those provoked by bacteria of the genus *Rhizobium* and strains of *Agrobacterium tumefaciens* as well as the fungal participants in mycorrhizae. Although nodules are induced on nonleguminous roots, presumably by actinomycetes, and on leaves, most interest has been centered on those caused by *Rhizobium*. These nodules are highly organized, well-differentiated structures that are self-limiting in size, and they usually benefit the host through the N_2 assimilated. Similar properties, of course, are possessed by nodules on nonleguminous spermatophytes. Galls or tumors resulting from excessive division of host cells are incited by many plant pathogens, like the fungus *Ustilago maydis*, but the so-called crown gall formed by *A. tumefaciens* on a multiplicity of dicotyledonous hosts is the only one that has been well characterized morphologically and biochemically. The bacterium initiates a series of events resulting in normal plant cells shifting to become tumor cells; once this change has been effected, the modified cells continue to grow and multiply, even in the absence of the causal bacterium, to give rise to a non-self-limiting, transferable gall in which the constituent cells proliferate at random and multiply excessively rather than following the well-regulated events of normal differentiation.

Numerous and diverse are the changes microorganisms bring about in the chemical composition of animals and plants. Pathogens clearly modify the chemistry of their hosts, and these alterations may be expressed in ways such as the following: increases or decreases in the activity of particular enzymes; shifts in the kinds and quantities of tissue proteins, amino acids, carbohydrates, and lipids; a rise or fall in sugar and organic acid levels; or the appearance of additional compounds of presumed or established host origin. Nonpathogens, too, influence the chemistry of organisms within or upon which they live; for example, the content of individual amino acids or vitamins and the amount of nitrogen in plants can be altered through microbial action, and comparative studies of germfree and conventional animals of a few species have shown that colonization affects the level of serum γ-globulin, the amount of free histamine excreted, and the metabolism of cholesterol and bile acids.

Critical physiological processes of higher organisms are stimulated or retarded as a consequence of colonization by parasites or free-living heterotrophs, or merely by the proximity of a metabolically active community. Depletion of O_2 by aerobes affects respiration and ion uptake of roots in soil and the functioning of animals resident in water, and intestinal bacteria modify the rate at which vertebrates consume O_2 and evolve CO_2. Some parasitic fungi and bacteria are known to reduce transpiration to such an extent that host plants wilt, while others diminish the rate of photosynthesis by destroying chlorophyll-containing tissue or by developing over leaf surfaces and thus interfering with light penetration. Moreover, toxins of gram-negative bacterial pathogens incite manifold physiological responses, such as stimulating the metabolism of sugar by mammalian cells. The permeability of cells, plant and animal, is partially or completely abolished by simple products or enzymes of parasites, and components of the damaged cells may then leak out; free-living heterotrophs living adjacent to roots can do the same kind of harm.

Of considerable interest and importance is the ability of selected populations to modify the reproductive behavior or contribute to sex determination of animals. For example, when deprived of the microsymbionts on which they rely, many insects are reproductively defective, and the ovaries of the symbiont-free insect may show distinct abnormalities or the females may no longer lay eggs. Sometimes eggs are laid but a factor essential for their development is lacking, a substance synthesized by the microsymbiont, and the eggs are nonviable. This is well illustrated by the ambrosia beetle, *Xyleborus ferrugineus*, the adults of which, though developing from germfree eggs, themselves lay solely nonviable eggs. Only when the females are colonized by the symbiotic fungus, *Fusarium solani*, are viable eggs made (Norris and Baker, 1967). Results of investigations with germfree lice and roaches also suggest that their symbionts synthesize compounds essential for reproduction of the insects. Furthermore, some algae infect marine copepods and prevent normal development of the gonads, and sterilization of females or castration of males by microbial parasites is not unheard of. Microsymbionts may be sex determinants as well, and the absence or presence of a definite microorganism may determine whether eggs develop into females or males (Buchner, 1954, 1955).

A well-recognized but still unexplained influence of microbial invasion is the production of fever. The disturbance in the body's temperature-regulating mechanism following infection by bacteria, fungi, and viruses seems to be attributable to a release of substances from the tissues that are injured. It is often argued that fever is a self-protective response to infection by pathogens, but this view is not supported by most available evidence. On the contrary, the elevation in body temperature may be quite deleterious (Bennett and Nicastri, 1960). Invasion by many microorganisms, however, is not a prelude to the immediate production of high fever.

MICROORGANISMS AS FOOD

Direct feeding on microbial cells is a prevalent habit. Marine and freshwater invertebrates and vertebrates, aboveground insects, subterranean soil animals, and major insect pests of trees are but a few of the microbe feeders. In some instances the sole significant nutrient source of the animal is the unique microbial population that is its symbiont, while in others the feeder consumes a broad range of microbial species and has no obligate relationships with any of them. Every major taxonomic group of microorganisms contains representatives that thus serve to maintain animals in nature: aquatic algae, fungi inhabiting woody tissue and soil, yeasts proliferating on decaying fruit, and bacteria and protozoa of soil and water. The associations about which most is known are aquatic invertebrates and vertebrates feeding on free-living algae and bacteria; aquatic animals that digest their algal symbionts; airborne insects consuming free-living bacteria, yeasts, and algae; members of the soil fauna devouring bacteria, fungi, actinomycetes, and protozoa; and beetles and ants that cultivate and gain nutrients from specific fungal populations.

Aquatic algae are the food for herbivorous invertebrates of oceans and lakes, species of fish, and ephydrid flies visiting the effluents of hot springs. A majority of marine and freshwater metazoa probably live not on dissolved nutrients but rather by filtering out and preying on particulate materials, much of which is in the form of living organisms. This grazing may be intense and occasionally leads to an abrupt decline in a phytoplankton bloom. The small algal biomass in waters that might be assumed to be highly productive seems to be determined by the rate of grazing, and this constant feeding may be so marked that the algal residents hardly appear to be growing. Should the grazing exceed the rate of algal multiplication, a patch with few algae may be created, whereas the absence of grazers a short distance away permits explosive algal development; situations of this sort, not uncommon in nature, reveal themselves in irregular distributions of metazoa and of the algae they devour. The perennial seasonal fluctuations in algal abundance often cause differences in grazing pressures with the time of year. Most viable algae that are eaten seem to be consumed in the surface waters of the oceans, but a portion of the algal community and partially decayed cellular matter sinks downward, only to become nutrients for herbivores in the water column or at the bottom. The enormous significance of algae as a food source is accepted by marine scientists, who believe that most oceanic algae die not from old age but instead meet their demise through activities of the grazers (Raymont, 1963).

Among the principle aquatic feeders on algae are copepods, cladocerans, euphausids, appendicularians, and the larvae of many metazoan orders (Fig. 18.2). Copepods, important components of the

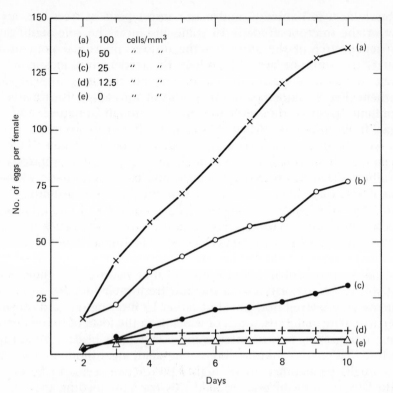

Figure 18.2 Egg laying by a copepod fed on different concentrations of *Skeletonema* (Marshall and Orr, 1964).

aquatic fauna, are selective grazers that consume immense numbers of diatoms and photosynthetic flagellates, not rarely causing the sudden termination of an algal bloom. Cladocerans are quite effective in devouring algae, too. Small crustaceans generally fare excellently on an algal diet, and they in turn may be consumed by aquatic carnivores. Scores of species of freshwater and marine fish, as well as insect and oyster larvae, eat filamentous or unicellular algae. On the other hand, the relative importance of aquatic bacteria as food for metazoa has not been thoroughly assessed, but they might well be preyed on by protozoa that in their turn are eaten by metazoa. In waste waters, however, bacteria as well as fungi and algae are reportedly utilized by the prevalent insects.

By contrast with free-living photoautotrophs, scant information has been collected regarding how algae entrapped and living symbiotically within invertebrates benefit their partners. On occasion, however, the algae die and are digested by their animal hosts; hence they are, in part at least, a food source. For example, freshwater sponges and the giant clam *Tridacna* destroy and digest algal cells contained within their bodies and presumably thereby gain considerably. The marine turbellarian *Convoluta roscoffensis* likewise assimilates constituents from those of its microsymbiotic algae that the animal kills (Fritsch, 1952; Yonge, 1944).

The innumerable soil populations have attracted great attention owing to their role in plant nutrition; yet, despite the extensive literature, the nutrition of animals grazing on the subterranean communities of microorganisms has been largely ignored. Nematodes are important terrestrial invertebrates, and they are able to devour bacteria, fungal hyphae and spores, protozoa, and algae. It has been proposed that nematodes in a hectare of land may consume each year 800 kg of bacterial cells (Overgaard Nielsen, 1949, 1961), grazing of a magnitude that undoubtedly is of consequence. Soils also are populated by large numbers of mites, collembolans, and enchytraeids that consume both fungi and bacteria, and although quantitative data are lacking, it is reasonable to assume that their feeding affects considerably the composition, density, and activity of soil microorganisms. Frequently invertebrates are seen to ingest humus or particulate plant remains, but the prime nutrient source may be the adhering microorganisms rather than the inanimate organic materials.

Insects of many kinds depend for their existence on heterotrophs. Species of *Drosophila*, for example, seek out and subsist on communities of yeasts and bacteria colonizing carbohydrate-rich plant materials, and mosquito and fly larvae similarly derive nutrients from bacteria and yeasts inhabiting the same regions. Furthermore, fascinating symbiotic relationships have developed between insects and the specific fungi on which they feed. Many of the insects concerned are wood inhabitants that damage living trees and felled timber. The ambrosia beetles, as one illustration, bore into wood but do not feed appreciably on it; instead they subsist largely or completely on the fungi the adults carry and introduce into the insect-built tunnels. Larvae as well as adult beetles consume the domesticated fungi, which include strains of *Ceratocystis, Leptographium, Botryodiplodia, Monilia,* and *Endomycopsis* (Baker, 1963). In the symbiosis involving leaf-cutting ants of the genus *Atta,* the insects shred and arrange leaves in heaps, on which distinctive fungi grow. The leaves have nutrients for the fungi, and the hyphae borne and cultivated by the insects proliferate and are utilized as food by the ants (Hartzell, 1967). Maintenance of such fungus farms apparently requires considerable skill, since, once abandoned, they become overgrown by other species and disappear.

METABOLITES INFLUENCING ANIMALS AND PLANTS

Microorganisms, as they multiply, synthesize compounds that may have an impact on plants and animals. The metabolites are sometimes intracellular, but a high percentage are excreted into the surroundings during active proliferation. Moreover, intracellular metabolites are released from the individual that produced them when the cell autolyzes or is attacked and the contents leak out. The compounds are numerous, vary in potency insofar as higher organisms are concerned, and range in complexity from the very simple to the remarkably complex. In the preceding section morphological and physiological changes induced by microorganisms were considered, but the specific products responsible for the responses discussed are generally not yet known. Still, through careful research, part of it old but much of recent date, a clear picture is gradually emerging of the types of microbiologically generated compounds that influence the growth, health, and vigor of macroorganisms. These products include toxins of several sorts, carbohydrates and organic acids or nitrogenous compounds that are carbon or nitrogen sources for animals, essential or stimulatory growth factors, plant growth regulators, estrogens, and substances detected by the sense of taste or smell.

Inasmuch as microorganisms acquired a reputation for doing harm long before the good they do was widely recognized, it is not surprising that considerable information has been collected on toxins they elaborate. These inhibitory metabolites may be formed by a population colonizing an animal or plant, as commonly occurs during certain diseases, but many are synthesized by organisms residing completely apart from the species harmed. Only the latter type of inhibitors is considered at the present time; those made by colonists are discussed at the end of this chapter. The microorganisms presently of concern do not invade the tissues of viable organisms and are not infectious. The products are made, usually by ubiquitous species, outside the plant or animal body when proper conditions exist for growth of the population and elaboration of the toxin, and the inhibitors have an effect only when they come into contact with a suscept or, with animals, when they are ingested. Many toxins of importance are products of noninfectious fungi and algae, and they not surprisingly are designated *mycotoxins* and *phycotoxins*. Noninfectious bacteria also are responsible for several highly potent poisons.

A list of species and genera synthesizing toxins effective against animals is given in Table 18.1. Not all isolates of each species or genus tabulated make the harmful metabolite, and closely related strains vary enormously in the quantity of the individual compounds they form. Susceptibility of the animal species differs considerably, and one may be wholly unaffected by a compound lethal to a second. The active principles include proteins, cyclic peptides, alkaloids, organic acids, lactones, and diverse aromatic compounds. Even H_2S can be lethal; the considerable amounts of H_2S appearing in flooded soil may kill terrestrial invertebrates, and that showing up in water during anaerobic decomposition of algal blooms or organic pollutants may be a major contributor to the death of fish. Massive mortalities of aquatic animals have also been attributed to the bacterial production of sulfuric acid in sulfur-rich coal seams and the entry of the acid mine drainage into adjacent streams and rivers. Although mushrooms like *Amanita* have been known for centuries as causes of acute poising, only recently have dissimilar fungi been recognized as sources of animal and human intoxication. Foodstuffs are colonized by fungi of genera not previously considered harmful, and feed prepared for domesticated animals, forage plants in the field, and, on rare occasions, commodities consumed by man thus become tainted with highly potent substances. The potential outcome of such fungal colonization is now appreciated by agricultural scientists, but much less so by physicians. One of the most dramatic instances of fungal intoxication took place in the Soviet Union during three winters of World War II, at which time species of *Fusarium* and *Cladosporium,* among others, grew on cereal grains remaining in the field unharvested owing to the hostilities. The toxin producers proliferated on the cereals despite the fact that snow covered the plant remains. Because of the near-starvation conditions, people in the affected district ate the moldy grain, and thousands developed symptoms of acute intoxication and many died (Joffe, 1965). Two species of bacteria, *Staphylococcus aureus* and *Clostridium botulinum,* multiply in food and there elaborate proteins that, in the latter instance, affect the nervous system or, with *S. aureus,* provoke inflammation of the stomach and intestinal lining.

Genera of blue-green algae such as *Anabaena* and *Microcystis,* dinoflagellates like *Gymnodinium* and *Gonyaulax,* and Chrysophyta like *Prymnesium* contain species implicated in phycotoxin biosynthesis. The problem of algal toxins is not new, as witnessed by the

Table 18.1

Representative Species Producing Compounds Toxic to Animals

Microorganism	Habitat of Microorganism	Susceptible Animal	Type of Toxin	Reference
		Fungi		
Amanita	Soil	Man	Cyclic peptide	Wieland (1968)
Aspergillus	Peanut, cottonseed	Turkeys, pigs	Aflatoxin	Ciegler and Lillehoj (1968)
Aspergillus, Penicillium	Corn feed	Pigs, cattle	Several kinds	Bamburg et al. (1969)
Claviceps	Pasture grasses	Cattle, sheep	Ergot alkaloid	Brook and White (1966)
Fusarium, Cladosporium	Unharvested grain	Man	Steroidal compound	Joffe (1965); Bamburg et al., (1969)
Penicillium, Aspergillus	Stored rice	Man	Complex aromatic	Bamburg et al. (1969)
Pithomyces chartarum	Dead pasture plants	Sheep, cattle	Sporodesmin	Brook and White (1966)
Rhizoctonia leguminicola	Clover hay	Cattle	Indolizidine	Aust et al. (1968)
Rhizopus nigricans	Millet	Man	—	Narasimhan et al. (1967)
		Algae		
Anabaena flosaquae	Fresh water	Ducks, gulls	—	Gorham (1964)
Gonyaulax	Marine water	Man	Purine derivative	Shilo (1967)
Gymnodinium	Marine water	Fish, mammals	—	Shilo (1967)
Microcystis aeruginosa	Fresh water	Waterfowl, livestock	Cyclic peptide	Gorham (1964)
Prymnesium parvum	Brackish water	Fish	Protein-containing	Shilo (1967)
		Bacteria		
Anaerobes	Soil	Nematodes	Butyric acid	Hollis and Rodriguez-Kabana (1966)
Anaerobes	Soil	Nematodes	H_2S	Rodriguez-Kabana et al. (1965)
Clostridium botulinum	Improperly canned food, raw meat	Man	Protein	Lamanna and Carr (1967)

statement in Exodus (7:20–21), "All the waters of the river were turned to blood. And the fish in the river died, the river stank, and the Egyptians could not drink the river water." Blue-greens flourishing in fresh water sometimes cause the death of birds, fish, and mammals, typically following drinking of water containing heavy algal growths. Dinoflagellates produce toxins that may cause massive mortalities of fish, dolphins, turtles, and other marine animals, and shellfish consuming *Gonyaulax* become tainted with the dinoflagellate toxin and, though themselves unaffected by the compound, pass it on to humans eating the shellfish, often with disastrous consequences (Gorham, 1964; Shilo, 1967).

Inhibitors derived from free-living heterotrophs may have a spectacular influence on higher plants, though, with the exception of H_2S and organic acids, the detrimental agents are quite different from those injuring animals. The proximity of rhizosphere bacteria and fungi to roots magnifies the potential impact, even when the toxin itself is readily biodegradable and hence has a short life. Indeed, several physiological disturbances exhibited by plants have been ascribed to low-molecular-weight metabolites made outside living plant tissues by noninvasive bacteria and fungi. Among the inorganic products H_2S stands out because this toxin not only is deleterious at low concentration but is produced abundantly in anaerobic regions supporting a sizable heterotrophic community. The roots of rice growing in flooded soils containing insufficient iron to bind all the H_2S as ferrous sulfide may thus be damaged (Takai and Kamura, 1966). Roots extending into waterlogged or poorly drained sites may likewise be exposed to detrimental levels of manganous ions or nitrite arising from anaerobic reduction of tetravalent manganese or nitrate. Nitrite, a potent toxin, may also accumulate in alkaline soils receiving ammonium fertilizers, because ammonium at high pH, or the free ammonia that exists under those conditions, selectively suppresses *Nitrobacter* and related bacteria that usually oxidize the nitrite generated by ammonium-oxidizing autotrophs, a toxicity also considered in Chapter 13. In addition, the precipitous fall in pH linked with the oxidation of sulfide ores destroys much of the vegetation along the banks and in streams receiving acid mine drainage.

An appreciation of the diversity of organic inhibitors that are made in nature is just now being gained. It has been known for some time that culture filtrates of common, noninvasive soil and rhizosphere fungi and bacteria contain toxins affecting rooted plants. Several of the compounds have been identified, and included in the list are fatty acids, simple alcohols, oxalic acid, phenolic substances like phenylacetic and 3-phenylpropionic acids, polysaccharides, and, infrequently, antibiotics. Yet even though a high percentage of the subterranean microflora is capable of generating these chemicals in vitro, a question always existed in the mind of the ecologist inasmuch as evidence for a role for such toxins in nature was lacking. Recent research has disclosed, however, that under certain conditions organic toxins are formed, and the compounds, though suitable as substrates for members of the indigenous community, persist for sufficiently long periods to make their presence felt. Insofar as present knowledge goes, the necessary conditions are met in three kinds of habitats: (a) flooded soils, where simple inhibitors accumulate and remain for some time on account of the prevailing anaerobiosis; (b) sites in immediate proximity to accumulations of decomposing plant litter, wherein the substrate level is high, the community dense, and the O_2 level low; and (c) regions where a particular crop or orchard plant has been cultivated continuously and yields remains that, on decay, are converted microbiologically to injurious products. The compounds so far identified in such soils include butyric, acetic, p-hydroxybenzoic, p-coumaric, vanillic, and protocatechuic acids, methanol, ethanol, the antibiotic patulin, and a few cyanogenic substances.

Organic compounds harmful to plants are important in many aspects of agriculture. Thus decomposition of crop residues shows up as a real problem when roots extend up to these remains and are therefore exposed to the excretions of large numbers of bacteria metabolizing the wealth of carbonaceous substrates. In fields where fruit trees have been grown for many years, progressive decreases in fruit production and difficulties in establishing seedlings are occasionally encountered. These problems are associated with toxins arising in the decomposition of plant constituents. A well-known case is the peach orchard. Peach trees are rich in amygdalin, and this innocuous compound is hydrolyzed jointly by members of the soil community and plant enzymes to two inhibitors, HCN and benzaldehyde, that contribute to root degeneration. A similar situation is encountered in apple orchards and in nurseries established on land where apples were previously cultivated; in this instance it is the phloridzin in the apple root that is degraded to toxic phenolic compounds (Patrick et al., 1964).

The microbial world teems with species of enormous benefit to animals and plants. Prominent among the helpful populations are those providing their neighbors with carbonaceous nutrients like carbohydrates and organic acids. The growing heterotrophic orchid, dependent for nutrients on the microbial partner, probably obtains

Table 18.2

Production of Volatile Fatty Acids in the Rumen [a]

Animal	Ration	Molar Percentage of Acids		
		Acetic	Propionic	Butyric
Cattle	Pasture	60.2	23.3	16.5
Cows, lactating	Hay and concentrate	63.3	20.1	16.6
Sheep	Hay	60.4	26.8	12.8
Steers	Hay, grain	59.5	19.1	21.4

[a] From Hungate (1966).

carbohydrates and other factors from its mycorrhizal fungus, and fungi probably supply sugars to permit germination of the orchid seeds. As stated previously, organic products made during photosynthesis by algae living symbiotically within invertebrates are assimilated by the animal symbionts. Glycerol stands out among the metabolites provided in this way to the invertebrate (Muscatine, 1967).

A delicate symbiosis is maintained between bacteria or protozoa, on the one hand, and numerous metazoa, on the other, one in which the microbial community functions essentially as the initial agent of animal digestion. Animals of various genera are themselves incapable of initiating the digestion of foods they ingest, relying instead on microorganisms to predigest the food and transform it to assimilable molecules (Table 18.2). Though the bulk of the diet of herbivores is composed of cellulose, hemicelluloses, and lignin, these plant constituents would be totally nonutilizable were it not for the anaerobic bacteria and protozoa borne in the animals' alimentary tracts. The herbivore is, from the vantage point of the microbiologist, a mobile fermentation vessel, an organism whose success in nature is due in no small way to the bacteria and protozoa depolymerizing the otherwise unavailable polysaccharides in the forage or wood.

Because domesticated ruminants play such a major part in agriculture, the symbiosis involving them has been thoroughly explored. The ruminants—cattle, sheep, goats, deer, antelopes, moose, and so on—get their food by grazing; yet they themselves possess few or none of the hydrolytic enzymes necessary to catalyze the conversion of plant polysaccharides into the simple kinds of molecules that can be absorbed. In the rumen, the most capacious of the four compartments of the ruminant's stomach, a heterogeneous and highly active community of bacteria and protozoa exists, one able to transform the polysaccharides to simpler, usable compounds. The end products of the fermentation are volatile fatty acids, CO_2, and CH_4. The gases are voided and the acids—in reality, wastes of the rumen anaerobes—are absorbed by the host. It has been estimated that about 70–85% of the feed, discounting its water content, is digested in the rumen, an activity attributable to microbial rather than animal enzymes (Gray, 1947). A portion of the volatile fatty acids is absorbed by way of the rumen wall and then enters the bloodstream, while a part passes into the omasum, another subdivision of the stomach, from which considerable assimilation also occurs.

Microorganisms live symbiotically not only in ruminants but also in other foraging animals. Here they serve an analogous function as initial digestive agents. In horses, elephants, and certain rodents, the cecum serves as the site of the microbial fermentation of polysaccharides. In field mice, hamsters, and some herbivorous marsupials it is the stomach that harbors the bacteria and protozoa (Hungate, 1966).

Flagellated protozoa residing in the alimentary tract of wood-eating termites behave in much the same way as the rumen anaerobes, cleaving cellulose to products of little or no use to the protozoa but crucial for the existence of the termite. The simple compounds released by the flagellates are absorbed and oxidized by the host. A termite can be freed of its symbionts by exposure to high O_2 tensions, but then the wood that is its dietary staple, though eaten, no longer supports it, and the termite must be fed simple, unnatural nutrients lest it perish in the midst of plenty. Reinfection of the defaunated insect with the flagellates restores the capacity to live on complex components of wood. Some roaches also consume wood, and their alimentary tracts likewise swarm with flagellates that hydrolyze cellulose into fragments assimilable by the insects (Buchner, 1965). In addition, anaerobes in the alimentary tracts of birds and beetle larvae may be initiators of cellulose digestion in an entirely analogous fashion. Birds of the genus *Indicator,* commonly called honeyguides, derive benefit from their specialized intestinal symbionts, but the microorganisms, strains of the bacterium *Micrococcus cerolyticus* and the yeast *Candida albicans,* in this instance aid the birds in their peculiar habit of feeding avidly on beeswax, a habit that would surely have led to indigestion were it not for the wax-cleaving powers of the microsymbionts (Friedmann, 1967). A few insects, moreover, maintain their symbionts externally, as with *Melittoma insulare,* which introduces into coconut wood a bacterium-yeast mixture that makes the plant tissues suitable as a nutrient for the insect (Brown, 1954).

The beneficial influence of the N_2 fixed in root nodules has already been considered. Heterotrophs in leaf nodules and blue-green algae in nodulated cycads, the fern *Azolla,* and the liverwort *Blasia* likewise must be useful to their hosts when the plants are growing in nitrogen-deficient soils, and the N_2-fixing symbiosis might give the plant a distinct competitive advantage over its non-nodulated neighbors in regions of nitrogen insufficiency. The N_2 fixed by free-living blue-greens in flooded paddy fields appears to become ultimately available, after its mineralization, for utilization by rice (Relwani and Subrahmanyan, 1963). As yet undefined is the ecological significance of the enormous numbers of *Beijerinckia* and *Azotobacter,* excellent N_2 fixers in the laboratory, that live on the foliage and roots of selected tropical plants, sites where the exudates may contain enough of the simple substrates needed by these bacteria for active N_2 fixation.

The microbial cell is crammed with vitamins, amino acids, and additional growth factors that are essential or stimulatory to higher organisms. These compounds either are excreted during the phase of active proliferation or are released upon death and lysis of the cell. The compounds thus liberated provoke favorable responses in many metazoa inhabited by microsymbionts, and plants, too, apparently derive value on occasion. An animal's nutritional needs increase as its capacity for biosynthesis of requisite metabolites diminishes, and animals living on a diet free of an essential growth factor they cannot synthesize quite often are colonized by microorganisms that overcome the inadequacies of the host. Rodents, birds, ruminants, and man thereby obtain a small or an appreciable number of essential compounds from the inhabitants of their stomachs or intestines. Bacteria colonizing different insects are well known to synthesize essential metabolites, an activity that is especially striking among insects feeding on wood and specific animal tissues, which are quite deficient insofar as the nutrition of the insect is concerned.

A summary of essential animal growth factors that are supplied by microorganisms is presented in Table 18.3. Several approaches have been used to gain evidence for these needs and to demonstrate the ability of communities in the gastrointestinal tract to supplement a deficient diet. A common procedure is to compare the requirements of germfree and conventional animals; for example, germfree rats and chickens or insects developing from sterile eggs have requirements not possessed by their conventional counterparts. Alternatively, the germfree but not the normal animal shows classical deficiency symptoms on a diet lacking in a particular vitamin or amino acid, but growth of the former is better and its symptoms disappear when the factor is added to the diet or when the germfree animal is inoculated

Table 18.3

Growth Factors Required by Animals and Synthesized by Inhabitants of the Alimentary Tract[a]

Growth Factor	Animal
Biotin	Rat, cow, insects
Folic acid	Rat, chicken, insects
Nicotinic acid	Cow, insects
Pantothenic acid	Cow, insects
Pyridoxine	Cow, insects
Riboflavin	Cow, insects
Thiamine	Cow, chicken, insects
Vitamin K	Rat
Carnitine	Insects
Choline	Insects
Sterol	Insects
Amino acids	Ruminant, rat, insects

[a] Based on Fraenkel (1952), Gustafsson (1959), Hungate (1966), Koch (1963), Luckey (1965), and Virtanen (1963).

either with a bacterium synthesizing that substance or with a mixture of populations. Sterile insect larvae often fail to develop normally on certain foods, but they are healthy if the diet is supplemented with the particular growth factor or with the appropriate protozoan, bacterial, fungal, or yeast symbiont. In some instances evidence for a microbial contribution is first suggested by the stimulation of growth when a microorganism is inoculated into germfree animals reared on rations deficient in a given vitamin. In nature these metazoan species seemingly do without external sources of the vitamins or amino acids because they are fortunate enough to have associated with them a community that makes up for their metabolic inadequacies. Surprisingly, blood has a suboptimal concentration of B vitamins for the well-being of some bloodsucking insects, but their microsymbionts compensate for the deficiencies in the insect's food source (Fraenkel, 1952). Man, too, apparently relies on the bacteria of his intestine for at least part of his requirement for vitamin K, biotin, pyridoxine, and vitamin B_{12}. Though little is known about the contribution of individual microbiologically synthesized amino acids, by contrast with B vitamins, the frequent finding that animals grow normally on diets deficient in amino acids they themselves are unable to make indicates a key role for the stomach or intestinal inhabitants.

Angiosperms may sometimes be unable to synthesize vitamins at an optimum rate; at least this is a reasonable explanation of data revealing that vitamin additions enhance the growth or a particular phase of plant development. Furthermore, studies with excised root cultures show that roots cannot form a number of B vitamins and amino acids, and it is not unreasonable to assume that natural roots may be deficient in compounds of these sorts. Moreover, plants can assimilate intact vitamin and amino acid molecules and are reported to benefit thereby. Growth factor excretion or release by populations of the rhizosphere, or even of leaf surfaces, might therefore have a considerable effect on rooted plants. Evidence for such a microbial function in nature, however, is wanting.

Plant growth regulators and related compounds are produced in vitro by free-living, parasitic, and mycorrhizal fungi as well as by bacteria isolated from soil, rhizosphere, phyllosphere, and diseased plant tissues. Indoleacetic acid, other auxin-like substances, gibberellins, and gibberellin-like compounds have now been isolated or their activities found in culture filtrates of a high percentage of heterotrophic species dwelling in environments where they might affect plant growth. Auxins are in fact present in soil, generated there by the indigenous microflora as they decompose carbonaceous matter. The absence of a body of experimental data establishing that growth-promoting substances occur in the rhizosphere or phyllosphere has not dampened speculation that such compounds are synthesized near roots or on foliage, there to increase growth, nutrient uptake, or crop yield. Still, simple organic molecules, some possibly precursors of the growth-promoting agents, are exuded from roots and leaves in nature, and these may be directly converted to auxins, gibberellins, or molecules with related effects; alternatively, the heterotrophic colonists on above- or below-ground tissues may cause the underlying plant cells to produce excessive quantities of growth regulators. Though the precise cause remains obscure, evidence exists that sterile plants contain less auxin than normal ones and that inoculation of the former results in a rise in auxin level (Libbert et al., 1969).

Invasive fungi are commonly believed to elaborate growth regulators in plant tissues they penetrate. Mycorrhizal fungi, for example, readily make indoleacetic acid in vitro, and as this compound causes histological modifications akin to those observed in mycorrhizae, it has been implicated in the induction of morphological as well as some physiological modifications attendant on mycorrhiza formation and function. Virulent pathogens frequently are potent producers of indoleacetic acid, and parasitized tissues not uncommonly contain more than nondiseased tissues. Yet, granted that the pathogenicity of these strains can often be correlated with, if not explained by, the auxin yield, the greater quantity of indoleacetic acid in infected tissues could be accounted for in an entirely different way, namely, by assuming that the pathogen excretes a chemical provoking the host to produce abnormally large amounts of its own growth promoters. A stronger argument can be made for the microbial biosynthesis of gibberellins in invaded tissues, particularly in the case of rice infected by *Gibberella fujikoroi*. Rice parasitized by this fungus exhibits all the symptoms of the presence of a gibberellin, and in fact it was from research on this very malady that the existence of gibberellins was first demonstrated (Wood, 1967).

Fungi proliferating on feed or on pasture plants elaborate metabolites that act as estrogens on the herbivore grazing on the moldy substance, and often swollen vulvas and mammary glands or abortion of the young may follow. Corn feed and forage legumes have been implicated in such disorders, and species of *Leptosphaerulina* and *Pseudopeziza* are among the fungal incitants. Moreover, plant pathogens may aggravate a problem already existing in forages possessing estrogenic constituents (Bamburg et al., 1969; Brook and White, 1966).

Numerous other compounds excreted by microorganisms or liberated upon cell lysis alter the morphology or physiology of plants and animals. For instance, an enzyme presumed to be characteristic of a higher organism may in reality be a product of its indigenous microflora, as with the urease found in association with the gastric mucosal tissue of rats and dogs (Delluva et al., 1968). Some enzymes, though indeed of host origin, are synthesized only in response to structural components of the microbial cell (Lyle and Jutila, 1968) or of its excretions. An attribute the practicing microbiologist encounters so frequently that he usually overlooks it must not be forgotten: microorganisms commonly generate odors. The smell of freshly turned earth is most likely the result of a *Streptomyces* metabolite, the odor of the human armpit seems to come in part from bacterial diphtheroids native to the skin, and the pungent aroma of putrefying vegetable matter and feces is assignable to the anaerobic bacteria contained therein. One of the objectionable features of eutrophication is surely the physical presence of the algae, but even more disagreeable are the smells emanating upon their death and subsequent anaerobic decomposition.

EFFECTS ON MINERAL NUTRITION
AND METABOLISM

Microbial communities regulate the availability, uptake, and at times the metabolism of inorganic nutrients by plants. There are clear lines of evidence pointing to a similar role for microorganisms in the metabolism of mineral nutrients by animals, and further inquiry will undoubtedly reveal additional ways by which the microflora or microfauna influences uptake and metabolism of inorganic nutrients by metazoa. Inasmuch as the major mechanisms by which subterranean communities govern the availability of soil-borne nutrients to plants were reviewed in the preceding two chapters, only facets pertaining directly to nutrient uptake will be considered here.

Mineralization and immobilization are among the principal means by which terrestrial populations regulate the availability of elements essential for plants. By degrading organic compounds and releasing the nitrogen as ammonium and nitrate, sulfur as sulfate, and phosphorus as phosphate, bacteria, fungi, and actinomycetes determine the extent of development of the vegetation since roots probably assimilate very little of these elements when they are bound in organic complexes. Conversely, inasmuch as each of these, as well as most plant nutrients, is essential for the underground heterotrophs, the assimilation of the elements by bacterial cells or fungal and actinomycete hyphae has detrimental and at times disastrous consequences for the vegetation. Such a competition between microorganisms and rooted plants for common needs is universal, but it is particularly critical in the case of nitrogen, and occasionally phosphorus. The competition is readily demonstrable in solution culture, in soil in the greenhouse, and in the field, revealing itself in a lower rate of uptake by roots in solution culture containing microorganisms than in a system supporting aseptically grown plants, provided that the nutrient level in the liquid is too low to satisfy the over-all demand (Barber, 1966), or in soil when the requirement of the microbial community for inorganic nutrients is raised as a result of the addition of carbonaceous substrates.

Bacteria and fungi, by mechanisms other than mineralization and immobilization, appreciably affect the uptake of inorganic ions by roots, but rarely is the biochemical basis for this influence established. The effect may be exerted on the nutrient element itself, as by an oxidation or reduction or by a change in its solubility, but frequently an alteration neither in the oxidation state nor in the solubility of the element can account for the results; rather, the heterotrophs seem to modify the physiology of the plant in some undefined way. In comparisons of seedlings grown axenically and in the presence of defined or unknown populations, the results obtained show that some bacteria and fungi enhance the uptake of phosphorus, potassium, and sulfur, while dissimilar heterotrophs diminish the absorption of calcium, phosphorus, rubidium, strontium, and zinc (Table 18.4). The responses vary with the microbial species, the identity of the plant, and environmental conditions.

Invasion of a plant by pathogens or symbionts also alters the pattern of uptake. A notable benefit following root penetration by mycorrhizal fungi is the enhanced assimilation of phosphate, in particular, but also of potassium and nitrogen, and though part of this increased absorption is ostensibly attributable to the greater surface area of mycorrhizal than of nonmycorrhizal roots, the former unquestionably are more efficient in ion uptake than the latter; in this manner the symbiotic fungi are uniquely important in the mineral nutrition of both trees and nonwoody plants (Harley and Lewis, 1969). Subterranean free-living, symbiotic, and parasitic microorganisms also modify translocation of the element assimilated. The formation of toxins in and around roots, the excretion of plant growth regulators, and the inhibition of root or root hair development have all been proposed as the basis for the microbiologically induced changes in nutrient absorption and translocation, but final explanations must await additional critical research.

When performed in the rhizosphere itself, several of the reactions considered in the preceding two chapters have a direct bearing on plant development and vigor. The dense rhizosphere community must place a drain on the O_2 supply in the root environs, and use of the O_2 and the concomitant lowering of the E_h increase the availability of nutrients like iron and manganese. Consumption of O_2 in this ecosystem simultaneously makes conditions suitable for denitrification, the responsible bacteria nearly always being numerous on the root surface, and, given the coincidence of an anaerobic microhabitat, root exudates, and nitrate, bacterial reduction of nitrate to N_2 will diminish the quantity of available nitrogen (Woldendorp, 1962). The CO_2 evolved by the rhizosphere community gives rise to carbonic acid; organic acids are presumably produced, too, and the carbonic and organic acids conceivably dissolve and make available insoluble nutrients such as phosphate and potassium. Con-

Table 18.4
Microbial Enhancement and Diminution of Ion Uptake by Plants

Ion	Effect on Up-take by Plant	Plant	Microorganism Used	Reference
Calcium	Reduced	Clover	Soil organisms	Trolldenier and Marckwordt (1962)
Phosphorus	Increased	Maple	Soil organisms	Akhromeiko and Shestakova (1958)
	Reduced	Tomato	Trichoderma	Subba-Rao et al. (1961)
Potassium	Increased	Wheat	Bacteria	Aleksandrov and Zak (1950)
Rubidium	Reduced	Clover	Soil organisms	Trolldenier and Marckwordt (1962)
Strontium	Reduced	Corn	Aspergillus	Frei (1963)
Sulfur	Increased	Maple	Soil organisms	Akhromeiko and Shestakova (1958)
Zinc	Reduced	Tomato	Trichoderma	Subba-Rao et al. (1961)
	Reduced	Corn	Soil bacterium	Ark (1936)

versely, manganese oxidizers are known to aggravate a soil deficiency for this element as they convert the soluble manganous ions to the unavailable, insoluble manganic oxides.

By assisting or causing calcification and decalcification, algae and bacteria affect the mineral composition of animals. The algae of concern, not themselves calcium accumulators, live symbiotically with the coelenterates characteristic of coral reefs in tropical marine waters. A prominent trait of these coelenterates is their calcareous skeleton, a structure that seems to be essential in protecting the reef from constant battering by waves. The zooxanthellae residing within the animals apparently increase, in some undefined manner, the calcification process and building of the skeletal matrix, and thus indirectly contribute to the deposition of vast amounts of calcium in the oceans (Goreau, 1961). The bacteria of concern are those participating in the first, crucial step in formation of dental caries, namely, the fermentation of simple carbohydrates within the microbiologically synthesized, polysaccharide-rich films adhering to the tooth surface. The acids generated in this fermentation cause decalcification and dissolution of the structural integrity of the tooth enamel.

CHANGES RESULTING FROM METABOLISM
OF O_2 AND CO_2

Metazoa and rooted plants respond to the O_2 produced by algae during photosynthesis and to consumption of the gas by aerobic populations. They also may be affected by the CO_2 evolved in heterotrophic metabolism and in the respiration of aquatic photoautotrophs. Spectacular responses of animals to O_2 generation and depletion are evident in lakes, rivers, and streams. In waters populated by reasonable numbers of algae, the dissolved O_2 concentration rises during the daytime, when conditions favor photosynthesis, and the water may even become supersaturated. Although algal respiration proceeds in sunlight, the amount of O_2 used in respiration is small by comparison with the concomitant release; upon the cessation of photosynthesis after sunset, however, algal consumption without evolution of O_2 brings about a decline in the concentration in water. Aerobic heterotrophs accompanying the photoautotrophs contribute to the O_2 diminution to a lesser or greater degree, depending on the age of the algal mat and its susceptibility to decay. In waters lacking algae but receiving organic pollutants and in soils where biological O_2 generation is usually inconsequential, microbial communities invariably bring about O_2 consumption, which is compensated for solely by diffusion from the overlying air.

Fish show quite dramatically what microbial metabolism of O_2 can do to vertebrates. Occasional fish kills have been ascribed to O_2 supersaturation accompanying the photosynthetic activity of dense algal masses, but far more often, especially in recent times with ever-increasing water pollution, the death of immense numbers of fish and other freshwater animals is prompted by the consumption of dissolved O_2 either by algal respiration or by heterotrophs decomposing an algal bloom or organic pollutants. Invertebrates able to tolerate low O_2 levels may assume dominance, but a drastic upheaval in the faunal community and mass mortalities are the inevitable outcome of total O_2 depletion.

Whenever algae grow in O_2-deficient waters, they may be helpful to their animal neighbors. Free-living algae, by providing this gas, may allow for growth where survival was previously not feasible, or they may permit an organism to endure until the resumption of favorable conditions. Such beneficence is not limited, and ciliates and coelenterates containing algal endosymbionts possibly share in the wealth so distributed. Plants rooted in soil subjected to low O_2 tension may profit, too, but because algae need both water and light in order to carry out photosynthesis, the spermatophytes that obtain the advantage are largely those in flooded soils, which often bear luxuriant algal blooms; here the chief plant of economic concern that stands to gain from the oxygenation is rice. Conversely, rooted plants do not tolerate the anaerobiosis resulting from the aerobes' deoxygenation of alga-free marshy or wet areas, and excessive deoxygenation is a prelude to eliminating higher plants, the roots of which demand O_2. The deoxygenation need not be complete, for microbial consumption of a goodly part of the O_2 in soil pores reduces ion uptake by roots, and the aboveground portions may then contain less nitrogen, phosphorus, potassium, calcium, and magnesium than they would have had were aeration adequate. At the same time the shift from an aerated to an anaerobic status favors the microbial biosynthesis of H_2S, and phytotoxic fatty acids and phenolic compounds may be generated as the organic materials undergo anaerobic decay.

Carbon dioxide, like O_2, is a mixed blessing in terms of its influence on plants. In small amounts the CO_2 evolved by a soil community does indeed accelerate the development of certain plant species. This improved growth may, in part, be caused by a solubilization of nutrient-containing minerals or a change in soil pH. A large amount of CO_2, conversely, inhibits ion absorption, diminishes water uptake, and otherwise damages roots, and there is no doubt that the hundreds of millions of bacteria in the rhizosphere are constantly generating a stream of CO_2 from the simple compounds exuded by the roots. The situation is aggravated in poorly drained or compacted land following periods of heavy rainfall or in the presence of accumulations of carbonaceous substrates, because the CO_2 cannot be readily dissipated or is produced in abundance. The problem of whether excess CO_2, anaerobiosis, an inorganic toxin, or some organic inhibitor is the sole or prime cause of root death has not been resolved, but no controversy remains on the identity of the culprits, namely, facultative and obligate anaerobes.

PROTECTION AGAINST AND PREDISPOSITION TO DISEASE

In addition to serving as actual incitants of disease, microorganisms are capable of modifying the course of disease and favoring, delaying, or preventing the initiation of attack on susceptible tissues by pathogens. This kind of detrimental or protective action by species that themselves are noninvasive is effected in several ways, and increasing attention has been given recently to research in this area because of the obvious importance in control of human, animal, and plant diseases. Populations that protect against or promote disease may do so either by directly influencing the pathogen or, alternatively, by provoking a modification in the physiology of the potential host.

Evidence has been accumulating for many years that bacteria, fungi, and actinomycetes in the rhizosphere very effectively ward off soil-borne plant pathogens. Almost invariably a pathogenic fungus or bacterium introduced into natural soil does much less harm and acts far more slowly than the same species when added to sterile soil supporting susceptible plants. Not infrequently the parasite fails to proliferate and soon is eliminated, though it would have thrived were it not for the many dissimilar populations already present. Furthermore, the root zone of healthy plants teems with organisms that can, by themselves, readily invade and do extensive damage, but they are held in check by the legions of innocuous residents of the same underground ecosystem.

A very similar situation exists in the throat and gastrointestinal system of man and animals, sites inhabited by communities containing potential disease-producing bacteria that are not allowed by their neighbors to invade the adjacent tissues. These same communities wholly eliminate certain other disease incitants that successfully reach the throat or alimentary tract. Upset of the normal community by the administration of large doses of antibiotics commonly destroys such microbiological buffers against human or animal disease, and a resident species heretofore occupying no better than a minor position or an alien to the ecosystem finds the new circumstances favorable, exploits them, and invades the nearby tissues (Table 18.5). This is the apparent reason for the rise in incidence of human infections by opportunistic fungi and occasional cases of acute staphylococcal enteritis following treatment with broad-spectrum antibiotics. As might be expected, germfree animals are notoriously susceptible to very small inocula of pathogenic bacteria, like *Bacillus anthracis,* at cell densities having no effect on conventional animals. In addition, germfree mammals are often killed when exposed to species, such as *Escherichia coli,* usually considered to be nonpathogenic or weakly so, and the sudden transfer of germfree rodents to a colony of conventional animals leads to their rapidly succumbing to agents carried by the well-colonized sister individuals, suggesting that microorgan-

Table 18.5

Predisposition to Establishment of *Candida albicans* in Intestines of Mice
Receiving Orally Administered Antibiotics [a]

Antibiotic Used	No. of Mice Showing C. albicans	Total No. of Mice	Mice Positive for C. albicans (%)
None	7	87	8
Chlortetracycline	25	89	28
Chloramphenicol	24	30	80
Oxytetracycline	25	30	83
Dihydrostreptomycin	21	29	72
Carbomycin	12	30	40

[a] From Huppert et al. (1955).

isms ineffective in the climax communities do in fact have potent
pathogenic capabilities (Luckey, 1965).

The basis for the protection afforded by the resident community
inhabiting both plants and animals is generally uncertain, but the
existing, albeit sparse, data indicate that it is usually the result of one
or more of four interactions previously cited—competition, amensalism, parasitism, or predation—or of a microbiologically induced
response by the pathogen's potential host. Competition between indigenous species and the parasite is probably the most common basis
for protection, but parasitism, including lysis, may be quite common
in soil. On the other hand, the generation by microorganisms of simple metabolic products or antibiotics may explain the protection of
hosts against selected disease agents. A likely site for antibiotic biosynthesis in nature is the rhizosphere, because of its nearly uninter-
rupted supply of simple organic substrates, and though no such
metabolites have yet been detected there, they could conceivably be
formed and in turn influence the rhizosphere community or be translocated to the aerial portion of the plant. The latter contention is not
implausible, as many deliberately applied antibiotics are taken up
through the roots and move into the stem and leaves. Induction in
the host of a biochemical response that seems to aid in warding off an
invader is well illustrated by the appearance of bactericidal substances
in the blood of germfree animals inoculated with single bacterial
cultures. Indeed, in comparison with conventional individuals, the
germfree animal has poorly developed defense mechanisms, presum-

ably because of the absence of microorganisms eliciting the formation of these defenses. That some of these mechanisms of resistance to infection are nonspecific, by contrast with the specificity of antibodies made in response to microbial invasion, is shown by the fact that a host becomes more resistant not just to a particular bacterium but also to antigenically unrelated organisms (Whitby et al., 1961).

Microbial activities may have the opposite effect and predispose a host to infection or overcome its innate resistance. Tissues through which penetration is accomplished are usually exposed to enzymes and products not only of the invasive population but also of species not normally capable of colonizing the tissues, and the enzymes and organic compounds elaborated by the saprobe might well favor penetration and pathogenesis by the parasite. Experimental observations have revealed that this does indeed occur; for example, *Entamoeba histolytica* neither invades nor causes lesions in germfree guinea pigs, but amebic lesions are readily discernible when the protozoa are introduced into conventional animals or germfree guinea pigs simultaneously inoculated with *Escherichia coli* (Phillips and Wolfe, 1959). Predisposition to infection is also evident in roots injured by microbial products, such as phenolic compounds, released during the degradation of organic remains up to which the roots have grown (Linderman and Toussoun, 1968). The barriers to parasitism erected by a plant that is normally disease-resistant can be destroyed or an avirulent strain can be provoked to invade healthy plant tissues by activities of nearby nonpathogens. It is not clear how resistance is broken or how one species promotes the invasiveness of a second. One might speculate, however, that a reduction in E_h and consumption of O_2 by nonpathogens might allow spores of a pathogenic strain of the obligately anaerobic *Clostridium* to germinate and its vegetative cells to multiply in human wounds or that enzymes or low-molecular-weight products of saprobes may alter host cell permeability so that penetration is facilitated or products stimulatory to fungal germination and growth leak out.

EFFECTS ASSOCIATED WITH PATHOGENICITY

The habitat of a pathogen, when it is living parasitically, is the host—or, more aptly, a particular tissue of its host. The diseased animal or plant is an environment for microorganisms, as much as is water or soil, although study of the diseased host is uniquely complicated by environmental feedback or defense mechanisms of the invaded organism. Conversely, research dealing with this kind of ecosystem is often simplified because the community of interest is frequently monospecific. In its parasitic phase the pathogen obtains food from its host and reacts to the environmental conditions within the surrounding tissues while at the same time doing harm. Should the habitat at that place and time be unsuitable, establishment or multiplication of the population and its ill effects will be prevented; for this reason the outcome of the interaction between host and parasite is determined by where and when contact is made and where and when the two participants interact. Host specificity, a characteristic of parasitism involving plant and animal as well as microbial hosts, is essentially a reflection of the uniqueness of the fit between microorganism and environment: very few viable environments satisfy the nutritional needs and tolerances to deleterious circumstances of species with narrow host specificities, while many are suitable for those with wide specificities.

Parasites need not be pathogenic, that is, be causative agents of disease, although in nature it is likely that one individual living at the expense of constituents of another nearly always does some injury, even if not readily discernible. The degree of pathogenicity is expressed in terms of *virulence,* and strains of a single species vary in their virulence for a particular animal or plant, from the completely innocuous to the highly destructive. Though it is actively growing, the potential of a virulent strain for doing harm may not be expressed owing to conditions prevailing in the microenvironment where the pathogen is lodged, a situation in animal and human pathology frequently designated subclinical infection. Whether an infection is subclinical or symptoms become grossly evident is controlled by the reaction between microorganism and habitat and is, therefore, within the province of ecology. A parallel between pathology and ecology is also evident in the phenomenon of *invasiveness,* the ability of a pathogen to spread from its initial site of encounter, for this is merely a special case of colonization in the face of environmental resistance.

Details of how a pathogen affects its environment—the invaded cell, tissue, or organism—are numerous, but generalizations are few. Emphasis in pathology has largely been, justifiably, on a disease by disease basis, and hence, with notable exceptions, progress has been slow in establishing the mechanisms underlying microbial modification of their viable surroundings, whether the mechanisms of interest are those responsible for invasiveness or for injury to host tissues. Nevertheless, except in rare instances, it is surely not the mere physical presence of the pathogen that does damage, because the mass of cells is negligible by comparison with the weight of the host or of the affected tissue; rather, it is something these cells do. Granted that occasionally a physical blockage of some portion of the infected host may be a basis for injury, pathogens influence their hosts far more often by means of cellular constituents and products they generate in situ.

Among the microbial products synthesized within the host ecosystem that are apparently responsible, directly or indirectly, for pathogenicity in animals and plants are a variety of enzymes, proteins with no recognized enzymatic activity, polypeptides, polysaccharides, lipids, and some astonishingly simple compounds. Because of their unrelated chemical structures, it is not surprising that these products show widely varying kinds of biological activity and elicit entirely different sorts of host responses. Only a small percentage of these substances themselves do the damage; instead the microbial constituent or product usually provokes a response in the surrounding tissues, which in turn react in a manner resulting ultimately in a deleterious change. A few of the metabolites are remarkably selective and act solely on a narrow range of animal or plant species or on a limited variety of tissue types, whereas others affect an array of diverse species or tissues. On occasion a remarkable correlation exists between a microbial characteristic, such as the synthesis of a specific enzyme or toxin, and the virulence of the culture under test, and it is therefore natural that these traits are postulated as being determinants of the deleterious effects noted in disease; however, the high correlation may be misleading and the hypothesis invalid. Even if the assumption later proves to be wrong, the property exhibiting this correlation nevertheless serves as a convenient virulence marker.

Major determinants of virulence are the enzymes elaborated by the invasive heterotroph, in particular enzymes that hydrolyze macromolecules concerned in structural integrity of the tissue surrounding the parasite. Simple or complex molecules with no enzymatic function but with toxic properties constitute a second group of critical factors involved in virulence. Enzymes of bacterial pathogens are sometimes designated toxins, but such usage will be avoided here for the sake of convenience. In general, although pathogenicity is on rare occasions attributable to a single microbial metabolite, as with the toxin of Clostridium tetani, usually the determinants of virulence and the ability to cause injury depend on a few or a multitude of substances and cellular characteristics. In addition to possessing the capacity to grow within a host, the invader must shield itself against host defense reactions, and the responsible protective factors or metabolites—which apparently are different from those involved in disease—may aid the parasite in its initial establishment and tissue colonization or in subsequent extension of the primary infection.

One of the principal problems in understanding the microorganism-host interaction results from the difference in the parasite's behavior in vitro and in vivo. The ecologist should not be surprised at this, since the environment of the culture flask or petri dish bears little semblance to that of the tissues of a living organism, and phenotypic alterations, as well as possibly the emergence of new genotypes, ought to be expected as a free-living population successfully colonizes or penetrates an animal or plant. Virulence factors thus may or may not be expressed in laboratory media. Adding to the difficulties in elucidating the mechanisms of pathogenicity or identifying the responsible metabolites is the progressive change in the tissue under attack, and alteration of the physicochemical environment undoubtedly calls forth metabolic responses by the microorganism (Smith, 1968).

In view of the vast assemblage of products synthesized in vitro, it is not surprising that many will have a deleterious effect on a test organism and sometimes provoke, at least in plants, symptoms essentially identical with those noted in natural infections, but only a scant number of this large collection have been unequivocally shown to be synthesized in vivo and to be determinants of virulence. The dilemma arises, as stated above, from the high probability that a compound formed in culture media may not be elaborated in the host environment, or not in sufficient quantities. However, a few toxins made by bacteria and fungi are remarkably specific, harming only the microorganism's particular host, and these could well be responsible for host specificity of the pathogens. Still, many toxins are injurious to tissues or activities of animals and plants that are never infected by the particular parasite in nature, and this probably is explained by the inability of the microorganism to make contact with or penetrate the appropriate tissues or its failure to synthesize the toxic metabolite in vivo. Ecological factors governing the activity and distribution of parasites when they are away from the host as well as their invasiveness must always be considered together with potential toxigenicity.

Toxins of significance in pathology fall into two categories. Some of those synthesized by bacteria and plant-pathogenic fungi are designated *exotoxins* because they are excreted by actively growing populations with no necessity for autolysis. By contrast, the *endotoxins* of gram-negative bacteria, such as *Salmonella*, are closely associated with the organisms and are released only when the cells autolyze or are otherwise disrupted. A number of toxins implicated in damage to hosts are listed in Table 18.6, from which it is evident that the effective substances range from simple ones like nitrite and oxalic acid to the highly complex proteins of *Clostridium tetani* and *Corynebacterium diphtheriae*. Fungal exotoxins and endotoxins of gram-negative bacteria, the latter being complex lipopolysaccharides having a polysaccharide and a phospholipid as integral units, are far less potent than the toxins of *Clostridium* and *Corynebacterium,* and consequently higher cell densities are necessary for pathologically significant levels to be attained. The relative simplicity of the toxins of plant-invading fungi, in comparison with the complexity of the protein and lipopolysaccharide toxins of bacteria parasitizing animals and man, may conceivably have been dictated by the need for a simple molecule to diffuse through tissues of plants, which do not have the developed circulatory system of metazoa.

For the majority of toxigenic heterotrophs, either the role of the poisons elaborated is as yet undefined or the compounds are of consequence but are not the sole factors associated with disease. However, the toxins of several microbial species quite clearly account for their pathogenicity. *Clostridium tetani* and *Corynebacterium diphtheriae* stand out in this regard among the bacteria affecting man, as do *Pseudomonas tabaci, Helminthosporium victoriae,* and *Periconia circinata* among the plant pathogens. The tetanus bacillus, for example, enters the body through wounds and has little or none of the physiological attributes essential for invasiveness, but once established in mechanically injured tissues that are devoid of O_2, it generates just a minute amount of its unique toxic protein. Though this anaerobe does not spread appreciably, ostensibly because it cannot, the trace quantity of poison is enough to cause death. That toxigenicity alone accounts for the ill done by *C. tetani* is indicated by observations that the purified protein gives all the symptoms of natural infection and that protective immunization against the toxin prevents tetanus. Observations of an identical nature support the view that *C. diphtheriae,* a human throat inhabitant capable of no more than minor invasion of superficial tissue, owes its reputation for evil to its potent exotoxin. Both toxins, however, are carried from the locale of their formation to remote sites of the body. The toxins of *P. tabaci, H. victoriae,* and *P. circinata* also produce symptoms of the respective diseases they cause in plants.

Two of the exotoxins listed have deceptively simple structures. One, nitrite, is formed from nitrate present in forage ingested by ruminants, and rumen bacteria may generate sufficient nitrite, as they

Table 18.6

Microbial Products Implicated in Injury to Hosts

Pathogen	Host	Disease	Active Principle	Reference
		Bacteria		
Anaerobes	Cow	Methemoglobinemia	Nitrite	Hungate (1966)
Bacillus cereus	Sawfly	—	Lecithinase	Lysenko (1967)
Clostridium perfringens	Man	Gas gangrene	Lecithinase	MacLennan (1962)
Clostridium tetani	Man	Tetanus	Protein exotoxin	MacLeod and Bernheimer (1965)
Corynebacterium diphtheriae	Man	Diphtheria	Protein exotoxin	MacLeod and Bernheimer (1965)
Erwinia carotovora	Celery	Rot	Pectic enzymes	Goodman et al. (1967)
Pseudomonas solanacearum	Tomato	Wilt	Cellulase	Kelman and Cowling (1965)
Pseudomonas tabaci	Tobacco	Wildfire	Methionine antimetabolite	Wood (1967)
		Fungus		
Fusarium oxysporum	Many plants	Wilt	Pectic enzymes	Wood (1967)
Helminthosporium victoriae	Oat	Victoria blight	Polypeptide	Goodman et al. (1967)
Penicillium expansum	Apple	Soft rot	Pectic enzymes	Goodman et al. (1967)
Periconia circinata	Sorghum	Milo disease	Polypeptide	Scheffer and Pringle (1961)
Sclerotium rolfsii	Bean	Blight	Oxalic acid	Bateman and Beer (1965)
Thielaviopsis basicola	Bean	Root rot	Lecithinase	Lumsden and Bateman (1968)

479

reduce the nitrate entering their anaerobic habitat, to poison the animal through a reaction between nitrite reaching the blood and its hemoglobin. The second, oxalic acid, is a common product of fungal metabolism, and it is formed in plants infected by *Sclerotium rolfsii* in quantities large enough to be a major contributory factor in the destruction of parasitized tissues.

The release of extracellular enzymes is a property of numerous disease-inciting bacteria, fungi, and protozoa, and the invasiveness or pathogenicity of the populations so endowed has been ascribed, usually on the basis of equivocal evidence, to the activity of these enzymes on host tissues or cells. The enzymes allegedly assist the microorganism in colonization or in facilitating its spread. Alternatively, the enzyme may catalyze the degradation of polymers to monomers that the invader can use as carbon sources for growth. The microbial enzymes frequently implicated in animal disease are lecithinase, hyaluronidase, coagulase, collagenase, and fibrinolysin, while those in plant pathology are the pectic enzymes, lecithinase, and cellulase. The importance of these enzymes to the colonizing population or to tissue integrity becomes obvious from a consideration of the identity of their substrates. Lecithinase hydrolyzes lecithin, a key component of the cell membrane, and cells so altered die because they cannot tolerate disruption of the membrane. Hyaluronidase depolymerizes the hyaluronic acid of the ground substance of metazoan connective tissue, in effect dissolving the cementing substance and thereby facilitating the spread of microorganisms through the tissues. Coagulase causes the clotting of blood plasma of man and animals. Collagenase digests the collagen in infected mammalian muscle or other tissues and thus eliminates the collagen barrier to microbial spread, while fibrinolysin acts on the defensive fibrin the animal makes and destroys the fibrin of blood clots. The pectic substances and cellulose of spermatophytes have a special significance, because invading fungi or bacteria must penetrate between adjacent cells or break directly through cell walls; thus the pectic substances that are key polysaccharides binding cells together and the cellulose that is a backbone of the plant cell wall will, when hydrolyzed by extracellular pectic enzymes and cellulase of the pathogens, permit microbial proliferation through the tissue and disruption of the affected region.

The evidence for a functional role for a few of these enzymes in colonization and pathogenesis is unshakable, but the data in regard to others remain unconvincing. As set forth in Table 18.6, pectic enzymes and cellulase have been implicated in occasional infections, and lecithinase is involved in gas gangrene caused by *Clostridium perfringens* and in the destruction of bean cells by *Thielaviopsis basicola,* for example. Though pectic enzymes are excreted in culture by a broad spectrum of plant inhabitants and crude preparations containing these enzymes elicit responses in suscepts that are similar to or identical with those noted in natural infections, only in rots and vascular wilts is the weight of evidence at the present time sufficiently convincing to argue for a functional role. The collagenases of *Clostridium histolyticum* and *C. perfringens* also seem to have an amply documented importance. On the other hand, despite the great attention given to the hyaluronidase of *Streptococcus* and *Clostridium,* the coagulase of *Staphylococcus,* and the fibrinolysin of streptococci and staphylococci, and admitting, too, the striking correlation between coagulase production and pathogenicity among *Staphylococcus* strains, the data regarding facilitation of bacterial invasion and spread by these enzymes are still equivocal. In addition to the enzymes mentioned above, an ancillary function in the colonization of their ultimate habitat must exist for other enzymes or toxins of bacterial and fungal pathogens, but limited information is available on such products.

Parasites have a number of additional effects on animals and plants. The activity or amount of host-synthesized enzymes may be altered, the rate of formation of normal constituents by the animal or plant can be modified, and occasionally novel metabolites are generated in response to infection. Protozoa bearing specialized attachment organelles sometimes bring about extensive physical injury to metazoa, and the consumption of countless blood cells or destruction of the intestinal mucosa by protozoa undoubtedly directly damages the physiological processes of mammals. The ultimate in physical harm to an animal, without question, is that done by predacious Moniliales and Zoopagales, fungi that ensnare and utterly destroy nematodes.

Microorganisms may disrupt the normal functioning of their host by mechanisms associated with mechanical blockage. At one time death resulting from *Bacillus anthracis* infections was ascribed to a physical interference with the flow of blood by the tremendous number of bacterial cells in the capillaries, but this view is now discredited. Yet a common outcome of specific plant diseases is a wilting linked directly with impaired water flow. The parasite responsible for some vascular wilts is indeed confined principally to large xylem elements in early stages of infection, and it appears that products of the invader, such as polysaccharides, or substances formed as enzymes of the parasite modify host tissues do in fact obstruct the flow of fluids through the vascular system and thus injure the plant (Wood, 1967).

References

REVIEWS

Ciegler, A., and Lillehoj, E. B. 1968. Mycotoxins. *Advan. Appl. Microbiol.*, **10**, 155–219.

Goodman, R. N., Kiraly, Z., and Zaitlin, M. 1967. *The Biochemistry and Physiology of Infectious Plant Disease.* Van Nostrand, Princeton.

Hungate, R. E. 1966. *The Rumen and Its Microbes.* Academic Press, New York.

Luckey, T. D. 1965. Effects of microbes on germfree animals. *Advan. Appl. Microbiol.*, **7**, 169–223.

MacLeod, C. M., and Bernheimer, A. W. 1965. Pathogenic properties of bacteria. *In* R. J. Dubos and J. G. Hirsch, Eds., *Bacterial and Mycotic Infections of Man.* Lippincott, Philadelphia. pp. 146–169.

Patrick, Z. A., Toussoun, T. A., and Koch, L. W. 1964. Effect of crop-residue decomposition products on plant roots. *Ann. Rev. Phytopathol.*, **2**, 267–292.

Raymont, J. E. G. 1963. *Plankton and Productivity in the Oceans.* Pergamon Press, Oxford.

Shilo, M. 1967. Formation and mode of action of algal toxins. *Bacteriol. Rev.*, **31**, 180–193.

Smith, H. 1968. Biochemical challenge of microbial pathogenicity. *Bacteriol. Rev.*, **32**, 164–184.

Wood, R. K. S. 1967. *Physiological Plant Pathology.* Blackwell, Oxford.

OTHER LITERATURE CITED

Akhromeiko, A. I., and Shestakova, V. A. 1958. *Mikrobiologiya*, **27**, 67–74.

Aleksandrov, V. G., and Zak, G. A. 1950. *Mikrobiologiya*, **19**, 97–104.

Ark, P. A. 1936. *Proc. Am. Soc. Hort. Sci.*, **34**, 216–221.

Aust, S. D., Broquist, H. P., and Rinehart, K. L. 1968. *Biotechnol. Bioeng.*, **10**, 403–412.

482

Baker, J. M. 1963. *In* P. S. Nutman and B. Mosse, Eds., *Symbiotic Associations.* Cambridge University Press, London. pp. 232–265.

Bamburg, J. R., Strong, F. M., and Smalley, E. B. 1969. *J. Agr. Food Chem.,* **17,** 443–450.

Barber, D. A. 1966. *Nature,* **212,** 638–640.

Bateman, D. F., and Beer, S. V. 1965. *Phytopathology,* **55,** 204–211.

Bennett, I. L., and Nicastri, A. 1960. *Bacteriol. Rev.,* **24,** 16–34.

Brook, P. J., and White, E. P. 1966. *Ann. Rev. Phytopathol.,* **4,** 171–194.

Brown, E. S. 1954. *Bull. Entomol. Res.,* **45,** 1–66.

Buchner, P. 1954. *Z. Morphol. Oekol. Tiere,* **42,** 550–633.

Buchner, P. 1955. *Z. Morphol. Oekol. Tiere,* **43,** 397–424.

Buchner, P. 1965. *Endosymbiosis of Animals with Plant Microorganisms.* Interscience Publishers, New York.

Delluva, A. M., Markley, K., and Davies, R. E. 1968. *Biochim. Biophys. Acta,* **151,** 646–650.

Fraenkel, G. 1952. *Tijdschr. Entomol.,* **95,** 183–195.

Frei, P. 1963. *Ber. Schweiz. Bot. Ges.,* **73,** 21–57.

Friedmann, H. 1967. *In* S. M. Henry, Ed., *Symbiosis,* Vol. 2. Academic Press, New York. pp. 291–316.

Fritsch, F. E. 1952. *Proc. Roy. Soc. (London), Ser. B,* **139,** 185–192.

Goreau, T. F. 1961. *In* H. M. Lenhoff and W. F. Loomis, Eds., *The Biology of Hydra.* University of Miami Press, Miami. pp. 269–282.

Gorham, P. R. 1964. *In* D. F. Jackson, Ed., *Algae and Man.* Plenum Press, New York. pp. 307–336.

Gray, F. V. 1947. *J. Exptl. Biol.,* **24,** 15–19.

Gustafsson, B. E. 1959. *Ann. N.Y. Acad. Sci.,* **78,** 166–174.

Harley, J. L., and Lewis, D. H. 1969. *Advan. Microbial Physiol.,* **3,** 53–81.

Hartzell, A. 1967. *In* S. M. Henry, Ed., *Symbiosis,* Vol. 2. Academic Press, New York. pp. 107–140.

Hollis, J. P., and Rodriguez-Kabana, R. 1966. *Phytopathology,* **56,** 1015–1019.

Huppert, M., Cazin, J., and Smith, H. 1955. *J. Bacteriol.,* **70,** 440–447.

Joffe, A. Z. 1965. *In* G. N. Wogan, Ed., *Mycotoxins in Foodstuffs.* M. I. T. Press, Cambridge, Mass. pp. 77–85.

Kelman, A., and Cowling, E. B. 1965. *Phytopathology,* **55,** 148–155.

Koaze, Y., Sakai, H., and Arima, K. 1957. *J. Agr. Chem. Soc. Japan,* **31,** 338–349.

Koch, A. 1963. *In* N. E. Gibbons, Ed., *Recent Progress in Microbiology.* University of Toronto Press, Toronto. pp. 151–161.

Lamanna, C., and Carr, C. J. 1967. *Clin. Pharmacol. Therap.,* **8,** 286–332.

Libbert, E., Kaiser, W., and Kunert, R. 1969. *Physiol. Plantarum,* **22,** 432–439.

Linderman, R. G., and Toussoun, T. A. 1968. *Phytopathology,* **58,** 1571–1574.

Lumsden, R. D., and Bateman, D. F. 1968. *Phytopathology*, **58**, 219–227.

Lyle, L. R., and Jutila, J. W. 1968. *J. Bacteriol.*, **96**, 606–608.

Lysenko, O. 1967. *In* P. A. Van Der Laan, Ed., *Insect Pathology and Microbial Control*. North Holland Publishing Co., Amsterdam. pp. 219–237.

MacLennan, J. D. 1962. *Bacteriol. Rev.*, **26**, 177–274.

Marshall, S. M., and Orr, A. P. 1964. *In* D. J. Crisp, Ed., *Grazing in Terrestrial and Marine Environments*. Blackwell, Oxford. pp. 227–238.

Muscatine, L. 1967. *Science*, **156**, 516–519.

Narasimhan, M. J., Ganla, V. G., Deodhar, N. S., and Sule, C. R. 1967. *Lancet*, 760–761.

Norris, D. M., and Baker, J. K. 1967. *Science*, **156**, 1120–1122.

Overgaard Nielsen, C. 1949. *Natura Jutlandica*, **2**, 1–131.

Overgaard Nielsen, C. 1961. *Oikos*, **12**, 17–35.

Phillips, B. P., and Wolfe, P. A. 1959. *Ann. N.Y. Acad. Sci.*, **78**, 308–314.

Relwani, L. L., and Subrahmanyan, R. 1963. *Current Sci. (India)*, **32**, 441–443.

Rodriguez-Kabana, R., Jordan, J. W., and Hollis, J. P. 1965. *Science*, **148**, 524–526.

Scheffer, R. P., and Pringle, R. B. 1961. *Nature*, **191**, 912–913.

Subba-Rao, N. S., Bidwell, R. G. S., and Bailey, D. L. 1961. *Can. J. Bot.*, **39**, 1759–1764.

Takai, Y., and Kamura, T. 1966. *Folia Microbiol.*, **11**, 304–313.

Trolldenier, G., and Marckwordt, U. 1962. *Arch. Mikrobiol.*, **43**, 148–151.

Virtanen, A. I. 1963. *Biochem. Z.*, **338**, 443–453.

Whitby, J. L., Michael, J. G., Woods, M. W., and Landy, M. 1961. *Bacteriol. Rev.*, **25**, 437–446.

Wieland, T. 1968. *Science*, **159**, 946–952.

Woldendorp, J. W. 1962. *Plant Soil*, **17**, 267–270.

Yonge, C. M. 1944. *Biol. Rev.*, **19**, 68–80.

Appendix

MICROORGANISMS REFERRED TO IN TEXT

Class: Schizomycetes (bacteria)
  Order: Pseudomonadales
    Suborder: Rhodobacteriineae
      Chlorobium, Chromatium, Rhodospirillum
    Suborder: Pseudomonadineae
      Acetobacter, Aeromonas, Bdellovibrio, Caulobacter, Desulfovibrio, Ferrobacillus, Gallionella, Halobacterium, Hydrogenomonas, Methanobacterium, Nitrobacter, Nitrosomonas, Pseudomonas, Thiobacillus, Vibrio, Xanthomonas
  Order: Chlamydobacteriales
    Sphaerotilus
  Order: Hyphomicrobiales
    Hyphomicrobium, Metallogenium
  Order: Eubacteriales
    Achromobacter, Actinobacillus, Aerobacter, Agarbacterium, Agrobacterium, Alcaligenes, Arthrobacter, Azotobacter, Bacillus, Bacteroides, Beijerinckia, Bordetella, Brevibacterium, Brucella, Chromobacterium, Clostridium, Corynebacterium, Diplococcus, Erwinia, Escherichia, Flavobacterium, Fusobacterium, Hemophilus, Klebsiella, Lactobacillus, Microbacterium, Micrococcus, Neisseria, Pasteurella, Pediococcus, Proteus, Rhizobium, Salmonella, Sarcina, Serratia, Shigella, Staphylococcus, Streptococcus, Veillonella
  Order: Actinomycetales (includes the actinomycetes)
    Actinomyces, Micromonospora, Mycobacterium, Streptomyces
  Order: Beggiatoales
    Beggiatoa

Order: Myxobacterales
   Archangium, Chondrococcus, Cytophaga, Myxococcus, Soran-gium
Order: Spirochaetales
   Borrelia, Treponema
Order: Mycoplasmatales
   Mycoplasma
Class: Microtatobiotes
Order: Rickettsiales (rickettsiae)
   Coxiella, Rickettsia

Fungi

Acrasiales (cellular slime molds)
   Dictyostelium
Myxomycetes (slime molds)
Phycomycetes
Order: Plasmodiophorales
   Spongospora
Order: Chytridiales (chytrids)
   Chytridium, Chytriomyces, Nucleophaga, Olpidium, Podo-chytrium, Rhizidium, Rhizophlyctis, Rhizophydium, Septo-sperma, Sphaerita, Zygorhizidium
Order: Blastocladiales
   Allomyces
Order: Saprolegniales
   Achlya, Aphanomyces, Ectrogella
Order: Lagenidiales
   Olpidiopsis
Order: Peronosporales
   Peronospora, Phytophthora, Pythium
Order: Mucorales
   Dispira, Endogone, Gilbertella, Mortierella, Mucor, Pilobolus, Piptocephalis, Rhizopus, Sporodinia, Zygorhynchus
Order: Entomophthorales
   Entomophthora
Order: Zoopagales
Uncertain affinity
   Coccidioides
Ascomycetes
Order: Endomycetales (contains many yeasts)

Debaromyces, Endomycopsis, Nematospora, Saccharomyces, Schizosaccharomyces
Order: Erysiphales
Uncinula
Order: Hypocreales
Balansia, Claviceps, Gibberella
Order: Sphaeriales
Ceratocystis, Chaetomium, Endothia, Glomerella, Leptosphaerulina, Neurospora, Ophiobolus, Physalospora, Venturia
Order: Phacidiales
Bifusella, Hypodermella
Order: Helotiales
Cyttaria, Pseudopeziza, Sclerotinia
Order: Pezizales
Dasyobolus
Basidiomycetes
Order: Ustilaginales
Thecaphora, Tilletia, Urocystis, Ustilago
Order: Uredinales
Hemileia, Puccinia
Order: Agaricales
Agaricus, Amanita, Armillaria, Boletus, Fomes, Lactarius, Polyporus, Polystictus, Tricholoma
Order: Nidulariales
Sphaerobolus
Fungi Imperfecti (Deuteromycetes)
Order: Sphaeropsidales
Botryodiplodia, Coniothyrium, Darluca, Diplodia, Rhodosticta
Order: Melanconiales
Colletotrichum
Order: Moniliales (contains some yeasts)
Acontium, Alternaria, Aspergillus, Blastomyces, Botrytis, Calcarisporium, Candida, Cephalosporium, Cephalothecium, Chalaropsis, Cladosporium, Cryptococcus, Epidermophyton, Fusarium, Gliomastix, Gonatobotryum, Graphium, Helminthosporium, Histoplasma, Leptographium, Microsporum, Monilia, Mycogone, Oospora, Penicillium, Periconia, Phymatotrichum, Pithomyces, Pityrosporum, Rhodotorula, Sporotrichum, Thielaviopsis, Trichoderma, Trichophyton, Trichosporon, Trichothecium, Tuberculina, Verticillium
Order: Mycelia Sterilia
Cenococcum, Rhizoctonia, Sclerotium

Algae

Division Cyanophyta (blue-green algae)
  Order: Chroococcales
    Anacystis, Microcystis, Synechococcus
  Order: Nostocales
    Anabaena, Anabaenopsis, Arthrospira, Aulosira, Calothrix, Cylindrospermum, Lyngbya, Microcoleus, Nostoc, Oscillatoria, Phormidium, Plectonema, Schizothrix, Scytonema, Tolypothrix, Trichodesmium
  Order: Stigonematales
    Mastigocladus
Division Rhodophyta (red algae)
  Order: Bangiales
    Porphyra
  Order: Nemalionales
    Asparagopsis
  Order: Ceramiales
    Delesseria
Division Cryptophyta
  Order: Cryptomonadales
    Cyanidium
Division Pyrrophyta (predominantly dinoflagellates)
  Order: Gymnodiniales
    Gymnodinium, Noctiluca, Oxyrrhis, Paulseniella
  Order: Peridiniales
    Ceratium, Gonyaulax
Division Bacillariophyta (diatoms)
  Class: Centrobacillariophyceae
    Biddulphia, Melosira, Planktoniella, Rhizosolenia, Skeletonema, Thalassiosira
  Class: Pennatibacillariophyceae
    Asterionella, Nitzschia, Pinnularia, Thalassiothrix
Division Phaeophyta (brown algae)
  Order: Laminariales
    Laminaria
  Order: Fucales
    Fucus
Division Chrysophyta
  Order: Ochromonadales
    Coccolithus, Dinobryon, Ochromonas, Phaeocystis

Order: Isochrysidales
    Prymnesium
Division Xanthophyta
  Order: Chloramoebales
    Olisthodiscus
  Order: Tribonematales
    Heterococcus
Division Euglenophyta
  Order: Euglenales
    Euglena, Peranema
Division Chlorophyta (green algae)
  Order: Volvocales
    Chlamydomonas, Dunaliella, Haematococcus, Pandorina, Volvox
  Order: Chlorococcales
    Chlorella, Chlorococcum, Scenedesmus, Trebouxia
  Order: Ulotrichales
    Trentepohlia
  Order: Ulvales
    Enteromorpha, Ulva
  Order: Zygnematales
    Closterium, Spirogyra
  Order: Siphonales
    Tydemania
Division Charophyta
  Order: Charales
    Chara

Protozoa

Class: Mastigophora (flagellates)
  Order: Chrysomonadida (see Algae, Division Chrysophyta)
    Oikomonas
  Order: Cryptomonadida (see Algae, Division Cryptophyta)
  Order: Phytomonadida (see Algae, Division Chlorophyta)
  Order: Euglenoidida (see Algae, Division Euglenophyta)
  Order: Dinoflagellida (see Algae, Division Pyrrophyta)
  Order: Rhizomastigida
    Dimorpha, Histomonas
  Order: Protomonadida
    Bodo, Cercomonas, Crithidia, Leishmania, Trypanosoma
  Order: Trichomonadida
    Trichomonas

Order: Hypermastigida
    Trichonympha
Class: Sarcodina (amebae, etc.)
  Order: Amoebida
    Acanthamoeba, Amoeba, Dimastigamoeba, Entamoeba, Mayorella
  Order: Testacida
    Corythion
Class: Sporozoa
  Order: Haemosporida
    Plasmodium
  Order: Haplosporida
    Toxoplasma
Class: Cnidosporidia
Class: Ciliata (ciliates)
  Order: Gymnostomatida
    Actinobolina, Didinium, Nassula, Perispira
  Order: Trichostomatida
    Colpoda, Isotricha, Woodruffia
  Order: Hymenostomatida
    Colpidium, Glaucoma, Leucophrys, Paramecium, Tetrahymena
  Order: Heterotrichida
    Metopus, Stentor
  Order: Oligotrichida
    Diplodinium, Entodinium, Halteria
  Order: Entodiniomorphida
    Polyplastron
  Order: Hypotrichida
    Diophrys, Pleurotricha, Stylonychia, Uroleptus, Uronychia
  Order: Peritrichida
    Opercularia, Trichodina
  Order: Opalinida
    Opalina, Zelleriella
Class: Suctoria
  Order: Suctorida
    Allantosoma

The taxonomic scheme is based on the following sources:

Ainsworth, G. C. 1961. *Dictionary of the Fungi.* Commonwealth Mycological Institute, Kew, Surrey, England.

Breed, R. D. S., Murray, E. G. D., and Smith, N. R. 1957. *Bergey's Manual of Determinative Bacteriology*. Williams & Wilkins, Baltimore.

Kudo, R. R. 1966. *Protozoology*. Thomas, Springfield, Ill.

Silva, P. C. 1962. *In* R. A. Lewin, Ed., *Physiology and Biochemistry of Algae*. Academic Press, New York. pp. 827–837.

Index